Reciprocity, Spatial Mapping and
Time Reversal in Electromagnetics

DEVELOPMENTS IN
ELECTROMAGNETIC THEORY AND APPLICATIONS

VOLUME 9

The titles published in this series are listed at the end of this volume.

Reciprocity, Spatial Mapping and Time Reversal in Electromagnetics

by

C. Altman
Department of Physics,
Technion – Israel Institute of Technology,
Haifa, Israel

and

K. Suchy
Institute for Theoretical Physics,
University of Düsseldorf, Germany

KLUWER ACADEMIC PUBLISHERS
DORDRECHT / BOSTON / LONDON

Library of Congress Cataloging-in-Publication Data

Altman, C.
Reciprocity, spatial mapping and time reversal in electromagnetics / C. Altman, K. Suchy.
p. cm. -- (Developments in electromagnetic theory and applications ; v. 9)
Includes bibliographical references and index.
ISBN 0-7923-1339-9 (HB : acid free paper)
1. Electromagnetism. 2. Reciprocity theorems. 3. Time reversal. I. Suchy, K. II. Title. III. Series: Developments in electromagnetic theory and applications ; 9.
QC760.A48 1991
537--dc20 91-20105

ISBN 0-7923-1339-9

Published by Kluwer Academic Publishers,
P.O. Box 17, 3300 AA Dordrecht, The Netherlands.

Kluwer Academic Publishers incorporates
the publishing programmes of
D. Reidel, Martinus Nijhoff, Dr W. Junk and MTP Press.

Sold and distributed in the U.S.A. and Canada
by Kluwer Academic Publishers,
101 Philip Drive, Norwell, MA 02061, U.S.A.

In all other countries, sold and distributed
by Kluwer Academic Publishers Group,
P.O. Box 322, 3300 AH Dordrecht, The Netherlands.

Printed on acid-free paper

Printed in the Netherlands

OBITUARY

Professor John Heading

Professor John Heading who died on 3 January was the editor of the book series Developments in Electromagnetic Theory and Application. He was also Professor of Applied Mathematics at the University College of Wales from 1968 until his retirement in September 1989.

Professor Heading was educated at the City of Norwich School and St Catherine's College, Cambridge. From 1943 to 1946 he was on war service working on the installation and maintenance of continental telephone repeater stations. He went to Cambridge in 1946, with an Open Exhibition in Mathematics and took a First Class in all three parts of the Mathematical Tripos.

He then spent three years as a research student in the Cavendish Laboratory and was awarded the degree of PhD in 1953 for his work in the field of theoretical ionospheric radio propagation.

Before going to Aberystwyth, Dr Heading held teaching posts in the University of London (at West Ham College of Technology) and at the University of Southampton as Reader in Applied Mathematics. He was a Fellow of the Cambridge Philosophical Society and a member of the London Mathematical Society.

Professor Heading was head of the Department of Applied Mathematics until its merger with the departments of Pure Mathematics and Statistics in 1986.

He will be succeeded as series editor by Professor Gary Roach.

TABLE OF CONTENTS

PREFACE

The choice of topics in this book may seem somewhat arbitrary, even though we have attempted to organize them in a logical structure. The contents reflect in fact the path of 'search and discovery' followed by us, on and off, for the last twenty years. In the winter of 1970-71 one of the authors (C.A.), on sabbatical leave with L.R.O. Storey's research team at the Groupe de Recherches Ionosphériques at Saint-Maur in France, had been finding *almost* exact symmetries in the computed reflection and transmission matrices for plane-stratified magnetoplasmas when symmetrically related directions of incidence were compared. At the suggestion of the other author (K.S., also on leave at the same institute), the *complex conjugate* wave fields, used to construct the eigenmode amplitudes via the mean Poynting flux densities, were replaced by the *adjoint* wave fields that would propagate in a medium with transposed constitutve tensors, *et voilà*, a scattering theorem—'reciprocity in **k**-space'—was found in the computer output. To prove the result analytically one had to investigate the properties of the adjoint Maxwell system, and the two independent proofs that followed, in 1975 and 1979, proceeded respectively via the matrizant method and the thin-layer scattering-matrix method for solving the scattering problem, according to the personal preferences of each of the authors. The proof given in Chap. 2 of this book, based on the hindsight provided by our later results, is simpler and much more concise.

Further investigation revealed that the 'conjugate' problem, in which the scattering matrix was the transpose of that in the given problem, was no more than a reflection mapping of the the adjoint problem (i.e. of the original problem with transposed constitutive tensors). Later, when media with bianisotropic constitutive tensors were investigated, it was found that conjugate (reciprocal) media and wave fields could be formed by any orthogonal spatial mapping of those in the original problem after media and fields were *reversed in time*. The result was quite general and not limited to stratified systems.

The second line of development was to find the link between 'reciprocity in **k**-space' and Lorentz reciprocity, involving currents and sources in physical space. This was done for plane-stratified media by applying the scattering theorem to the plane-wave spectrum of eigenmodes radiated by one current source and reaching the second source. The reverse linkage, from Lorentz reciprocity to reciprocity in **k**-space, had already been found by Kerns (1976).

Application of *restricted* time reversal as a means to obtain Lorentz reciprocity relations was the immediate generalization. (Dissipative processes are not 'time reversed', and so the time reversal is 'restricted'.)

The relation between time reversal and reciprocity is not new. It has often been discussed in the scientific literature. In the context of Lorentz reciprocity it has been applied by Deschamps and Kesler (1967), and possibly by others. We believe however that this is the first time that time reversal has been presented in a systematic and mathematically well-defined procedure to serve as a tool for solving problems of reciprocity and scattering symmetries (reciprocity in **k**-space). The use of time reversal gives rise to problems of causality when sources are present, but when the interaction between *two* systems is involved (Lorentz reciprocity) the non-causal effects are irrelevant.

The insight gained during these investigations has enabled us to present many of the earlier theorems and results, both our own and those of other workers, in a compact and unified approach. Much of the material is new. The generalization of Kerns' theorem in Chap. 7, for instance, had not yet been published at the time of writing of this book. We would like to hope that these ideas may prove stimulating to other workers in the field.

In conclusion, one of the authors (C.A.) would like to express his indebtedness to Professor H. Cory and Dr. E. Fijalkow for their contributions in developing the computer programs that revealed scattering theorems in the computer printouts, in the heroic days when programs were still punched on paper tapes and corrections inserted with scissors and glue. He is also indebted to Dr. A. Schatzberg for his important contribution in bridging the gap between reciprocity in **k**-space and Lorentz reciprocity in physical space.

The authors would like to thank the Deutsche Forschungsgemeinschaft, the Heinrich-Hertz-Stiftung, the Technion–Israel Institute of Technology and the University of Düsseldorf, for their generous financial support during the many years of cooperation between us.

As this manuscript being prepared for press we learnt of the untimely death of Professor John Heading, the first managing editor of the series 'Developments in Electromagnetic Theory and Applications', and of this book in particular. We are indebted to Professor Heading for having invited us to contribute to the series and for his friendly encouragement and advice at all stages of this work. His contribution to the subject of wave propagation and reciprocity was considerable, and has influenced numerous workers in the field.

C. Altman and K. Suchy

Haifa – Düsseldorf
January 1991.

Introduction: scope and aims

There are two underlying themes in this book. The first concerns the use and application of the formally adjoint and the Lorentz-adjoint Maxwell system of equations, the latter so-called since it leads to the Lorentz reciprocity theorem. The (formally) adjoint Maxwell system is shown to play an essential role in the derivation of scattering theorems. With the aid of the adjoint wave fields in anisotropic and possibly absorbing media, one obtains a *bilinear concomitant* vector $\mathbf{P}$, having zero divergence, which in the case of the Maxwell system turns out to be no more than a generalization of the Poynting vector and reduces to the time-averaged Poynting vector in loss-free media. Making use of normalized adjoint eigenmodes we may decompose an arbitrary wave field into its component eigenmodes, and a complex amplitude a_α and its adjoint $\bar{a}_\alpha$ may be defined so that the algebraic sum of the products $\bar{a}_\alpha a_\alpha$ for all eigenmodes equals the component flux density P_z of the generalized Poynting vector in a specified direction $\hat{\mathbf{z}}$ (normal to the stratification, for instance, in stratified media). These results are used in the derivation of scattering theorems ('reciprocity in **k**-space') for plane-stratified and curved-stratified anisotropic media (Chaps. 2 and 3), and in the generalization of a reciprocity theorem involving scattering from an arbitrary object immersed in a homogeneous or plane-stratified anisotropic absorbing medium (Chap. 7). In media with sources (currents) use is made of the Lorentz-adjoint, rather than the (non-physical) formally adjoint Maxwell system, so that the fields that obey Maxwell's equations in a hypothetical Lorentz-adjoint medium are 'physical' and obey a radiation condition at infinity. The Lorentz-adjoint fields and currents thus obtained are related to those in the given system by a Lorentz reciprocity relation, and the results are applied to a variety of media, anisotropic and bianisotropic, including cold magnetoplasmas, moving media, chiral media and compressible magnetoplasmas.

The second underlying theme concerns the relationship between the time-reversed and the Lorentz-adjoint Maxwell systems. In loss-free media the two systems are found to be identical: a Lorentz-adjoint medium is a time-reversed medium. In both cases the direction of a fixed magnetic field (in a magnetoplasma, for instance) is reversed, and a moving medium will move in the opposite direction. In absorbing (dissipative) media the two systems are again identical if by 'time reversal' we understand ***restricted time reversal***, in which all quantities are reversed in time *except* the dissipative processes. Collisional losses are thus unaffected by restricted time reversal. The time-reversed medium is ***reciprocal*** to the given medium, in the sense that currents and fields subject to Maxwell's equations in such media obey a Lorentz reciprocity relation with respect to currents and fields in the given medium. The fact that time-reversed wave fields obey Maxwell's equations in the time-reversed medium, leads to immediate and intuitively self-evident applications. If a ray path, i.e. the trajectory of a wave packet, is known in a given medium, then the time-reversed ray path, i.e. the reverse trajectory, will be a valid solution in the time-reversed (Lorentz-adjoint) medium. In scattering problems, time reversal interchanges incoming and outgoing wave fields, which is the basic idea behind the derivation of the 'reciprocity in **k**-space' scattering theorems.

If sources are present the situation is not so simple, for one then finds that even though the Lorentz-adjoint and the time-reversed governing equations are identical, the solutions are different! The one is causal, being derived from retarded potentials or retarded Green's functions, the other is non-causal, being derived from advanced potentials. However, this does not necessarily imply that non-causal time-reversed solutions are of no practical interest. For if one considers the reaction of the field of one current distribution (antenna) on another, and then compares it with that in the reciprocal situation (in which the roles of transmitter and receiver are interchanged and the medium is replaced by a Lorentz-adjoint medium), the same result is found in the two cases! The receiver in the time-reversed situation sees incoming fields converging on it from infinity, as well as the ray(s) arriving from the transmitter. In the Lorentz-adjoint, reciprocal situation the only incoming wave fields are those that emanate from the transmitter. The reaction of the transmitter on the receiver, which is mediated in both cases by the same connecting ray (or rays), is thus identical too, even though the process in the case of (restricted) time reversal is manifestly non-causal. Time reversal thus provides an intuitively simple prescription for visualizing or deriving the solution to a reciprocal problem (time-reversed media, time-reversed currents and fields) if the solution to the given problem is known.

Historically, the problem of reciprocity in electromagnetics developed in two different directions. The classic work of Lorentz (1896), and its development by Sommerfeld (1925), Dällenbach (1942), Rumsey (1954), Cohen (1955), Harrington and Villeneuve (1958), Kong and Cheng (1970), – the list is only partial – dealt with the interchangeability of receiving and transmitting antennas in a given and in a 'transposed', 'complementary' or 'adjoint' medium. A parallel, and seemingly unrelated, line of development initiated by Budden (1954), Barron and Budden (1959) and developed by Pitteway and Jespersen (1966), Heading (1971), Suchy and Altman (1975), treated 'reciprocity in **k**-space', which dealt with the symmetry properties of scattering matrices in cold plane-stratified magnetoplasmas. These results were extended by Altman, Schatzberg and Suchy (1981) to more general anisotropic or bianisotropic media, and by Suchy and Altman (1989) to curved-stratified media. The gap between the two approaches was bridged partially by Kerns (1976) who demonstrated that the Lorentz reciprocity theorem could be used to derive an 'adjoint reciprocity (scattering) theorem' involving an arbitrary scattering body in free space, and by Schatzberg and Altman (1981) who showed that the scattering theorem in plane-stratified magnetoplasmas led to Lorentz reciprocity between fields and sources in that medium. The relationship between the two types of reciprocity was concealed by the fact that whereas the Lorentz reciprocity theorem interrelates fields and currents in a given and in a Lorentz-adjoint (time-reversed) medium, the scattering theorems generally interrelate incoming and outgoing waves in the same medium for different, symmetrically related directions of incidence. The problem was clarified by Altman, Schatzberg and Suchy (1984) who used the concept of a 'conjugate' medium, which was no more than a spatial mapping of the Lorentz-adjoint medium. Suppose that a plane-stratified magnetoplasma is 'time reversed'. This means that the direction of the external magnetic field, which is odd under time reversal, is reversed. If a reflection mapping is now applied to the medium, viz. reflection with respect to a symmetry plane containing the normal to the stratification and the external magnetic field, the plane-stratified medium is unaffected but the external magnetic field, being an axial vector, is again reversed in direction, and we have thereby recovered the original medium. The given medium is thus 'self-conjugate'. This means that if a scattering problem is solved by some distribution of incoming and outgoing wave fields, another solution will be given by time reversing all wave fields and then reflecting (mapping) them with respect to the symmetry plane. The relations between these given and conjugate (reflected, time-reversed) wave fields in the same medium is the substance of the scattering theorems in plane-stratified media discussed in Chaps. 2 and 3.

The structure of the book is as follows. In Chap. 1 the refractive indices and wave polarizations for a cold electron magnetoplasma are derived, as well as the form of the eigenmodes in a plane-stratified plasma. The equations governing their propagation in the stratified medium are developed (the Clemmow-

Heading coupled wave equations) and the numerical techniques for solving them (Budden's full-wave numerical methods, the Pitteway method of uncoupled penetrating and non-penetrating modes, the Altman-Cory thin-layer scattering-matrix method, the matrizant methods). This survey is far from exhaustive but is a description of each numerical approach that led to the derivation of a scattering theorem, or that revealed scattering (reciprocity) relations in the computed outputs, with each new scattering theorem reflecting the computer program from which it derived. The survey is intended as a basis for Chaps. 2 and 3, in which many of the earlier scattering ('reciprocity in k-space') theorems are rederived within a unified framework with the hindsight provided by the later work on the adjoint system, spatial mapping and time reversal.

In Chap. 4, in the discussion of media with sources, the formally adjoint Maxwell system is modified to give the Lorentz-adjoint system, and the Lorentz-adjoint fields and currents are shown to be related to those in the given medium by the Lorentz reciprocity theorem. Particular attention is paid to media that are self-conjugate, i.e. that reduce to their original form after time reversal and an orthogonal mapping (e.g. reflection) are applied to them. The resulting Lorentz-type reciprocity theorem then involves reflected (or other orthogonally mapped) currents and fields in the reciprocal problem.

In Chap. 5 the plane-wave spectrum of eigenmodes generated by an arbitrary current distribution in a plane-stratified medium is developed, and a Green's function derived in terms of elements, reflection and transmission, of the scattering matrix. With its aid the 'reaction' of the field of one current distribution on another is found, and application of the scattering theorem for plane-stratified media then yields a 'conjugate' Green's function and a resulting Lorentz-type reciprocity theorem relating the original currents and fields to reflected currents and fields in the same (self-conjugate) medium. Reciprocity in k-space is thereby shown to lead to Lorentz reciprocity in physical space, but including a reflected set of currents and fields in the reciprocal situation.

The orthogonal mapping of vector and tensor fields is discussed in Chap. 6, including the mapping of the tensor Green's functions, the constitutive tensors and the scattering matrices. Two types of transformation or mapping are considered. The one is linked to a fixed (usually cartesian) coordinate system, whereas the other is *coordinate free*. In both cases we consider only *active* transformations of the vector and tensor fields that are mapped from one region of space to another. The first type of transformation (coordinate linked) is best suited for use as a *passive* transformation, in which the transformed quantities are fixed in space while the coordinate axes are transformed. This is mathematically equivalent to the active transformations but conceptually quite different.

In Chap. 7 the mathematical formalism of ***restricted*** time reversal is developed. The time-reversed and Lorentz-adjoint Maxwell systems are found to be identical. To test whether this finding has greater universality, we consider a compressible magnetoplasma. This medium is of interest, from our point of view, mainly because the 10-element electromagnetic-acoustic wave field that it supports consists of four different constituents, the electric, the magnetic, the velocity and the pressure fields. Each has its characteristic behaviour under reflection (polar- or axial-vectors) and under time reversal (odd or even). It is found that two possible Lorentz-adjoint systems may be constructed mathematically to satisfy a Lagrange identity and yield a reciprocity relation, but only one of them, corresponding to the time-reversed system, turns out to be physically relevant.

Application of the properties of the formally adjoint system (from which one derives modal biorthogonality and constructs a zero-divergence bilinear concomitant vector) and of the Lorentz-adjoint system, in which the incoming (outgoing) eigenmodes have been time reversed into outgoing (incoming) modes, leads to a generalization of a scattering theorem due to Kerns, in which an anisotropic scattering object is now imbedded in a homogeneous or plane-stratified, anisotropic, and possibly absorbing medium. To complete the discussion of the behaviour of the various types of media under spatial mapping, time reversal and reciprocity, we consider optically active chiral media which are interesting in that they exhibit spatial, rather than frequency dispersion. The analysis of Lorentz reciprocity in chiral media leads to a useful formulation in which each eigenmode transmission channel is exhibited explicitly. In the final section of Chap. 7 the non-physicality of the (non-causal) time-reversed wave fields is demonstrated when they are generated by current sources. However, when determining the reaction of the field of one current source on another in a given and in the reciprocal situation, one finds that reaction of the field of the second current on the first in the reciprocal situation is precisely that predicted by the time-reversal analysis, the non-physical time-reversed fields that converge inwards from infinity playing no role in the interaction. Here too time reversal gives a correct description of the reciprocal interaction between two current sources.

Chapter 1

Wave propagation in a cold magnetoplasma

1.1 Conductivity and permittivity tensors of a weakly-ionized cold magnetoplasma

A plasma permeated by an external magnetic field is a birefringent medium for the propagation of electromagnetic waves. In this sense it is similar to uniaxial crystals, but has some important differences. While the symmetry axes in crystals are represented by polar vectors, the magnetic field $\mathbf{b}$ is an axial vector, an expression of which is the gyration of the charges q_s in a plasma about the magnetic field lines. Thus a magnetoplasma is called a *gyrotropic medium*.

To relate the mean perturbed velocities $\boldsymbol{v}_s$ of the charged plasma species s to the electric field $\mathbf{E}$ of the wave and the external magnetic field $\mathbf{b}$, we use the momentum balance for each species in a weakly ionized plasma in the (approximate) form

$$m_s n_{s0}\left(\frac{\partial \boldsymbol{v}_s}{\partial t}+\nu_s \boldsymbol{v}_s\right)+\nabla p_s-q_s n_{s0}\left(\mathbf{E}+\boldsymbol{v}_s \times \mathbf{b}\right)=\mathbf{f}_s \tag{1.1}$$

with m_s and q_s representing the particle mass and charge respectively, n_{s0} the equilibrium number density, ν_s the momentum-transport collision frequency with neutral particles, p_s the small-amplitude pressure perturbation and $\mathbf{f}_s$ the force-density source. The magnetic wave field being of order $\mathbf{E}/c$, where $c=1/\sqrt{\varepsilon_0\mu_0}$ is the speed of light in vacuo, can be neglected in comparison with the external magnetic field $\mathbf{b}$.

Because of the small-amplitude pressure perturbation p_s, we need the (approximate) energy balance [116, eq. (94.6)]

$$\left(\frac{\partial}{\partial t}+\alpha_s\nu_s\right)p_s+\gamma p_{s0}\nabla\cdot\boldsymbol{v}_s=\pi_s \tag{1.2}$$

to relate p_s to the mean perturbed velocity $\boldsymbol{v}_s$. Here γ=5/3 is the translational specific-heat ratio, π_s is the power-density source and

$$p_{s0}:=n_{s0}\mathcal{K}T \tag{1.3}$$

is the unperturbed partial pressure; $\mathcal{K}$ is Boltzmann's constant and T the temperature. The factor

$$\alpha_s:=\frac{2\mu_{sn}}{m_n}\qquad\text{with}\qquad\alpha_e\ll 1\,,\qquad\alpha_i\approx 1 \tag{1.4}$$

is twice the ratio of the reduced mass μ_{sn} to the mass m_n of a neutral particle [116, eq. (94.7)].

To derive a dispersion equation and polarization ratios we consider the plasma to be homogeneous and stationary. Far from the sources we may assume plane electromagnetic waves with wave vectors $\mathbf{k}$ and angular frequency ω, with wave fields $\mathbf{E}$, $\mathbf{H}$, $\boldsymbol{v}_s$ and p_s varying as

$$\mathbf{E},\,\mathbf{H},\,\boldsymbol{v}_s,\,p_s\sim\exp[i(\omega t-\mathbf{k}\cdot\mathbf{r})]$$

For media varying slowly in space and time this plane-wave ansatz can be generalized to an *eikonal ansatz*

$$\mathbf{E},\,\mathbf{H},\,\boldsymbol{v}_s,\,p_s\sim\exp[-i\phi(\mathbf{r},t)]\qquad\text{with}\qquad\mathbf{k}:=\nabla\phi,\qquad\omega:=-\frac{\partial\phi}{\partial t} \tag{1.5}$$

which is valid in the so-called *geometric optics approximation.*

The momentum balance equation (1.1) becomes

$$m_sn_{s0}\left(i\omega+\nu_s\right)\boldsymbol{v}_s-i\mathbf{k}p_s-q_sn_{s0}\left(\mathbf{E}+\boldsymbol{v}_s\times\frac{\omega_{cs}\hat{\mathbf{b}}}{q_s/|q_s|}\right)=\mathbf{f}_s \tag{1.6}$$

with

$$\hat{\mathbf{b}}:=\frac{\mathbf{b}}{|\mathbf{b}|} \tag{1.7}$$

denoting the unit vector in the direction of the magnetic field $\mathbf{b}$, and

$$\omega_{cs}:=\frac{|q_s|}{m_s}|\mathbf{b}| \tag{1.8}$$

denoting the *gyro- (cyclotron) frequency*. The energy balance equation (1.2) becomes

$$(i\omega + \alpha_s \nu_s)\, p_s - i\gamma p_{s0}\mathbf{k} \cdot \boldsymbol{v}_s = \pi_s \tag{1.9}$$

In the *cold plasma approximaton* we neglect $\mathbf{k}p_s$ in comparison with the term $\omega m_s n_{s0} \boldsymbol{v}_s$ in (1.6). With the small-amplitude pressure perturbation p_s proportional to the square of the mean perturbed velocity v_s, i.e.

$$p_s \approx m_s n_{s0} v_s{}^2 \tag{1.10}$$

this approximation means that the mean perturbed speed v_s is much smaller than the phase velocity ω/k:

$$v_s \ll \frac{\omega}{k} \tag{1.11}$$

and therefore requires high phase velocities. In the energy balance (1.9) a high phase velocity ω/k is tantamount to the vanishing of the temperature $T \sim p_{s0}$ (1.3). This is the reason for the connotation 'cold plasma approximation'. All plasmas in this book will be considered to be 'cold', except the compressible magnetoplasma discussed in Sec. 7.4.

Another approximation has already been implied in the momentum balance equation (1.6) for electrons ($s = e$), viz. the small-collision approximation [117, eq. (3.31)]

$$\omega \gg \nu_e \ll |\omega - \omega_{ce}| \tag{1.12}$$

If it is violated, the momentum balance (1.6) for the electrons should be replaced by an expression for the electron mobility derived from kinetic theory [117, eqs. (19.8d),(32.9),(32.17)].

In the cold plasma approximation eq. (1.6) may be written (neglecting $\mathbf{k}p_s$) as

$$\boldsymbol{\eta}_s (q_s n_{s0})^2 \boldsymbol{v}_s - q_s n_{s0}\mathbf{E} = \mathbf{f}_s \tag{1.13}$$

in which the *resistivity tensor* $\boldsymbol{\eta}_s$ is given by

$$\boldsymbol{\eta}_s(\hat{\mathbf{b}}) := \frac{m_s}{q_s{}^2 n_{s0}} \left[(i\omega + \nu_s)\mathsf{I} + \operatorname{sgn}(q_s)\omega_{cs}\hat{\mathbf{b}} \times \mathsf{I}\right] = \boldsymbol{\eta}_s{}^T(-\hat{\mathbf{b}}) \tag{1.14}$$

where I denotes the unit tensor and $\boldsymbol{\eta}_s{}^T$ the transpose of $\boldsymbol{\eta}_s$. The *mobility tensor* $\boldsymbol{\mu}_s$ is defined by the relation

$$\boldsymbol{v}_s =: \boldsymbol{\mu}_s \mathbf{E} \qquad \text{with} \qquad \boldsymbol{\mu}_s = \frac{1}{q_s n_{s0}} \boldsymbol{\eta}_s{}^{-1} \tag{1.15}$$

and the *conductivity tensor* $\boldsymbol{\sigma}$ in Ohm's law for the electric (conduction) current density, defined by

$$\mathsf{j} = \sum_s q_s n_{s0} \boldsymbol{v}_s =: \boldsymbol{\sigma}\mathbf{E} \tag{1.16}$$

is the sum of the reciprocal resistivity tensors,

$$\boldsymbol{\sigma}(\hat{\mathbf{b}}) = \sum_s q_s n_{s0} \boldsymbol{\mu}_s = \sum_s \boldsymbol{\eta}_s^{-1}(\hat{\mathbf{b}}) = \boldsymbol{\sigma}^T(-\hat{\mathbf{b}}) \tag{1.17}$$

For the inversion of the resistivity tensor $\boldsymbol{\eta}_s$ (1.14), the (complete) system of three orthogonal projectors

$$\mathcal{P}_{\pm 1} := \frac{1}{2}\left(\mathbf{I} - \hat{\mathbf{b}}\hat{\mathbf{b}}^T \pm i\hat{\mathbf{b}} \times \mathbf{I}\right), \qquad \mathcal{P}_0 := \hat{\mathbf{b}}\hat{\mathbf{b}}^T \tag{1.18}$$

satisfying the completeness and orthogonality relations

$$\mathcal{P}_1 + \mathcal{P}_{-1} + \mathcal{P}_0 = \mathbf{I} \tag{1.19}$$

and

$$\mathcal{P}_i \mathcal{P}_j = \delta_{ij} \mathcal{P}_j \tag{1.20}$$

is introduced. It is shown in Appendix A.2 that any gyrotropic tensor $\mathbf{A}$ whose general form is

$$\begin{aligned} \mathbf{A}(\hat{\mathbf{b}}) &= a_\perp(\mathbf{I} - \hat{\mathbf{b}}\hat{\mathbf{b}}^T) + i a_\times(\hat{\mathbf{b}} \times \mathbf{I}) + a_\parallel \hat{\mathbf{b}}\hat{\mathbf{b}}^T \\ &= \mathbf{A}^T(-\hat{\mathbf{b}}) \end{aligned} \tag{1.21}$$

may be decomposed into a linear combination of such projectors

$$\mathbf{A} = \lambda_1 \mathcal{P}_1 + \lambda_{-1} \mathcal{P}_{-1} + \lambda_0 \mathcal{P}_0 \tag{1.22}$$

where λ_1, λ_{-1} and λ_0 are the three distinct eigenvalues of $\mathbf{A}$, given by

$$\lambda_1 = a_\perp + a_\times, \qquad \lambda_{-1} = a_\perp - a_\times, \qquad \lambda_0 = a_\parallel \tag{1.23}$$

The inverse tensor $\mathbf{A}^{-1}$ is then given by, cf. (A.11),

$$\mathbf{A}^{-1} = \frac{1}{\lambda_1}\mathcal{P}_1 + \frac{1}{\lambda_{-1}}\mathcal{P}_{-1} + \frac{1}{\lambda_0}\mathcal{P}_0 \tag{1.24}$$

Decomposing the resistivity tensor $\boldsymbol{\eta}_s$ (1.14) in this manner

$$\boldsymbol{\eta}_s = \eta_{s,1}\mathcal{P}_1 + \eta_{s,-1}\mathcal{P}_{-1} + \eta_{s,0}\mathcal{P}_0 \tag{1.25}$$

with $\eta_i \to \lambda_i$ representing the eigenvalues of $\boldsymbol{\eta}_s$, we may identify the eigenvalues as

$$\eta_{s,\pm 1} = \frac{i[\omega \mp \operatorname{sgn}(q_s)\omega_{cs}] + \nu_s}{q_s^2 n_{s0}/m_s} \qquad \eta_{s,0} = \frac{i\omega + \nu_s}{q_s^2 n_{s0}/m_s} \tag{1.26}$$

The reciprocal tensor $\boldsymbol{\eta}_s^{-1}$, cf. (1.24), is then

$$\boldsymbol{\eta}_s^{-1} = \frac{1}{\eta_{s,1}}\boldsymbol{P}_1 + \frac{1}{\eta_{s,-1}}\boldsymbol{P}_{-1} + \frac{1}{\eta_{s,0}}\boldsymbol{P}_0 \tag{1.27}$$

and the mobility tensor $\boldsymbol{\mu}_s$ (1.15) becomes

$$\boldsymbol{\mu}_s = \mu_{sL}\boldsymbol{P}_1 + \mu_{sR}\boldsymbol{P}_{-1} + \mu_{sP}\boldsymbol{P}_0 \tag{1.28}$$

with eigenvalues

$$\mu_{sL} = \frac{q_s/m_s}{i[\omega - \operatorname{sgn}(q_s)\omega_{cs}] + \nu_s}, \qquad \mu_{sR} = \frac{q_s/m_s}{i[\omega + \operatorname{sgn}(q_s)\omega_{cs}] + \nu_s},$$

$$\mu_{sP} = \frac{q_s/m_s}{i\omega + \nu_s} \tag{1.29}$$

The reason for the subscripts R, L and P will become apparent later, following eq. (1.63). The conductivity tensor $\boldsymbol{\sigma}$ (1.17) is now

$$\boldsymbol{\sigma} = \sigma_L\boldsymbol{P}_1 + \sigma_R\boldsymbol{P}_{-1} + \sigma_P\boldsymbol{P}_0 \tag{1.30}$$

with the eigenvalues

$$\sigma_{R,L,P} = \sum_s q_s n_{s0}\mu_{sR,L,P} = \sum_s \frac{1}{\eta_{s1,-1,0}} \tag{1.31}$$

The first two of the following *constitutive relations*

$$\mathbf{D} = \varepsilon_0\mathbf{E}\,, \qquad \mathbf{j} = \boldsymbol{\sigma}(\hat{\mathbf{b}})\mathbf{E}\,, \qquad \mathbf{B} = \mu_0\mathbf{H} \tag{1.32}$$

are customarily combined as

$$\mathbf{D} - \frac{i}{\omega}\mathbf{j} = \boldsymbol{\varepsilon}\mathbf{E}, \quad \text{with} \quad \boldsymbol{\varepsilon}(\hat{\mathbf{b}}) = \varepsilon_0\mathsf{I} - \frac{i}{\omega}\boldsymbol{\sigma}(\hat{\mathbf{b}}) = \boldsymbol{\varepsilon}^T(-\hat{\mathbf{b}}) \tag{1.33}$$

defining thereby the *plasma permittivity* $\boldsymbol{\varepsilon}(\hat{\mathbf{b}})$. Its eigenvalues, in conventional notation [113, Sec. 1-2], are

$$\frac{\varepsilon_L}{\varepsilon_0} =: L = 1 - \frac{i\sigma_L}{\omega\varepsilon_0} = 1 - \frac{1}{\omega}\sum_s \frac{\omega_{ps}^2}{\omega - \operatorname{sgn}(q_s)\omega_{cs} - i\nu_s} \tag{1.34a}$$

$$\frac{\varepsilon_R}{\varepsilon_0} =: R = 1 - \frac{i\sigma_R}{\omega\varepsilon_0} = 1 - \frac{1}{\omega}\sum_s \frac{\omega_{ps}^2}{\omega + \operatorname{sgn}(q_s)\omega_{cs} - i\nu_s} \tag{1.34b}$$

$$\frac{\varepsilon_P}{\varepsilon_0} =: P = 1 - \frac{i\sigma_P}{\omega\varepsilon_0} = 1 - \frac{1}{\omega}\sum_s \frac{\omega_{ps}{}^2}{\omega - i\nu_s} \tag{1.34c}$$

with the ***plasma frequencies*** ω_{ps} given by

$$\omega_{ps}{}^2 := \frac{q_s{}^2 n_{s0}}{m_s\varepsilon_0} \tag{1.35}$$

The plasma permittivity $\boldsymbol{\varepsilon}$ may be written in terms of the orthogonal projectors, $\mathcal{P}_{\pm 1}$ and $\mathcal{P}_0$, as

$$\frac{\boldsymbol{\varepsilon}}{\varepsilon_0} = L\mathcal{P}_1 + R\mathcal{P}_{-1} + P\mathcal{P}_0 \tag{1.36}$$

as in (1.22), with $\lambda_1, \lambda_{-1}, \lambda_0 \to L, R, P$. Using conventional notation, cf. [113, Sec. 1-2],

$$S := \tfrac{1}{2}(R + L), \qquad D := \tfrac{1}{2}(R - L) \tag{1.37}$$

we obtain

$$\frac{\boldsymbol{\varepsilon}}{\varepsilon_0} = S(\mathsf{I} - \hat{\mathbf{b}}\hat{\mathbf{b}}^T) - iD\hat{\mathbf{b}} \times \mathsf{I} + P\hat{\mathbf{b}}\hat{\mathbf{b}}^T \tag{1.38}$$

as in (A.25) in the Appendix.

In ionospheric and magnetospheric physics the following notation for the frequency ratios is common:

$$X_s := \frac{\omega_{ps}{}^2}{\omega^2} \qquad Y_s := \frac{\omega_{cs}}{\omega} \qquad Z_s := \frac{\nu_s}{\omega} \tag{1.39}$$

In terms of them the eigenvalues (1.34) of the plasma permittivity $\boldsymbol{\varepsilon}/\varepsilon_0$ now read

$$L = 1 - \sum_s \frac{X_s}{1 - iZ_s - \operatorname{sgn}(q_s)Y_s}, \qquad R = 1 - \sum_s \frac{X_s}{1 - iZ_s + \operatorname{sgn}(q_s)Y_s},$$

$$P = 1 - \sum_s \frac{X_s}{1 - iZ_s} \tag{1.40}$$

and their linear combinations (1.37) are

$$S = 1 - \sum_s \frac{X_s}{(1 - iZ_s)^2 - Y_s{}^2}, \qquad D = -\sum_s \frac{X_s \operatorname{sgn}(q_s)Y_s}{(1 - iZ_s)^2 - Y_s{}^2} \tag{1.41}$$

In a cartesian frame, with unit vectors $\hat{\mathbf{x}}$, $\hat{\mathbf{y}}$ and $\hat{\mathbf{z}} = \hat{\mathbf{b}}$, the plasma permittivity $\boldsymbol{\varepsilon}/\varepsilon_0$ has the matrix representation

$$\frac{\boldsymbol{\varepsilon}}{\varepsilon_0} = \begin{bmatrix} S & iD & 0 \\ -iD & S & 0 \\ 0 & 0 & P \end{bmatrix} \tag{1.42}$$

Another cartesian frame, used in Sec. 2.4 with unit vectors $\hat{\mathbf{x}}'$, $\hat{\mathbf{y}}'$=$\hat{\mathbf{y}}$ and $\hat{\mathbf{z}}'$, is adapted to the Earth's magnetic field **b**. With $\hat{\mathbf{z}}'$ pointing vertically upwards, $y' = 0$ representing the magnetic meridian plane and I the magnetic inclination, we have

$$\hat{\mathbf{b}} = \hat{\mathbf{x}}' \cos I + \hat{\mathbf{z}}' \sin I \tag{1.43}$$

with the transformation

$$\begin{bmatrix} E_{x'} \\ E_{y'} \\ E_{z'} \end{bmatrix} = \begin{bmatrix} \sin I & 0 & \cos I \\ 0 & 1 & 0 \\ -\cos I & 0 & \sin I \end{bmatrix} \begin{bmatrix} E_x \\ E_y \\ E_z \end{bmatrix} \tag{1.44}$$

for the components of the vector **E**. In this frame the matrix representation of the plasma permittivity tensor $\boldsymbol{\varepsilon}/\varepsilon_0$ is given by

$$\begin{aligned} \frac{\boldsymbol{\varepsilon}}{\varepsilon_0} &= \begin{bmatrix} S\sin^2 I + P\cos^2 I & iD\sin I & (P-S)\sin I\cos I \\ -iD\sin I & S & iD\cos I \\ (P-S)\sin I\cos I & -iD\cos I & S\cos^2 I + P\sin^2 I \end{bmatrix} \\ &= \begin{bmatrix} S - C\hat{b}_{x'}^2 & iD\hat{b}_{z'} & -C\hat{b}_{x'}\hat{b}_{z'} \\ -iD\hat{b}_{z'} & S & iD\hat{b}_{x'} \\ -C\hat{b}_{x'}\hat{b}_{z'} & -iD\hat{b}_{x'} & S - C\hat{b}_{z'}^2 \end{bmatrix} \end{aligned} \tag{1.45}$$

where $C := S - P$, and $\hat{b}_{x'}=\cos I$, $\hat{b}_{y'}=0$ and $\hat{b}_{z'}=\sin I$ are the direction cosines of the Earth's magnetic field **b** in the magnetic meridian plane.

1.2 Dispersion equation and polarization ratios

To establish a dispersion equation for electromagnetic waves in cold magnetoplasmas, Maxwell's homogeneous equations

$$\nabla \times \mathbf{H} - \frac{\partial \mathbf{D}}{\partial t} - \mathbf{j} = 0, \qquad \nabla \times \mathbf{E} + \frac{\partial \mathbf{B}}{\partial t} = 0 \tag{1.46}$$

in the geometric optics approximation (1.5), viz.

$$i\mathbf{k} \times \mathbf{H} + i\omega \mathbf{D} + \mathbf{j} = 0, \qquad \mathbf{k} \times \mathbf{E} - \omega \mathbf{B} = 0 \tag{1.47}$$

have to be solved together with the the constitutive relations (1.33),

$$\mathbf{D} - \frac{i}{\omega}\mathbf{j} = \boldsymbol{\varepsilon}\mathbf{E}, \qquad \mathbf{B} = \mu_0 \mathbf{H} \tag{1.48}$$

The combination of Maxwell's two equations (1.47) together with $\mathbf{B}=\mu_0\mathbf{H}$ (1.48) leads to the algebraic wave equation

$$\mathbf{n}\times(\mathbf{n}\times\mathbf{E})+\frac{\boldsymbol{\varepsilon}}{\varepsilon_0}\mathbf{E}=\left[\mathbf{n}\mathbf{n}^T-n^2\mathsf{I}+\frac{\boldsymbol{\varepsilon}}{\varepsilon_0}\right]\mathbf{E}=0 \tag{1.49}$$

where n is the refractive index and $\mathbf{n}$ the *refractive index vector*

$$\mathbf{n}:=\frac{\mathbf{k}}{\omega\sqrt{\varepsilon_0\mu_0}}=\frac{c}{\omega}\mathbf{k}\,,\qquad |\mathbf{n}|=n \tag{1.50}$$

The *dispersion equation* is the solubility condition

$$\det\left[\mathbf{n}\mathbf{n}^T-n^2\mathsf{I}+\frac{\boldsymbol{\varepsilon}}{\varepsilon_0}\right]=0 \tag{1.51}$$

In a cartesian frame (x, y, z) with

$$\hat{\mathbf{z}}:=\hat{\mathbf{b}}\,,\qquad \hat{\mathbf{y}}:=\frac{\hat{\mathbf{b}}\times\hat{\mathbf{n}}}{\sin\theta}\,,\qquad \hat{\mathbf{x}}:=\hat{\mathbf{y}}\times\hat{\mathbf{z}}=\hat{\mathbf{n}}\cdot\frac{\mathsf{I}-\hat{\mathbf{b}}\hat{\mathbf{b}}^T}{\sin\theta} \tag{1.52}$$

where $\cos\theta=\hat{\mathbf{b}}\cdot\hat{\mathbf{n}}$, the algebraic wave equation (1.49), with the matrix representation (1.42) for $\boldsymbol{\varepsilon}$, becomes [113, Sec. 1-3, eq. (20)]

$$\begin{bmatrix} S-n^2\cos^2\theta & iD & n^2\sin\theta\cos\theta \\ -iD & S-n^2 & 0 \\ n^2\cos\theta\sin\theta & 0 & P-n^2\sin^2\theta \end{bmatrix}\begin{bmatrix} E_x \\ E_y \\ E_z \end{bmatrix}=0 \tag{1.53}$$

The dispersion equation (1.51) reads [113, Sec. 1-3, eqs. (21)–(24)]

$$n^4(S\sin^2\theta+P\cos^2\theta)-n^2\left[(RL+SP)\sin^2\theta+2SP\cos^2\theta\right]+RLP=0 \tag{1.54a}$$

where the identity, cf. (1.37),

$$S^2-D^2=RL$$

has been used. This may be rearranged in the form

$$\tan^2\theta=-\frac{P(n^2-R)(n^2-L)}{(n^2S-RL)(n^2-P)} \tag{1.54b}$$

from which we obtain the dispersion relations for $\theta=0=\tan\theta$, and $\theta=\pi/2$, $\tan\theta\to\infty$:

$$\begin{aligned} \theta=0: &\qquad n^2=R\,,\quad n^2=L\,,\quad P=0 \\ \theta=\frac{\pi}{2}: &\qquad n^2=P\,,\qquad n^2=\frac{RL}{S} \end{aligned} \tag{1.55}$$

The cofactors of the top row of the square matrix in (1.53) give the electric wave-field *polarization ratios* for the different modes [1, eq. (4.1)]

$$E_x : E_y : E_z = (n^2 - S)(n^2 \sin^2\theta - P) : -iD(n^2 \sin^2\theta - P) : (n^2 - S)n^2 \cos\theta \sin\theta \quad (1.56)$$

The solutions of the dispersion equation (1.54) are [113, Sec. 1-3, eqs. (26),(27)]

$$n^2 = \frac{(RL + SP)\sin^2\theta + 2SP\cos^2\theta \pm \sqrt{(RL - SP)^2 \sin^4\theta + 4D^2P^2\cos^2\theta}}{2(S\sin^2\theta + P\cos^2\theta)} \quad (1.57)$$

They are generally named the *cold plasma modes* or, in the ionospheric literature, the *magnetoionic modes.* For a (one-species $s \to e$) electron plasma the dispersion formula (1.57) can be rearranged (with the corresponding expressions (1.40) and (1.41) for L, R, P, S and D) in the form

$$n^2 = 1 - \frac{X}{1 - iZ - \dfrac{Y^2 \sin^2\theta}{2(1 - X - iZ)} \pm \sqrt{\dfrac{Y^4 \sin^4\theta}{4(1 - X - iZ)^2} + Y^2 \cos^2\theta}} \quad (1.58)$$

This equation was first derived by Lassen (1927) [89, eq. 18] and independently, without collisions, by Appleton (1928 and 1932) [16,17], Goldstein(1928) [60] and Hartree (1931) [64]. It is usually referred to as the Appleton-Hartree formula in the ionospheric literature, but historically the more appropriate name seems to be the Appleton-Lassen formula, as used for instance by Budden [34, Sec. 3],[33, Secs. 3.12 and 4.6].

At low frequencies, $\omega_c/\omega =: Y > 1$, when the term ${\omega_p}^2/\omega^2 =: X$ becomes very large because of its dependence on ω^{-2} [we have dropped the subscript s=e in ω_{cs} and ω_{ps}], the so called *quasi-longitudinal approximation* holds [104, eq. 8.1.3]

$$\frac{Y^4 \cos^4\theta}{4|(1 - X - iZ)^2|} << Y^2 \cos^2\theta \quad (1.59)$$

provided only that the direction of propagation is not too transverse (with respect to the magnetic field **b**), i.e. provided that θ does not approach $\pi/2$. Eq. (1.58) then reduces to

$$n^2 \approx 1 - \frac{X}{1 - iZ \pm Y\cos\theta} \quad (1.60)$$

If $Y\cos\theta > 1$, a propagating mode—the so called *whistler mode* [114]—is obtained with the lower sign in (1.60), which has interesting properties in the usual conditions of VLF ionospheric or magnetospheric propagation, with

$$1 << Y\cos\theta >> Z\,, \qquad X >> Y$$

and (1.60) then takes the form

$$n^2_{whistler} \approx 1 - \frac{X}{1 - iZ - Y\cos\theta} \rightarrow \frac{X}{Y\cos\theta} \tag{1.61}$$

For propagation parallel to the external magnetic field **b**, the dispersion equation (1.54) has three solutions

$$n^2 = R, \qquad n^2 = L, \qquad P = 0 \tag{1.62}$$

with the corresponding polarization ratios (1.56)

$$E_x : E_y : E_z = \overbrace{(1 : -i : 0)}^{R}, \overbrace{(1 : i : 0)}^{L}, \overbrace{(0 : 0 : finite)}^{P} \tag{1.63}$$

The first two (propagating) modes have circular transverse polarizations with opposite helicity, viz.

$$\begin{aligned} \mathbf{E}_{R,L} &= (\hat{\mathbf{x}} \mp i\hat{\mathbf{y}})\exp(i\omega t) \\ &= (\hat{\mathbf{x}}\cos\omega t \pm \hat{\mathbf{y}}\sin\omega t) + i(\hat{\mathbf{x}}\sin\omega t \mp \hat{\mathbf{y}}\cos\omega t) \end{aligned}$$

Looking down the direction of propagation, $\mathbf{E}_R$ has *right-handed* polarization, $\mathbf{E}_L$ is *left-handed*, and hence the symbols R and L; the polarization of the mode P, on the other hand, is *parallel* (longitudinal).

In the absence of collisions ($Z_s{:=}\nu_s/\omega{=}0$) the eigenvalues R, L and P, (1.34) and (1.40), of the plasma permittivity $\boldsymbol{\varepsilon}/\varepsilon_0$ (1.33) are real, and so too is the square of the the refractive index n (1.57). For negative values of n^2 no propagation is possible. The corresponding mode has a *cutoff* at $n^2 = 0$. Such cutoffs occur, as can be seen from the dispersion equation (1.54), for vanishing RLP, i.e. for

$$R = 0, \qquad L = 0 \quad \text{or} \quad P = 0 \tag{1.64}$$

or, with the aid of (1.40), for

$$\sum_s \frac{X_s}{1 + \mathrm{sgn}(q_s)Y_s} = 1, \quad \sum_s \frac{X_s}{1 - \mathrm{sgn}(q_s)Y_s} = 1 \quad \text{or} \quad \sum_s X_s = 1 \tag{1.65}$$

In the case of an electron plasma in particular, with $\mathrm{sgn}(q_e) = -1$, the cutoffs occur when

$$X = 1 - Y, \qquad X = 1 + Y \quad \text{or} \quad X = 1 \tag{1.66}$$

In Fig. 1.1 the dependence of n^2 (1.58) on $X =: {\omega_p}^2/\omega^2$ is shown, for $Y = 2$, $Z = 0$, with $\theta = \arccos(\hat{\mathbf{b}}\cdot\hat{\mathbf{n}})$ as the parameter. The wave polarizations, right-

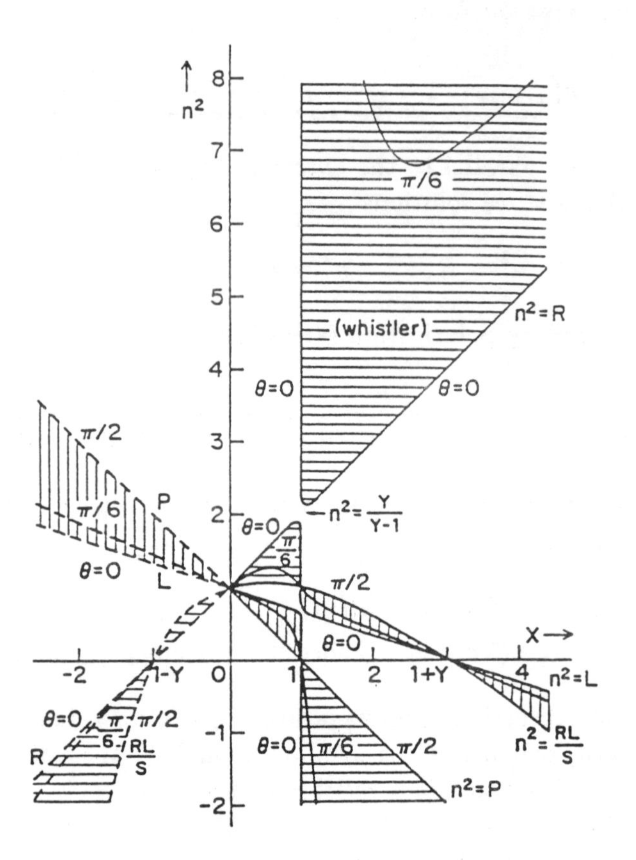

Figure 1.1: Square of refractive index n^2 versus $X:=\omega_p{}^2/\omega^2$ (1.58) for a cold electron magnetoplasma. $Y:=\omega_c/\omega=2$, $Z:=\nu/\omega=0$. Three values of the parameter $\theta:=\arccos(\hat{\mathbf{b}}\cdot\hat{\mathbf{n}})$ are shown, 0, $\pi/6$ and $\pi/2$, see eq. (1.55). Curves for left-handed polarizations lie in the vertically-hatched areas, right-handed polarizations in the horizontal hatching.

or left-handed, in each region are indicated, and the region in which there is whistler-type propagation is marked.

Denoting the two solutions (1.57) of the dispersion equation (1.54) by $n_\alpha{}^2$ and $n_\beta{}^2$ and the corresponding electric wave fields, as determined by the polarization ratios in (1.56), by $\mathbf{E}_\alpha$ and $\mathbf{E}_\beta$, we may write the algebraic wave equation (1.49), with the unit vector $\hat{\mathbf{n}}:=\mathbf{n}/n$, as an eigenvalue equation

$$\left[\frac{\boldsymbol{\varepsilon}(\mathbf{b})}{\varepsilon_0} - n_\alpha{}^2(\mathbf{I}-\hat{\mathbf{n}}\hat{\mathbf{n}}^T)\right]\mathbf{E}_\alpha = 0, \qquad \det\left[\frac{\boldsymbol{\varepsilon}(\mathbf{b})}{\varepsilon_0} - n_\alpha{}^2(\mathbf{I}-\hat{\mathbf{n}}\hat{\mathbf{n}}^T)\right] = 0 \quad (1.67)$$

with the two cold plasma modal wave fields $\mathbf{E}_{\alpha,\beta}$ as (right) eigenvectors. Consider next the eigenmode equation for the reciprocal (left) eigenvectors, cf. (A.15) in Appendix A.1,

$$\left[\frac{\boldsymbol{\varepsilon}(\mathbf{b})^T}{\varepsilon_0} - \bar{n}_\beta{}^2(\mathsf{I} - \hat{\mathbf{n}}\hat{\mathbf{n}}^T)\right]\overline{\mathbf{E}}_\beta = 0 \tag{1.68}$$

or, transposing,

$$\overline{\mathbf{E}}_\beta{}^T\left[\frac{\boldsymbol{\varepsilon}(\mathbf{b})}{\varepsilon_0} - \bar{n}_\beta{}^2(\mathsf{I} - \hat{\mathbf{n}}\hat{\mathbf{n}}^T)\right] = 0 \qquad \det\left[\frac{\boldsymbol{\varepsilon}(\mathbf{b})}{\varepsilon_0} - \bar{n}_\beta{}^2(\mathsf{I} - \hat{\mathbf{n}}\hat{\mathbf{n}}^T)\right] = 0 \tag{1.69}$$

The eigenvalue equations for $\bar{n}_\beta^2$ and n_α^2, (1.67) and (1.69), are identical and so clearly

$$\bar{n}_\beta = n_\beta \tag{1.70}$$

However, the direction of the refractive index vector, cf. (1.50),

$$\bar{\mathbf{n}}_\beta := \frac{\bar{\mathbf{k}}_\beta}{\omega\sqrt{\varepsilon_0\mu_0}} = \frac{c}{\omega}\bar{\mathbf{k}}_\beta\,, \qquad |\bar{\mathbf{n}}_\beta| =: n_\beta \tag{1.71}$$

is not necessarily the same as that of $\mathbf{n}_\beta$. In fact if we write the first of Maxwell's equations (1.47), with the aid of (1.48), in the form

$$\left[\omega\frac{\boldsymbol{\varepsilon}}{\varepsilon_0}\,,\ \mathbf{k}_\alpha \times \mathsf{I}\right]\begin{bmatrix}\mathbf{E}_\alpha\\ \mathbf{H}_\alpha\end{bmatrix} = 0$$

and the corresponding equation for the reciprocal (left) eigenvector in the form

$$\left[\omega\frac{\boldsymbol{\varepsilon}^T}{\varepsilon_0}\,,\ -\mathbf{k}_\alpha \times \mathsf{I}\right]\begin{bmatrix}\overline{\mathbf{E}}_\alpha\\ \overline{\mathbf{H}}_\alpha\end{bmatrix} = \left[\omega\frac{\boldsymbol{\varepsilon}^T}{\varepsilon_0}\,,\ \bar{\mathbf{k}}_\alpha \times \mathsf{I}\right]\begin{bmatrix}\overline{\mathbf{E}}_\alpha\\ \overline{\mathbf{H}}_\alpha\end{bmatrix} = 0$$

using the fact that

$$[\mathbf{k} \times \mathsf{I}]^T = -\mathbf{k} \times \mathsf{I}$$

we see that

$$\bar{\mathbf{k}}_\alpha = -\mathbf{k}_\alpha\,, \qquad \bar{\mathbf{n}}_\alpha = -\mathbf{n}_\alpha \tag{1.72}$$

This result implies that if the modal wave field $(\mathbf{E}_\alpha\,,\mathbf{H}_\alpha)$ propagates with a wave vector $\mathbf{k}_\alpha$ in the medium $\boldsymbol{\varepsilon}(\mathbf{b})$, then a 'reciprocal wave field' $(\overline{\mathbf{E}}_\alpha\,,\overline{\mathbf{H}}_\alpha)$ propagates with a wave vector $\bar{\mathbf{k}}_\alpha = -\mathbf{k}_\alpha$ in the medium $\boldsymbol{\varepsilon}(-\mathbf{b}) = \boldsymbol{\varepsilon}^T(\mathbf{b})$ (1.33). Recalling that $\bar{n}_\alpha{}^2 = n_\alpha{}^2$ (1.70), the polarization ratios of the electric wave fields are equal in the two cases and we may write

$$\overline{\mathbf{E}}_\alpha(-\hat{\mathbf{b}}, -\hat{\mathbf{n}}) = \mathbf{E}_\alpha(\hat{\mathbf{b}}, \hat{\mathbf{n}})\,, \qquad \overline{\mathbf{H}}_\alpha(-\hat{\mathbf{b}}, -\hat{\mathbf{n}}) = -\mathbf{H}_\alpha(\hat{\mathbf{b}}, \hat{\mathbf{n}}) \tag{1.73}$$

since, from (1.47) and (1.48),

$$\mathbf{H}_\alpha = \frac{1}{\omega\mu_0}\mathbf{k}_\alpha \times \mathbf{E}_\alpha = -\frac{1}{\omega\mu_0}\overline{\mathbf{k}}_\alpha \times \overline{\mathbf{E}}_\alpha = -\overline{\mathbf{H}}_\alpha$$

The terms 'reciprocal medium' and 'reciprocal wave fields' are used in a different context in the chapters that follow, but mathematically they are closely related if not identical to the ones used here. We may anticipate some concepts developed in later chapters by noting that the 'reciprocal medium' and 'reciprocal wave fields' are obtained by *time reversing* the original medium $\boldsymbol{\varepsilon}(\mathbf{b})$ and wave vectors $\mathbf{k}$ which, in the case of the cold magnetoplasma, means reversing the external magnetic field $\mathbf{b}$ of the medium and the magnetic wave field $\mathbf{H}$, and through it the direction of the propagation vector $\mathbf{k} \sim \mathbf{E} \times \mathbf{H}$.

Multiplication of (1.67) from the left with $\overline{\mathbf{E}}_\beta{}^T$ and of (1.69) from the right with $\mathbf{E}_\alpha$, with subsequent subtraction, yields

$$(n_\alpha{}^2 - n_\beta{}^2)\overline{\mathbf{E}}_\beta{}^T[\mathbf{I} - \hat{\mathbf{n}}\hat{\mathbf{n}}^T]\mathbf{E}_\alpha = 0 \tag{1.74}$$

This is a biorthogonality relation

$$\overline{\mathbf{E}}_{t\beta} \cdot \mathbf{E}_{t\alpha} = \delta_{\alpha\beta}\overline{\mathbf{E}}_{t\alpha}{}^T\mathbf{E}_{t\alpha} \tag{1.75}$$

for the transverse (to $\mathbf{k}$) components of $\mathbf{E}_\alpha$ and $\overline{\mathbf{E}}_\beta$:

$$\mathbf{E}_{t\alpha} := [\mathbf{I} - \hat{\mathbf{n}}\hat{\mathbf{n}}^T]\mathbf{E}_\alpha\,, \qquad \overline{\mathbf{E}}_{t\beta}{}^T := \overline{\mathbf{E}}_\beta{}^T[\mathbf{I} - \hat{\mathbf{n}}\hat{\mathbf{n}}^T] \tag{1.76}$$

since

$$\mathbf{I} - \hat{\mathbf{n}}\hat{\mathbf{n}}^T = [\mathbf{I} - \hat{\mathbf{n}}\hat{\mathbf{n}}^T][\mathbf{I} - \hat{\mathbf{n}}\hat{\mathbf{n}}^T]$$

is a projector onto the plane transverse to the wave-normal direction $\hat{\mathbf{n}}$.

In a cartesian frame (ξ, η, ς), with

$$\hat{\boldsymbol{\varsigma}} := \hat{\mathbf{n}}\,, \qquad \hat{\boldsymbol{\eta}} := \frac{\hat{\mathbf{b}} \times \hat{\mathbf{n}}}{\sin\theta} = \hat{\mathbf{y}}\,, \qquad \hat{\boldsymbol{\xi}} := \hat{\boldsymbol{\eta}} \times \hat{\boldsymbol{\varsigma}} = \frac{\hat{\mathbf{n}}\hat{\mathbf{n}}^T - \mathbf{I}}{\sin\theta} \cdot \hat{\mathbf{b}} \tag{1.77}$$

we have

$$\overline{E}_\varsigma = -E_\varsigma\,, \qquad \overline{E}_\eta = E_\eta\,, \qquad \overline{E}_\xi = -E_\xi \tag{1.78}$$

and therefore the biorthogonality relation (1.75) takes the form

$$\overline{E}_{\xi\beta}E_{\xi\alpha} + \overline{E}_{\eta\beta}E_{\eta\alpha} = E_{\eta\beta}E_{\eta\alpha} - E_{\xi\beta}E_{\xi\alpha} = 0$$

yielding the products of the *transverse polarizations*, ρ_α and ρ_β,

$$\rho_\alpha\rho_\beta := \left(\frac{E_\eta}{E_\xi}\right)_\alpha \left(\frac{E_\eta}{E_\xi}\right)_\beta = 1\,, \qquad \alpha \neq \beta \tag{1.79}$$

The components of $\mathbf{E}$ in the two cartesian frames $\hat{\mathbf{x}}, \hat{\mathbf{y}}, \hat{\mathbf{z}}$ and $\hat{\boldsymbol{\xi}}, \hat{\boldsymbol{\eta}}, \hat{\boldsymbol{\varsigma}}$ (1.52) and (1.77), are related by the transformation

$$\begin{bmatrix} E_\xi \\ E_\eta \\ E_\varsigma \end{bmatrix} = \begin{bmatrix} \cos\theta & 0 & -\sin\theta \\ 0 & 1 & 0 \\ \sin\theta & 0 & \cos\theta \end{bmatrix} \begin{bmatrix} E_x \\ E_y \\ E_z \end{bmatrix} \tag{1.80}$$

Using the polarization ratios (1.56) in the (x, y, z) frame, we obtain the corresponding ratios in the (ξ, η, ς) frame [1, eq. (4.2)] in the form

$$E_\xi : E_\eta : E_\varsigma = (n^2 - S)P\cos\theta \;:\; iD(n^2\sin^2\theta - P) \;:\; (n^2 - S)(n^2 - P)\sin\theta \tag{1.81}$$

In Chap. 2 we shall need to express the composite 6-element wave-field eigenvector in the form

$$\mathbf{e}_\alpha := (\mathbf{E}_\alpha, \mathcal{H}_\alpha) = (E_\xi\,,\, E_\eta\,,\, E_\varsigma\,;\, \mathcal{H}_\xi\,,\, \mathcal{H}_\eta\,,\, \mathcal{H}_\varsigma\,) \tag{1.82}$$

in which we have introduced the normalized magnetic wave-field vector, cf. [32, Sec. 2.10],

$$\mathcal{H} := \sqrt{\frac{\mu_0}{\varepsilon_0}}\mathbf{H} \tag{1.83}$$

From (1.47) and (1.48) we have

$$\mathbf{k} \times \mathbf{E} = \omega\mu_0\mathbf{H} = \omega\sqrt{\varepsilon_0\mu_0}\mathcal{H} = k_0\mathcal{H}\,, \qquad k_0 := \omega\sqrt{\varepsilon_0\mu_0}$$

which yields, with $\mathbf{n}=n\hat{\boldsymbol{\varsigma}}=\mathbf{k}/k_0$,

$$\mathbf{n} \times \mathbf{E} = \mathcal{H}\,, \qquad \mathcal{H}_\xi = -nE_\eta\,, \quad \mathcal{H}_\eta = nE_\xi\,, \qquad \rho := \frac{E_\eta}{E_\xi} = -\frac{\mathcal{H}_\xi}{\mathcal{H}_\eta} \tag{1.84}$$

The eigenmode wave field for the mode α then becomes

$$\mathbf{e}_\alpha = (1\,,\, \rho_\alpha\,,\, \sigma_\alpha\,;\, -\rho_\alpha n_\alpha\,,\, n_\alpha\,,\, 0)\, E_{\xi,\alpha} \tag{1.85}$$

with the *longitudinal polarization* $\sigma := E_\varsigma / E_\xi$.

1.3 Dispersion equation for a stratified magnetoplasma

If the magnetoplasma is plane stratified and its variation in the normal direction, say $\hat{\mathbf{z}}$, is so rapid that the geometrical optics approximation (1.5) is no longer valid, we must revert to Maxwell's differential equations (1.46) for time-harmonic waves

$$\mathbf{E}\,,\,\mathbf{H} \sim \exp(i\omega t) \tag{1.86}$$

in the form

$$\nabla\times\mathbf{H} - i\omega\boldsymbol{\varepsilon}\mathbf{E} = \mathbf{J}_{\mathrm{e}}\,, \qquad \nabla\times\mathbf{E} + i\omega\mu_0\mathbf{H} = -\mathbf{J}_{\mathrm{m}} \tag{1.87}$$

where $\mathbf{J}_{\mathrm{e}}$ and $\mathbf{J}_{\mathrm{m}}$ are the electric and equivalent magnetic current densities and $\boldsymbol{\varepsilon}:=\varepsilon_0\mathbf{I} - i\boldsymbol{\sigma}/\omega$ (1.33) is the plasma permittivity. With an ansatz for the separation of variables

$$\mathbf{E}(x,y,z) \sim \xi(x)\eta(y)\mathbf{E}(z)\,, \qquad \mathbf{H}(x,y,z) \sim \xi(x)\eta(y)\mathbf{H}(z) \tag{1.88}$$

the separation constants k_x, k_y are given by

$$\frac{1}{\xi}\frac{d\xi}{dx} = -ik_x\,, \qquad \frac{1}{\eta}\frac{d\eta}{dy} = -ik_y \tag{1.89}$$

Hence, together with the expression (1.86) for time-harmonic waves, we have

$$\begin{aligned}\mathbf{E}(\mathbf{r},t) &\sim \mathbf{E}(z)\exp[i(\omega t - k_x x - k_y y)] =: \mathbf{E}(z)\exp[i(\omega t - \mathbf{k}_t\cdot\mathbf{r})]\\ \mathbf{H}(\mathbf{r},t) &\sim \mathbf{H}(z)\exp[i(\omega t - k_x x - k_y y)] =: \mathbf{H}(z)\exp[i(\omega t - \mathbf{k}_t\cdot\mathbf{r})]\end{aligned} \tag{1.90}$$

the constancy of the tangential component $\mathbf{k}_t$ of the wave vector $\mathbf{k}:=\nabla\phi$ (1.5) expressing Snell's law.

Maxwell's partial differential equations (1.87) become

$$\begin{aligned}-i\mathbf{k}_t\times\mathbf{H} + \hat{\mathbf{z}}\times\frac{d\mathbf{H}}{dz} - i\omega\boldsymbol{\varepsilon}\mathbf{E} &= \mathbf{J}_{\mathrm{e}}\\ -i\mathbf{k}_t\times\mathbf{E} + \hat{\mathbf{z}}\times\frac{d\mathbf{E}}{dz} + i\omega\mu_0\mathbf{H} &= -\mathbf{J}_{\mathrm{m}}\end{aligned} \tag{1.91}$$

the common factor $\exp[i(\omega t - \mathbf{k}_t\cdot\mathbf{r})]$ having been dropped. With the aid of the unit tensor $\mathbf{I}$ this system of coupled ordinary differential equations may be written as

$$\begin{bmatrix} i\omega\boldsymbol{\varepsilon} & i\mathbf{k}_t\times\mathbf{I} - \hat{\mathbf{z}}\times\mathbf{I}\dfrac{d}{dz}\\ -i\mathbf{k}_t\times\mathbf{I} + \hat{\mathbf{z}}\times\mathbf{I}\dfrac{d}{dz} & i\omega\mu_0\mathbf{I}\end{bmatrix}\begin{bmatrix}\mathbf{E}\\ \mathbf{H}\end{bmatrix} = \begin{bmatrix}-\mathbf{J}_{\mathrm{e}}\\ -\mathbf{J}_{\mathrm{m}}\end{bmatrix} \tag{1.92}$$

or, more compactly, as

$$\left[i\omega\mathbf{K} - ik_x\mathbf{U}_x - ik_y\mathbf{U}_y + \mathbf{U}_z\frac{d}{dx}\right]\mathbf{e} = -\mathbf{j} \tag{1.93}$$

in terms of the 6×6 tensors

$$\mathbf{K} := \begin{bmatrix} \boldsymbol{\varepsilon} & \mathbf{0} \\ \mathbf{0} & \mu_0 \mathbf{I} \end{bmatrix} \tag{1.94}$$

$$\mathbf{U}_x := \begin{bmatrix} \mathbf{0} & -\hat{\mathbf{x}} \times \mathbf{I} \\ \hat{\mathbf{x}} \times \mathbf{I} & \mathbf{0} \end{bmatrix} = \mathbf{U}_x{}^T, \quad \mathbf{U}_y := \begin{bmatrix} \mathbf{0} & -\hat{\mathbf{y}} \times \mathbf{I} \\ \hat{\mathbf{y}} \times \mathbf{I} & \mathbf{0} \end{bmatrix} = \mathbf{U}_y{}^T$$

$$\mathbf{U}_z := \begin{bmatrix} \mathbf{0} & -\hat{\mathbf{z}} \times \mathbf{I} \\ \hat{\mathbf{z}} \times \mathbf{I} & \mathbf{0} \end{bmatrix} = \mathbf{U}_z{}^T = \left[\begin{array}{ccc:ccc} & & & 0 & 1 & 0 \\ & \mathbf{0} & & -1 & 0 & 0 \\ & & & 0 & 0 & 0 \\ \hdashline 0 & -1 & 0 & & & \\ 1 & 0 & 0 & & \mathbf{0} & \\ 0 & 0 & 0 & & & \end{array}\right] \tag{1.95}$$

and the 6×1 column vectors

$$\mathbf{e} := \begin{bmatrix} \mathbf{E} \\ \mathbf{H} \end{bmatrix}, \qquad \mathbf{j} := \begin{bmatrix} \mathbf{J}_{\mathrm{e}} \\ \mathbf{J}_{\mathrm{m}} \end{bmatrix} \tag{1.96}$$

With $\mathbf{K}$, $\mathbf{U}_x$ and $\mathbf{U}_y$ combined into a single 6×6 tensor $\mathbf{C}$,

$$\mathbf{C} := \omega \mathbf{K} - k_x \mathbf{U}_x - k_y \mathbf{U}_y \tag{1.97}$$

eq. (1.93) becomes

$$\left[i\mathbf{C} + \mathbf{U}_z \frac{d}{dz} \right] \mathbf{e} = -\mathbf{j} \tag{1.98}$$

Note that the differential operator $\mathbf{U}_z d/dz$ acts only on the tangential components $\mathbf{E}_t$ and $\mathbf{H}_t$. The remaining two equations are algebraic relations between the normal components $\mathbf{E}_z$, $\mathbf{H}_z$ and the the tangential components $\mathbf{E}_t$ $\mathbf{H}_t$.

In the following section (1.4) we shall separate differential and algebraic equations by elimination of the normal components $\mathbf{E}_z$ and $\mathbf{H}_z$ from the differential equations. Here we decompose the wave field $\mathbf{e}$ into modes by means of the formal substitution

$$\frac{d}{dz} \rightarrow -i\kappa(z) \tag{1.99}$$

in the source-free 6×6 system (1.98), which leads to the purely algebraic 6×6 system

$$[\mathbf{C} - \kappa \mathbf{U}_z]\mathbf{e} = 0 \tag{1.100}$$

This is an eigenvalue problem in which the eigenvalues κ_α and corresponding right and left (reciprocal) eigenvectors $\mathbf{e}_\alpha$ and $\bar{\mathbf{e}}_\beta$ solve the respective equations

$$[\mathbf{C} - \kappa_\alpha \mathbf{U}_z]\mathbf{e}_\alpha = 0, \qquad \bar{\mathbf{e}}_\beta{}^T [\mathbf{C} - \kappa_\beta \mathbf{U}_z] = 0 \tag{1.101}$$

Premultiplying the first equation in (1.101) with $\bar{\mathbf{e}}_\beta{}^T$, postmultiplying the second with $\mathbf{e}_\alpha$ and subtracting, we obtain

$$(\kappa_\alpha - \kappa_\beta)\bar{\mathbf{e}}_\beta{}^T \mathbf{U}_z \mathbf{e}_\alpha = 0 \tag{1.102}$$

establishing the biorthogonality of $\bar{\mathbf{e}}_\beta$ and $\mathbf{e}_\alpha$ with respect to the 'mixed Poynting product', viz.

$$\bar{\mathbf{e}}_\beta{}^T \mathbf{U}_z \mathbf{e}_\alpha = \left(\overline{\mathbf{E}}_\beta \times \mathbf{H}_\alpha + \mathbf{E}_\alpha \times \overline{\mathbf{H}}_\beta\right) \cdot \hat{\mathbf{z}} = \delta_{\alpha\beta}\left(\overline{\mathbf{E}}_\alpha \times \mathbf{H}_\alpha + \mathbf{E}_\alpha \times \overline{\mathbf{H}}_\alpha\right) \cdot \hat{\mathbf{z}} \tag{1.103}$$

If we substitute $d/dz \to -i\kappa$ (1.99) in (1.92) and compare with Maxwell's equations (1.47), with the constitutive relations (1.48) inserted,

$$\begin{bmatrix} i\omega\varepsilon & i\mathbf{k}\times\mathbf{I} \\ -i\mathbf{k}\times\mathbf{I} & i\omega\mu_0\mathbf{I} \end{bmatrix} \begin{bmatrix} \mathbf{E} \\ \mathbf{H} \end{bmatrix} = 0 \tag{1.104}$$

for a slowly varying plasma, the two systems coincide when we put

$$\mathbf{k} = \mathbf{k}_t + \kappa\hat{\mathbf{z}} \tag{1.105}$$

In terms of the refractive index vector $\mathbf{n}:=c\mathbf{k}/\omega$ (1.50) this reads

$$\mathbf{n} = \mathbf{n}_t + q\hat{\mathbf{z}}, \qquad n^2 = n_t{}^2 + q^2 \qquad \text{with} \qquad q := \frac{c\kappa}{\omega} \tag{1.106}$$

and the dispersion equation

$$\det\left[\frac{c}{\omega}\mathbf{C} - q\mathbf{U}_z\right] = 0 \tag{1.107}$$

i.e. the solubility condition for the 6×6 system (1.101) may therefore be deduced from the dispersion equation, (1.51) and (1.54), if we put

$$\begin{aligned} n^2\cos^2\theta &= (\mathbf{n}\cdot\hat{\mathbf{b}})^2 = (\mathbf{n}_t\cdot\hat{\mathbf{b}} + q\hat{b}_z)^2 \\ &= (\mathbf{n}_t\cdot\hat{\mathbf{b}})^2 + 2q(\mathbf{n}_t\cdot\hat{\mathbf{b}})\hat{b}_z + q^2\hat{b}_z{}^2 \end{aligned} \tag{1.108}$$

The resulting equation is the dispersion quartic [33, eqs. (6.15) and (6.17)]:

$$\begin{aligned} & q^4\left[S + (P-S)\hat{b}_z{}^2\right] \\ & +2q^3(P-S)\hat{b}_z(\mathbf{n}_t\cdot\hat{\mathbf{b}}) \\ & +q^2\left\{2Sn_t{}^2 + (P-S)(\mathbf{n}_t\cdot\hat{\mathbf{b}})^2 + \left[RL - SP + (P-S)n_t{}^2\right]\hat{b}_z{}^2 - (RL+SP)\right\} \\ & +2q\left[RL - SP + (P-S)n_t{}^2\right]\hat{b}_z(\mathbf{n}_t\cdot\hat{\mathbf{b}}) \\ & +Sn_t{}^4 - (RL+SP)n_t{}^2 + \left[RL - SP + (P-S)n_t{}^2\right](\mathbf{n}_t\cdot\hat{\mathbf{b}})^2 + RLP \\ & = 0 \end{aligned} \tag{1.109}$$

which, for a (one-species) electron plasma, is just the quartic equation derived by Booker (1936) [22, eq. 7].

1.4 Coupled differential equations for the tangential components of the wave field

To obtain a 4×4 system of differential equations for the four tangential components $\mathbf{E}_t$ and $\mathbf{H}_t$ of the wave field, we eliminate the two normal components E_z and H_z from the 6×6 system (1.92). Scalar multiplication of the second equality in (1.91) with $\hat{\mathbf{z}}$ yields

$$-i\hat{\mathbf{z}} \cdot \mathbf{k}_t \times \mathbf{E} + i\omega\mu_0 H_z = -J_{mz}$$

We introduce the normalized magnetic wave-field vector $\mathcal{H} := \sqrt{\mu_0/\varepsilon_0}\,\mathbf{H}$, as in (1.83). Then, with the refractive index vector $\hat{\mathbf{n}}{:=}c\mathbf{k}/\omega$ (1.50), we get

$$\mathcal{H}_z = (\hat{\mathbf{z}} \times \mathbf{n}_t) \cdot \mathbf{E}_t + i\frac{c}{\omega} J_{mz} \tag{1.110}$$

We now partition the permittivity tensor $\boldsymbol{\varepsilon}$ (1.33),

$$\boldsymbol{\varepsilon} = \begin{bmatrix} \varepsilon_{tt} & : & \vec{\varepsilon}_{tz} \\ \cdots & \cdot & \cdots \\ \vec{\varepsilon}_{zt}^{\,T} & : & \varepsilon_{zz} \end{bmatrix} \tag{1.111}$$

cf. [53, eq. 8.2.(2c)], with

$$\varepsilon_{tt} := \begin{bmatrix} \varepsilon_{xx} & \varepsilon_{xy} \\ \varepsilon_{yx} & \varepsilon_{yy} \end{bmatrix}, \quad \vec{\varepsilon}_{tz} := \begin{bmatrix} \varepsilon_{xz} \\ \varepsilon_{yz} \end{bmatrix}, \quad \vec{\varepsilon}_{zt}^{\,T} := \begin{bmatrix} \varepsilon_{zx} \\ \varepsilon_{zy} \end{bmatrix}^T = [\varepsilon_{zx}\,,\,\varepsilon_{zy}] \tag{1.112}$$

For the sake of clarity in the subsequent discussion we write these column or square matrices in their explicit vector-dyadic representation:

$$\begin{aligned} \vec{\varepsilon}_{tz} &:= (\hat{\mathbf{x}}\varepsilon_{xz} + \hat{\mathbf{y}}\varepsilon_{yz})\,, \qquad \vec{\varepsilon}_{zt} := (\varepsilon_{zx}\hat{\mathbf{x}} + \varepsilon_{zy}\hat{\mathbf{y}}) \\ \varepsilon_{tt} &:= \hat{\mathbf{x}}\varepsilon_{xx}\hat{\mathbf{x}}^T + \hat{\mathbf{x}}\varepsilon_{xy}\hat{\mathbf{y}}^T + \hat{\mathbf{y}}\varepsilon_{yx}\hat{\mathbf{x}}^T + \hat{\mathbf{y}}\varepsilon_{yy}\hat{\mathbf{y}}^T \end{aligned} \tag{1.113}$$

in order to unravel expressions we shall encounter presently containing terms like $\hat{\mathbf{z}} \times \vec{\varepsilon}_{tz}\vec{\varepsilon}_{zt}^{\,T}$ or $\hat{\mathbf{z}} \times \varepsilon_{tt}$,

$$\begin{aligned} \hat{\mathbf{z}} \times \vec{\varepsilon}_{tz}\vec{\varepsilon}_{zt}^{\,T} &= \hat{\mathbf{z}} \times (\hat{\mathbf{x}}\varepsilon_{xz} + \hat{\mathbf{y}}\varepsilon_{yz})\left(\varepsilon_{zx}\hat{\mathbf{x}}^T + \varepsilon_{zy}\hat{\mathbf{y}}^T\right) \\ &= (\hat{\mathbf{y}}\varepsilon_{xz} - \hat{\mathbf{x}}\varepsilon_{yz})\left(\varepsilon_{zx}\hat{\mathbf{x}}^T + \varepsilon_{zy}\hat{\mathbf{y}}^T\right) \\ &\rightarrow \begin{bmatrix} -\varepsilon_{yz} \\ \varepsilon_{xz} \end{bmatrix} [\varepsilon_{zx}\,,\,\varepsilon_{zy}] = \begin{bmatrix} -\varepsilon_{yz}\varepsilon_{zx} & -\varepsilon_{yz}\varepsilon_{zy} \\ \varepsilon_{xz}\varepsilon_{zx} & \varepsilon_{xz}\varepsilon_{zy} \end{bmatrix} \end{aligned} \tag{1.114}$$

or

$$\begin{aligned}\hat{\mathbf{z}} \times \boldsymbol{\varepsilon}_{tt} &= \hat{\mathbf{z}} \times \left(\hat{\mathbf{x}}\varepsilon_{xx}\hat{\mathbf{x}}^T + \hat{\mathbf{x}}\varepsilon_{xy}\hat{\mathbf{y}}^T + \hat{\mathbf{y}}\varepsilon_{yx}\hat{\mathbf{x}}^T + \hat{\mathbf{y}}\varepsilon_{yy}\hat{\mathbf{y}}^T\right) \\ &= \hat{\mathbf{y}}\varepsilon_{xx}\hat{\mathbf{x}}^T + \hat{\mathbf{y}}\varepsilon_{xy}\hat{\mathbf{y}}^T - \hat{\mathbf{x}}\varepsilon_{yx}\hat{\mathbf{x}}^T - \hat{\mathbf{x}}\varepsilon_{yy}\hat{\mathbf{y}}^T \\ &\to \begin{bmatrix} -\varepsilon_{yx} & -\varepsilon_{yy} \\ \varepsilon_{xx} & \varepsilon_{xy} \end{bmatrix}\end{aligned} \tag{1.115}$$

in which we have given the matrix representation in the last line in each case.

Dot multiplication of the first equation in (1.91) with $\sqrt{\mu_0/\varepsilon_0}\,\hat{\mathbf{z}}$ leads to

$$-i\hat{\mathbf{z}} \cdot (\mathbf{k}_t \times \mathcal{H}) - i\frac{\omega}{c\varepsilon_0}(\vec{\varepsilon}_{zt} \cdot \mathbf{E}_t + \varepsilon_{zz}E_z) = \sqrt{\mu_0/\varepsilon_0}\, J_{ez}$$

and solving for E_z yields, cf. [53, eq. 8.2.(3a)],

$$E_z = -\frac{1}{\varepsilon_{zz}}\left(\vec{\varepsilon}_{zt} \cdot \mathbf{E}_t + \varepsilon_0\hat{\mathbf{z}} \times \mathbf{n}_t \cdot \mathcal{H}_t\right) + \frac{i}{\omega\varepsilon_{zz}}J_{ez} \tag{1.116}$$

Having expressed the normal components, E_z and $\mathcal{H}_z$, in terms of the tangential components, $\mathbf{E}_t$ and $\mathcal{H}_t$, we should like to transform the differential operator $(\hat{\mathbf{z}} \times \mathbf{I}\, d/dz)$ into the operator $(\mathbf{I} - \hat{\mathbf{z}}\hat{\mathbf{z}}^T)d/dz$, using the transverse projector (see the equations following (A.20) in Appendix A.2)

$$-(\hat{\mathbf{z}} \times \mathbf{I})(\hat{\mathbf{z}} \times \mathbf{I}) = \mathbf{I} - \hat{\mathbf{z}}\hat{\mathbf{z}}^T$$

where, typically

$$\left[\mathbf{I} - \hat{\mathbf{z}}\hat{\mathbf{z}}^T\right]\mathcal{H} = \mathcal{H}_t$$

This can be achieved by cross multiplication of the equations in (1.91) with $-\hat{\mathbf{z}}$, and substitution of $\mathcal{H}=\sqrt{\mu_0/\varepsilon_0}\,\mathbf{H}$, to give

$$\begin{aligned}\frac{d\mathcal{H}_t}{dz} + i\mathbf{k}_t\mathcal{H}_z + i\frac{\omega}{c\varepsilon_0}\left\{(\hat{\mathbf{z}} \times \boldsymbol{\varepsilon}_{tt}) \cdot \mathbf{E}_t + \hat{\mathbf{z}} \times \vec{\varepsilon}_{tz}E_z\right\} &= -\hat{\mathbf{z}} \times Z_0\,\mathbf{J}_{et} \\ \frac{d\mathbf{E}_t}{dz} + i\mathbf{k}_tE_z - i\frac{\omega}{c}\hat{\mathbf{z}} \times \mathcal{H}_t &= \hat{\mathbf{z}} \times \mathbf{J}_{mt}\end{aligned} \tag{1.117}$$

with $Z_0{:=}\sqrt{\mu_0/\varepsilon_0}$. Substituting the expressions (1.110) and (1.116) for the normal components, E_z and $\mathcal{H}_z$, into these equations we obtain, cf. [53, problem 8.1],

$$\frac{d\mathbf{E}_t}{dz} - i\frac{\omega}{c}\mathbf{n}_t\frac{\vec{\varepsilon}_{zt}^{\,T}}{\varepsilon_{zz}}\mathbf{E}_t - i\frac{\omega}{c}\left(\frac{\varepsilon_0}{\varepsilon_{zz}}\mathbf{n}_t(\hat{\mathbf{z}} \times \mathbf{n}_t)^T + \hat{\mathbf{z}} \times \mathbf{I}\right)\mathcal{H}_t - \frac{\mathbf{n}_t}{c\varepsilon_{zz}}J_{ez} = \hat{\mathbf{z}} \times \mathbf{J}_{mt} \tag{1.118}$$

$$\frac{d\mathcal{H}_t}{dz} + i\frac{\omega}{c}\left\{\mathbf{n}_t(\hat{\mathbf{z}}\times\mathbf{n}_t)^T + \hat{\mathbf{z}}\times\left(\frac{\varepsilon_{tt}}{\varepsilon_0} - \frac{\vec{\varepsilon}_{tz}\vec{\varepsilon}_{zt}^{\,T}}{\varepsilon_0\varepsilon_{zz}}\right)\right\}\mathbf{E}_t - i\frac{\omega}{c}\frac{\hat{\mathbf{z}}\times\vec{\varepsilon}_{tz}}{\varepsilon_{zz}}(\hat{\mathbf{z}}\times\mathbf{n}_t)^T\mathcal{H}_t$$
$$-\mathbf{n}_t J_{mz} = -\hat{\mathbf{z}}\times Z_0\mathbf{J}_{et} \tag{1.119}$$

This can be written as a 4×4 system for the four tangential components, $\mathbf{E}_t$ and $\mathcal{H}_t$, in the form [39]

$$\frac{d}{dz}\begin{bmatrix}\mathbf{E}_t\\ \mathcal{H}_t\end{bmatrix} + i\frac{\omega}{c}\mathsf{N}^{(4)}\begin{bmatrix}\mathbf{E}_t\\ \mathcal{H}_t\end{bmatrix} = \begin{bmatrix}0 & \hat{\mathbf{z}}\times\mathsf{I}\\ -\hat{\mathbf{z}}\times\mathsf{I} & 0\end{bmatrix}\begin{bmatrix}Z_0\mathbf{J}_{et}\\ \mathbf{J}_{mt}\end{bmatrix} + \begin{bmatrix}\frac{\mathbf{n}_t}{c\varepsilon_{zz}}J_{ez}\\ \mathbf{n}_t J_{mz}\end{bmatrix} \tag{1.120}$$

with the 4×4 matrix $\mathsf{N}^{(4)}$ given by

$$\mathsf{N}^{(4)} := \begin{bmatrix} -\mathbf{n}_t\frac{\vec{\varepsilon}_{zt}}{\varepsilon_{zz}} & -\hat{\mathbf{z}}\times\mathsf{I} + \frac{\varepsilon_0}{\varepsilon_{zz}}\mathbf{n}_t(\mathbf{n}_t\times\hat{\mathbf{z}})^T \\ -\hat{\mathbf{z}}\times\left(\frac{\vec{\varepsilon}_{tz}\vec{\varepsilon}_{zt}^{\,T}}{\varepsilon_0\varepsilon_{zz}} - \frac{\varepsilon_{tt}}{\varepsilon_0}\right) - \mathbf{n}_t(\mathbf{n}_t\times\hat{\mathbf{z}})^T & \frac{(\hat{\mathbf{z}}\times\vec{\varepsilon}_{tz})(\mathbf{n}_t\times\hat{\mathbf{z}})^T}{\varepsilon_{zz}} \end{bmatrix} \tag{1.121}$$

Eq. (1.121) simplifies if the x-axis is chosen to lie in the plane spanned by $\hat{\mathbf{z}}$ (normal to the stratification) and $\hat{\mathbf{n}}$, the direction of the wave vector ($\mathbf{k}=k\,\hat{\mathbf{n}}$), so that with Snell's law (1.90)

$$\mathbf{n}_t = s\,\hat{\mathbf{x}} \qquad \text{where} \qquad s := \sin\theta \tag{1.122}$$

With the aid of (1.114) and (1.115), eq. (1.121) becomes

$$\mathsf{N}^{(4)} = \frac{1}{\varepsilon_{zz}}\begin{bmatrix} -s\varepsilon_{zx} & -s\varepsilon_{zy} & 0 & \varepsilon_{zz} - s^2\varepsilon_0 \\ 0 & 0 & -\varepsilon_{zz} & 0 \\ \varepsilon_{yz}\varepsilon_{zx} - \varepsilon_{yx}\varepsilon_{zz} & -\varepsilon_{yy}\varepsilon_{zz} + \varepsilon_{yz}\varepsilon_{zy} + s^2\varepsilon_{zz} & 0 & s\varepsilon_{yz} \\ \varepsilon_{xx}\varepsilon_{zz} - \varepsilon_{xz}\varepsilon_{zx} & -\varepsilon_{xz}\varepsilon_{zy} + \varepsilon_{xy}\varepsilon_{zz} & 0 & -s\varepsilon_{xz} \end{bmatrix} \tag{1.123}$$

If the medium is source free, i.e. if the currents $\mathbf{J}_\mathrm{e}$ and $\mathbf{J}_\mathrm{m}$ are everywhere zero, then (1.120) becomes

$$\frac{d}{dz}\mathbf{e}^{(4)} + ik_0\mathsf{N}^{(4)}\mathbf{e}^{(4)} = 0\,, \qquad k_0 := \frac{\omega}{c} \tag{1.124}$$

with

$$\mathbf{e}^{(4)} := \begin{bmatrix}\mathbf{E}_t\\ \mathcal{H}_t\end{bmatrix}, \qquad \mathcal{H}_t := \sqrt{\frac{\mu_0}{\varepsilon_0}}\mathbf{H}_t \tag{1.125}$$

If the sign of any one of the four tangential components of $\mathbf{E}_t$, $\mathcal{H}_t$ in $\mathbf{e}^{(4)}$ is changed by multiplication of $\mathbf{e}^{(4)}$ by one of the four matrices

$$\mathsf{P}^{(4)} := \begin{bmatrix} -1 & 0 & 0 & 0 \\ 0 & 1 & 0 & 0 \\ 0 & 0 & 1 & 0 \\ 0 & 0 & 0 & 1 \end{bmatrix}, \begin{bmatrix} 1 & 0 & 0 & 0 \\ 0 & -1 & 0 & 0 \\ 0 & 0 & 1 & 0 \\ 0 & 0 & 0 & 1 \end{bmatrix}, \begin{bmatrix} 1 & 0 & 0 & 0 \\ 0 & 1 & 0 & 0 \\ 0 & 0 & -1 & 0 \\ 0 & 0 & 0 & 1 \end{bmatrix}$$

$$\text{or} \quad \begin{bmatrix} 1 & 0 & 0 & 0 \\ 0 & 1 & 0 & 0 \\ 0 & 0 & 1 & 0 \\ 0 & 0 & 0 & -1 \end{bmatrix} = \left[\mathsf{P}^{(4)}\right]^{-1} \tag{1.126}$$

then the signs of the corresponding row and column of the matrix $\mathsf{N}^{(4)}$, (1.121) or (1.123), must be changed accordingly. The resultant 4×1 system

$$\frac{d\mathbf{g}}{dz} + ik_0\mathsf{T}\mathbf{g} = 0\,, \qquad \mathbf{g} := \mathsf{P}^{(4)}\mathbf{e}^{(4)} = \begin{bmatrix} E_x \\ -E_y \\ \mathcal{H}_x \\ \mathcal{H}_y \end{bmatrix}, \qquad \mathsf{T} := \mathsf{P}^{(4)}\mathsf{N}^{(4)}\mathsf{P}^{(4)} \tag{1.127}$$

was first derived by Clemmow and Heading [39, eq. 16] using the second matrix in (1.126) for $\mathsf{P}^{(4)}$. The *propagation matrix* T, cf. $\mathsf{N}^{(4)}$ in (1.123), is then

$$\mathsf{T} = \frac{1}{\varepsilon_{zz}} \begin{bmatrix} -s\varepsilon_{zx} & s\varepsilon_{zy} & 0 & \varepsilon_{zz} - s^2\varepsilon_0 \\ 0 & 0 & \varepsilon_{zz} & 0 \\ \varepsilon_{yz}\varepsilon_{zx} - \varepsilon_{yx}\varepsilon_{zz} & \varepsilon_{yy}\varepsilon_{zz} - \varepsilon_{yz}\varepsilon_{zy} - s^2\varepsilon_{zz} & 0 & s\varepsilon_{yz} \\ \varepsilon_{xx}\varepsilon_{zz} - \varepsilon_{xz}\varepsilon_{zx} & \varepsilon_{xz}\varepsilon_{zy} - \varepsilon_{xy}\varepsilon_{zz} & 0 & -s\varepsilon_{xz} \end{bmatrix} \tag{1.128}$$

cf. [33, eq. (7.82)]. Eq. (1.127) was derived also by Rawer and Suchy [105, eq. (13.9)] using the third matrix in (1.126) for $\mathsf{P}^{(4)}$. All four propagation matrices T (1.127) are transposed with respect to their trailing diagonals when the direction of the external magnetic field is reversed, $\mathbf{b} \to -\mathbf{b}$, since $\boldsymbol{\varepsilon}^T(\mathbf{b}) = \boldsymbol{\varepsilon}(-\mathbf{b})$ (1.33). This result is used in Sec. 3.2.1 to prove the property of modal biorthogonality required in the derivation of the eigenmode scattering theorem.

Eigenvalues and eigenvectors of the propagation matrix T

Suppose we have solved the characteristic equation for T,

$$[\mathsf{T} - q_\alpha \mathsf{I}]\mathbf{g}_\alpha = 0\,, \qquad \det[\mathsf{T} - q_\alpha \mathsf{I}] = 0 \tag{1.129}$$

We construct the eigenmode matrix $\mathbf{G}$ from the four normalized eigenvectors $\hat{\mathbf{g}}_\alpha$:

$$\mathbf{G} := [\hat{\mathbf{g}}_1\; \hat{\mathbf{g}}_2\; \hat{\mathbf{g}}_{-1}\; \hat{\mathbf{g}}_{-2}] \tag{1.130}$$

in which we assume that the normalization of the eigenvectors has been performed in some systematic way, e.g. by equating the component $E_{\alpha x}$ in each $\mathbf{g}_\alpha$ to unity. If G is not singular then T is diagonalized by G,

$$\mathsf{G}^{-1}\mathsf{T}\mathsf{G} = \begin{bmatrix} q_1 & \cdot & \cdot & \cdot \\ \cdot & q_2 & \cdot & \cdot \\ \cdot & \cdot & q_{-1} & \cdot \\ \cdot & \cdot & \cdot & q_{-2} \end{bmatrix} =: \mathsf{Q} \tag{1.131}$$

Now an arbitrary wave field $\mathbf{g}$ may be decomposed into the four eigenvectors $\mathbf{g}_\alpha$ by means of the transformation

$$\boldsymbol{a} = \mathsf{G}^{-1}\mathbf{g}, \qquad \mathbf{g} = \mathsf{G}\boldsymbol{a} = \sum_\alpha a_\alpha \hat{\mathbf{g}}_\alpha = \sum_\alpha \mathbf{g}_\alpha \tag{1.132}$$

Substituting $\mathbf{g}$ from (1.132) into (1.127), and assuming that the medium is homogeneous, i.e. that T is constant, we get with the aid of (1.131)

$$\mathsf{G}\boldsymbol{a}' = -ik_0\mathsf{T}\mathsf{G}\boldsymbol{a}, \qquad \boldsymbol{a}' = -ik_0\mathsf{Q}\boldsymbol{a} \tag{1.133}$$

Using (1.131) and (1.132) we find that the solutions of (1.133) are indeed

$$a_\alpha \sim \exp(-ik_0 q_\alpha z), \quad \mathbf{g}_\alpha(z) = \mathbf{g}_\alpha(0)\exp(-ik_0 q_\alpha z), \quad \alpha = \pm1, \pm2 \tag{1.134}$$

The eigenvectors $\hat{\mathbf{g}}_\alpha$ of T thus represent the characteristic wave fields or eigenmodes of the medium, and a_α are the modal amplitudes in the eigenmode decomposition. The eigenvalues q_α are the roots of the Booker quartic, $\det[\mathsf{T} - q_\alpha \mathsf{I}] = 0$ (1.129), derived already in Sec. 1.3, eq. (1.109).

1.5 Numerical methods of solution

1.5.1 Motivation and background

Any numerical method that calculates the outgoing eigenmode wave fields or amplitudes produced by a set of incoming eigenmodes incident on a plane-stratified medium, is based on a set of governing equations. These may be a system of differential equations that will be integrated numerically through the medium, or a set of matrix relations that are recursively modified as additional layers are added to a plane-stratified slab until the entire medium is reconstructed. In either case the symmetries of the scattering relations between incoming and outgoing eigenmodes must be contained in the symmetries inherent in the governing equations. Such symmetries in the scattering relations, which we shall call *scattering theorems*, were found on several occasions in computer outputs, and the analytical proofs were then sought and found in the governing equations. Some such scattering theorems and their derivation from the governing equations are discussed in Chaps. 2 and 3. In this section those numerical methods are described that are relevant to, and serve as a basis for the later discussion. Our survey is far from exhaustive, but we shall attempt to present an overall view of the problems and the main lines of development in this field. An excellent summary of the numerical methods has been given by Budden [33, Chap. 18].

The need for reflection and transmission coefficients

Until the mid-fifties the main motivation for developing numerical methods for solving the equations governing radio-wave propagation in the plane-stratified ionospheric magnetoplasma, was to produce a set of reflection coefficients for plane waves having arbitrary directions of incidence on the ionosphere from below. The ionosphere was viewed primarily as a reflecting medium permitting radio communication between stations far beyond the line of sight. Ray tracing methods, refined possibly by phase integral calculations near zeros or branch points of the complex refractive indices (reflection or coupling points) where ray methods broke down, yielded satisfactory results at high frequencies ($\gtrsim$ 1MHz). In the low and very low frequency ranges, where the variation in ionospheric parameters within a wavelength in the medium was large, the methods of ray tracing were inappropriate and full-wave solutions were mandatory.

The pioneering work by Budden and his coworkers in the fifties [30,31,21], aimed at developing such techniques, was subject to the severe constraints imposed primarily by the limited computer memory available for storage of the program and intermediate results of computation. Pitteway's full-wave computer program [98], using his penetrating and non-penetrating modes, belongs to this category, and was written for a computer which had no more than 8 (eight!) kbytes of available storage for the program. These first generation computer codes may be regarded as models of carefully thought-out programming, designed to produce accurate results as fast as possible and with maximum economy in computer storage.

The concept of transmission coefficients (${}_{\parallel}T_{\parallel}$, ${}_{\parallel}T_{\perp}$ etc.) for linearly polarized radio waves, parallel ($\parallel$) or perpendicular ($\perp$) to the plane of incidence, that penetrated into 'free space above the ionosphere', was widespread during the 1950's and later. The work of Storey (1953) [114] on whistlers, cf. Secs. 1.2 and 2.1.3, indicated however that the ionosphere extended (unexpectedly) to heights of many thousands of kilometres, and that very low frequency signals could penetrate into the ionosphere, follow the curved geomagnetic field lines to great heights, and finally emerge into free space below the ionosphere in the conjugate hemisphere. To follow the behaviour of such waves one needed to know their transmission coefficients as a function of height *within the ionosphere*, in terms of their upgoing energy flux density (Poynting vector) which is a conserved (constant) quantity in a slowly varying, lossless, plane-stratified medium. This then was the motivation for developing computer programs, such as Pitteway's full-wave method, for the computation of transmission coefficients of whistler-type signals.

1.5.2 The full-wave methods of Budden and Pitteway. The problem of numerical swamping

The methods of Budden [30,31] and Pitteway [98] are based on the numerical integration of four first-order linear differential equations, the Clemmow-Heading coupled wave equations (1.127)

$$\mathrm{g}' = -ik_0\mathsf{T}\mathrm{g}\,, \qquad \mathrm{g} := (E_x\,, -E_y\,, \mathcal{H}_x\,, \mathcal{H}_y\,)\,, \qquad \mathcal{H} := \sqrt{\mu_0/\varepsilon_0}\,\mathrm{H} \quad (1.135)$$

The numerical integration proceeds downwards, starting with two independent upgoing wave fields well above the $X = 1+Y$ reflection level [see eq. (1.66) and Fig. 1.1], with $X := {\omega_p}^2/\omega^2$ and $Y := \omega_c/\omega$ denoting the respective electron plasma- and gyrofrequencies. One wave is essentially the upgoing whistler mode (Fig. 1.1), the wave frequencies considered being well below the electron gyrofrequency, $\omega << \omega_c$, and the other is the evanescent continuation of the mode reflected at the $X = 1+Y$ level. Each solution (wave field) is integrated independently until free space below the ionosphere is reached. The numerical integration is carried out by a Runge-Kutta method in which the derivative g' $(= -ik_0\mathsf{T}\mathrm{g})$ is calculated several times within each integration step δz, and is equivalent to representing the variation of g within the step by a fourth degree polynomial. Below the ionosphere the wave fields are decomposed into two independent sets of upgoing and downgoing (incident and reflected) waves, and suitable combinations of them then give the required reflection coefficients. We consider some of the details.

The starting solutions

Budden's method [30,31] was used initially at a frequency of 16kHz. Above the $X = 1 + Y$ reflection level, with $\omega_p\,, \omega_c >> \omega$, large values of the refractive index n_α are encountered, cf. (1.60) and (1.61), with a negligibly small collision frequency $\nu/\omega =: Z << 1$,

$$n_\alpha{}^2 \approx 1 - \frac{X}{1 \mp Y \cos\theta} \approx \pm \frac{\omega_p{}^2}{\omega\omega_c \cos\theta}, \qquad n_2 = i|n_1| \tag{1.136}$$

where $\cos\theta = \hat{\mathbf{n}} \cdot \hat{\mathbf{b}}$. The wave-normal directions by Snell's law are nearly vertical, so that the (x, y, z) coordinate system used above (see also Sec. 1.3) in which the z-axis is normal to the stratification, coincides with the (ξ, η, ς) system (1.77) in which the ς-axis is along the wave normal. The wave polarizations, (1.63) and (1.81), are given by

$$\rho_\alpha = \frac{E_y}{E_x} = -\frac{\mathcal{H}_x}{\mathcal{H}_y} \approx \mp i, \qquad \alpha = 1, 2 \tag{1.137}$$

so that

$$\begin{aligned} \mathbf{g}_1 &:= (E_x, -E_y, \mathcal{H}_x, \mathcal{H}_y)_{(\alpha=1)} = (1, i, in_1, n_1)E_{x,1} \\ \mathbf{g}_2 &:= (1, -i, -in_2, n_2)E_{x,2} \end{aligned} \tag{1.138}$$

Although $\mathbf{g}_2$ is not an accurate representation of the upgoing evanescent wave (in view of the approximations made), the required evanescent component will increase exponentially in the downward integration so that any unwanted initial downgoing components will become negligible.

The situation with the other initial solution $\mathbf{g}_1$, the *upgoing* whistler-type wave, is not so simple. Suppose that the initial value, calculated from (1.138), differs from the true value by an amount $\delta\mathbf{g}$. The governing equations automatically decompose any 'error field' $\delta\mathbf{g}$ in the downward integration into a linear superposition of eigenmodes $\delta\mathbf{g}_\alpha$ $(\alpha = \pm 1, \pm 2)$, cf. (1.132),

$$\delta\mathbf{g} = \sum_\alpha \delta\mathbf{g}_\alpha = \sum_\alpha \delta a_\alpha \hat{\mathbf{g}}_\alpha \tag{1.139}$$

where $\hat{\mathbf{g}}_\alpha$ represents the wave field of the normalized eigenmode α, and δa_α the elementary modal amplitude. Thus a *downgoing* whistler mode $\delta\mathbf{g}_{-1}$, inter alia, will appear in the initial solution, and will persist in the downward integration into free space below the ionosphere to give a spurious reflected whistler signal. Some *upgoing* evanescent mode $\delta\mathbf{g}_2$ is also introduced via the initial error field, and will grow exponentially in the the downward integration. However, if this solution grows only moderately, as will occur at very low frequencies, the two solutions $\mathbf{g}_1$ and $\mathbf{g}_2$ obtained below the ionosphere will still be linearly independent, and no difficulty is incurred from this source. This is not the case at higher freqencies, as will be seen presently.

A more accurate method for obtaining the starting solutions has been given by Pitteway [98]. The matrix $[\mathbf{T} - \gamma \mathbf{I}]$ is applied r times to an arbitrary wave field $\mathbf{g}$, where γ is a constant to be selected for each initial solution. (The computer is of course already programmed to calculate $\mathbf{Tg}$ for use in the numerical integration.) Since the wave field $\mathbf{g}$ is a superposition of the four eigenmodes (1.132),

$$\mathbf{g} = \sum_{\alpha} a_{\alpha} \hat{\mathbf{g}}_{\alpha}$$

the process yields, by virtue of the eigenmode equation (1.129),

$$[\mathbf{T} - \gamma \mathbf{I}]^r \mathbf{g} = \sum_{\alpha} (q_{\alpha} - \gamma)^r a_{\alpha} \hat{\mathbf{g}}_{\alpha} \tag{1.140}$$

The mode for which $|q_{\alpha} - \gamma|$ is the largest is the one that will survive repeated application of $[T - \gamma \mathbf{I}]$, yielding $\hat{\mathbf{g}}_{\alpha}$ and q_{α}. In the high ionosphere, at low and very low frequencies, the values of q_{α} ($\approx n_{\alpha}$), cf. (1.136), are given approximately by

$$q_1\,,\, q_2\,,\, q_{-1}\,,\, q_{-2} \approx (1\,,\, i\,,\, -1\,,\, -i) n_1\,, \qquad n_1 \approx \frac{\omega_p}{(\omega \omega_c \cos\theta)^{1/2}} \tag{1.141}$$

Thus if the eigenmode $\mathbf{g}_1$ for instance is required, γ will be chosen to lie on the negative real axis, $\gamma \approx -n_1$, so that $|q_{\alpha} - \gamma|$ will be the largest for $q_{\alpha} = q_1$. Similarly, $\mathbf{g}_2$ and q_2 will be generated when $\gamma \approx -in_1$.

The problem of numerical swamping

The error fields in the upgoing whistler-mode ('penetrating') solution that tends to blow up in the downward integration with exponential z-dependence, (1.136) and (1.138),

$$\exp\left(\frac{\omega}{c} n_1 |z - z_0|\right) \approx \exp\left(\frac{\omega^{1/2} \omega_p |z - z_0|}{c(\omega_c \cos\theta)^{1/2}}\right) \tag{1.142}$$

This factor, although not dominant at the very low frequencies (16 kHz) used in the earlier work, becomes a serious problem at frequencies of hundreds of kHz, due to the exponential dependence on $k \sim \lambda^{-1}$ (where λ is the wavelength in the medium) which, as we have just seen, is proportional to to $\omega^{1/2}$ at whistler frequencies. Even if the initial solution $\mathbf{g}_1(z_0)$ is computed exactly, the 'truncation errors' in the downward integration, due to the finite (say, 8-figure) accuracy of the computer, introduce the unwanted exponentially-growing 'error field'. The result is that during the downward integration the first, penetrating solution, which we denote $\mathbf{g}^{(1)}$—during the integration both solutions $\mathbf{g}^{(1)}$ and $\mathbf{g}^{(2)}$ become superpositions of all four eigenmodes $\mathbf{g}_\alpha$—tends to converge towards (to be 'swamped by') the second, evanescently increasing solution and to be lost as an independent solution.

Pitteway [98] solved the problem of numerical swamping by constraining the penetrating solution to be hermitian orthogonal to the originally evanescent, non-penetrating one by adding to it, at regular intervals, an appropriate fraction a of the non-penetrating solution:

$$\mathbf{g}^{(1)} \to \mathbf{g}^{(1)} + a\mathbf{g}^{(2)}, \qquad (\mathbf{g}^{(1)} + a\mathbf{g}^{(2)})^\star \cdot \mathbf{g}^{(2)} = 0, \qquad a^\star = -\frac{\mathbf{g}^{(1)\star} \cdot \mathbf{g}^{(2)}}{\mathbf{g}^{(2)\star} \cdot \mathbf{g}^{(2)}} \tag{1.143}$$

A linear combination of the two independent solutions $\mathbf{g}^{(1)}$ and $\mathbf{g}^{(2)}$ is also a solution, and so $\mathbf{g}^{(2)}$ and the repeatedly adjusted $\mathbf{g}^{(1)}$ remain independent below the ionosphere. Since the consecutive values of a are stored, the initial penetrating solution can be reconstructed after the integration has been completed.

Barron and Budden [21] modified Budden's earlier method described above, by using a 2×2 admittance matrix A, rather than the wave fields, as the independent variable. The elements of A gave the ratios of magnetic to electric wave-field components in the two independent solutions, and consequently yielded slowly varying ratios even for evanescent waves, thus overcoming the problem of numerical swamping. The method was faster than the previous one, but the intermediate values of A could not easily be related to the wave propagation processes within the ionosphere without additional computation. It has been successfully applied to warm plasmas by Budden and Jones [34] to calculate the angular width of the Z-coupling radio window through which the myriametric, non-thermal electromagnetic radiation, generated by intense upper-hybrid electroststic oscillations at the plasmapause, can escape into the magnetospheric cavity [74].

Calculation of reflection and transmission coefficients

In Budden's [30] treatment each solution, $\mathbf{g}^{(1)}$ and $\mathbf{g}^{(2)}$, below the ionosphere is decomposed into up- and downgoing modes, $\mathbf{u}$ and $\mathbf{d}$, parallel ($\parallel$) and perpendicular ($\perp$) to the plane of incidence:

$$\begin{aligned} \mathbf{g}^{(1)} &= \mathbf{u}^{(1)} + \mathbf{d}^{(1)}, \qquad & \mathbf{g}^{(2)} &= \mathbf{u}^{(2)} + \mathbf{d}^{(2)} \\ \mathbf{u} &= \mathbf{u}_\parallel + \mathbf{u}_\perp, \qquad & \mathbf{d} &= \mathbf{d}_\parallel + \mathbf{d}_\perp \end{aligned} \tag{1.144}$$

with z-dependence

$$\mathbf{u} \sim \exp(-ik_0 q_0 z), \qquad \mathbf{d} \sim \exp(ik_0 q_0 z) \tag{1.145}$$

The free-space wave vectors are $\mathbf{k}^{\pm}=k_0(s\,,0\,,\pm q_0)$ with $k_0:=\omega/c$, $s:=\sin\theta$, $q_0:=\cos\theta$ in terms of the angle of incidence θ. The corresponding normalized free-space eigenmodes $\hat{\mathrm{g}}_\alpha(\alpha=\pm1\,,\pm2)$ may then be grouped into a 4×4 modal matrix [33, eq. (11.44)]:

$$\mathbf{G} := [\hat{\mathrm{g}}_1\,\hat{\mathrm{g}}_2\,\hat{\mathrm{g}}_{-1}\,\hat{\mathrm{g}}_{-2}] \equiv \left[\hat{\mathbf{u}}_{\|}\;\hat{\mathbf{u}}_{\perp}\;\hat{\mathbf{d}}_{\|}\;\hat{\mathbf{d}}_{\perp}\right] = \begin{bmatrix} q_0 & 0 & -q_0 & 0 \\ 0 & -1 & 0 & -1 \\ 0 & -q_0 & 0 & q_0 \\ 1 & 0 & 1 & 0 \end{bmatrix} \tag{1.146}$$

remembering that $\mathrm{g}:=(E_x\,,-E_y\,,\mathcal{H}_x\,,\mathcal{H}_y)$ (1.135). The solutions $\mathrm{g}^{(1)}$ and $\mathrm{g}^{(2)}$ may now be decomposed into eigenmodes,

$$\mathrm{g}^{(1)} = \sum_{\alpha=\pm1,\pm2} a_\alpha^{(1)}\hat{\mathrm{g}}_\alpha =: \mathbf{G}a^{(1)}\,, \qquad \mathrm{g}^{(2)} = \mathbf{G}a^{(2)}\,, \qquad a := \begin{bmatrix} a_1 \\ a_2 \\ a_{-1} \\ a_{-2} \end{bmatrix} \equiv \begin{bmatrix} a_{\|+} \\ a_{\perp+} \\ a_{\|-} \\ a_{\perp-} \end{bmatrix} \tag{1.147}$$

with downgoing (reflected) eigenmode amplitudes, $a_{\|-}$ and $a_{\perp-}$, related to the upgoing (incident) amplitudes, $a_{\|+}$ and $a_{\perp+}$, through the reflection matrix $\mathbf{R}$:

$$\begin{bmatrix} a_{\|-}^{(1)} & a_{\|-}^{(2)} \\ a_{\perp-}^{(1)} & a_{\perp-}^{(2)} \end{bmatrix} =: \mathcal{D} =: \mathbf{R}\mathcal{U} \equiv \begin{bmatrix} {}_{\|}R_{\|} & {}_{\|}R_{\perp} \\ {}_{\perp}R_{\|} & {}_{\perp}R_{\perp} \end{bmatrix} \begin{bmatrix} a_{\|+}^{(1)} & a_{\|+}^{(2)} \\ a_{\perp+}^{(1)} & a_{\perp+}^{(2)} \end{bmatrix} \tag{1.148}$$

Since $\mathcal{U}$ is non-singular (the columns of $\mathcal{U}$ represent independent solutions), the reflection matrix $\mathbf{R}$ is then obtained through

$$\mathbf{R} = \mathcal{D}\mathcal{U}^{-1} \tag{1.149}$$

In Pitteway's method the decomposition of the two solutions, $\mathrm{g}^{(1)}$ and $\mathrm{g}^{(2)}$, below the ionosphere into up- and downgoing waves, $\mathbf{u}^{(i)}$ and $\mathbf{d}^{(i)}$, $i=1,2$, proceeds with aid of (1.145) as follows:

$$\begin{aligned} \mathrm{g}^{(i)} &= \mathbf{u}^{(i)} + \mathbf{d}^{(i)} \\ \frac{d\mathrm{g}^{(i)}}{dz} &= -ik_0q_0\left(\mathbf{u}^{(i)} - \mathbf{d}^{(i)}\right) \end{aligned} \tag{1.150}$$

in which $d\mathrm{g}^{(i)}/dz$ (1.135) is known from the numerical integration. $\mathbf{u}^{(i)}$ and $\mathbf{d}^{(i)}$ are then found simply in terms of $\mathrm{g}^{(i)}$ and $d\mathrm{g}^{(i)}/dz$.

Now the upgoing wave $\mathbf{u}^{(2)}$, which becomes the evanescent solution $\mathbf{g}^{(2)}$ in the high ionosphere, is unambiguously defined up to a multiplying constant, and is called the 'non-penetrating mode', n [98], $\mathbf{u}^{(2)} \rightarrow \mathbf{u}_n$, $\mathbf{d}^{(2)} \rightarrow \mathbf{d}_n$. The ratio of the z-components of the downgoing to upgoing (incident) energy flux densities yields the corresponding reflection coefficient. Either of the following expressions may be used

$$|R_n|^2 = \frac{\mathbf{d}_n^\star \cdot \mathbf{d}_n}{\mathbf{u}_n^\star \cdot \mathbf{u}_n} = \frac{(\mathbf{E}_n^-)^\star \cdot \mathbf{E}_n^-}{(\mathbf{E}_n^+)^\star \cdot \mathbf{E}_n^+} \tag{1.151}$$

where $\mathbf{E}_n^\pm$ are the up- and downgoing electric wave vectors, two components of which, $E_x^\pm$ and $E_y^\pm$, are specified in $\mathbf{u}_n$ and $\mathbf{d}_n$, and the third, $E_z^\pm$, is obtained from the orthogonality of $\mathbf{k}^\pm$ and $\mathbf{E}_n^\pm$ in free space:

$$\mathbf{k}^\pm \cdot \mathbf{E}_n^\pm = (s\,,\,0\,,\,q_0) \cdot (E_x^\pm\,,\,E_y^\pm\,,\,E_z^\pm) = 0 \tag{1.152}$$

The upgoing constituent $\mathbf{u}^{(1)}$ of the the solution $\mathbf{g}^{(1)}$, that has been constrained to become hermitian orthogonal to $\mathbf{g}^{(2)}$ at regular intervals, is largely arbitrary. Just as $\mathbf{u}_n \equiv \mathbf{u}^{(2)}$ gives the minimum upward energy flux at great heights, it is appropriate to construct a 'penetrating mode', $\mathbf{u}_p = \mathbf{u}^{(1)} + b\mathbf{u}_n$, below the ionosphere so as to maximize the energy transmission to great heights. In terms of electric wave fields, we seek a constant b such that

$$\mathbf{E}_p^+ \rightarrow \mathbf{E}^{(1)} + b\mathbf{E}_n^+\,, \qquad \text{or} \qquad \mathbf{u}_p \rightarrow \mathbf{u}^{(1)} + b\mathbf{u}_n \tag{1.153}$$

gives maximum energy transmission. This is achieved by imposing again a condition of hermitian orthogonality, cf. (1.143):

$$(\mathbf{E}_p^+)^\star \cdot \mathbf{E}_n^+ = (\mathbf{E}^{(1)^+} + b\mathbf{E}_n^+)^\star \cdot \mathbf{E}_n^+ = 0\,, \qquad b^\star = -\frac{(\mathbf{E}^{(1)^+})^\star \cdot \mathbf{E}_n^+}{(\mathbf{E}_n^+)^\star \cdot \mathbf{E}_n^+} \tag{1.154}$$

This can easily be confirmed by noting that the upward energy flux density of the 'penetrating mode' is proportional to $|\mathbf{E}_n^+|^2$. If we add some non-penetrating mode to it, i.e. $\mathbf{E}_p^+ \rightarrow \mathbf{E}_p^+ + c\mathbf{E}_n^+$, the energy flux at a height z in the high ionosphere is unaffected (since the non-penetrating component is not transmitted), but the upward energy flux at a height z_0 below the ionosphere is proportional to

$$(\mathbf{E}_p^+ + c\mathbf{E}_n^+)^\star \cdot (\mathbf{E}_p^+ + c\mathbf{E}_n^+) = |\mathbf{E}_p^+|^2 + |c\mathbf{E}_n^+|^2 + 2\mathcal{R}e\left[(\mathbf{E}_p^+)^\star \cdot c\mathbf{E}_n^+\right] \tag{1.155}$$

the last term being zero because of (1.154). This is clearly a minimum, (and hence the transmission coefficient, which we denote by $\tau_p(z, z_0)$, a maximum), when c=0. Thus $\tau_p(z, z_0)$ may be calculated from the ratio of the time-averaged Poynting flux densities $\langle S_z(z)\rangle$ and $\langle S_z(z_0)\rangle$ at heights z and z_0 respectively:

$$|\tau_p(z, z_0)|^2 = \frac{\langle S_z(z)\rangle}{\langle \mathrm{S}\rangle z_0}, \qquad \langle S_z(z)\rangle = \mathcal{R}e\,[E_x^{\,*}\mathcal{H}_y \;-\; E_y^{\,*}\mathcal{H}_x\,] \tag{1.156}$$

the wave-field components referring either to the upgoing waves in the penetrating mode at z_0 below the ionosphere or to the whistler mode at z in the high ionosphere.

A reciprocity theorem due to Pitteway and Jespersen [100], discussed in Sec. 3.2.4, shows that the transmission coefficient for the upgoing penetrating mode in a given incident direction equals the transmission coefficient for the downgoing whistler mode in a symmetrically related direction.

1.5.3 Methods using discrete homogeneous strata

In the 1960's, with the advent of faster computers with much larger storage capacities, the Runge-Kutta methods for numerical integration of the differential equations were replaced to a large extent by matrix multiplication techniques. In the method of Price [102], and the earlier and essentially equivalent method of Johler and Harper [73], the ionosphere was divided stepwise into thin homogeneous strata, and transfer or 'propagator' matrices were used to compute the *amplitudes* of eigenmodes at each successive interface by means of recursive matrix multiplication. A more efficient technique, the matrizant method (discussed in Sec. 1.5.5), developed by Keller and Keller [78] and Volland [128, 129, 130], permitted the use of a much larger step size by taking into account the inhomogeneity of each layer. Since eigenmode amplitudes rather than wave fields were the dependent variables in both methods, numerical swamping was less of a problem in that numerical truncation of the progressive wave amplitude no longer introduced errors that grew exponentially. In a matrix multiplication method due to Nagano et al. [95], on the other hand, in which the wave fields were the dependent variables, the ensuing swamping difficulty was solved by means of repeated Gram-Schmidt orthogonalization of the two solutions during the recursive matrix multiplication, as in Pitteway's method described in the previous section.

In the thin-layer scattering-matrix technique of Altman and Cory [3,4] in which the reflection and transmission coeficients (i.e. the ratio of the wave fields) were the dependent variables, numerical swamping was avoided in the same way as with the admittance matrix method of Barron and Budden [21].

The methods using discrete homogeneous strata, and specifically the propagator (transfer matrix) and the scattering matrix techniques, had a number of features in common. The eigenvalues q_α (roots of the Booker quartic) and eigenvectors $\mathbf{g}_\alpha$ (the characteristic modal polarizations) were calculated in each layer, and continuity of the tangential wave-field components across each interface then yielded the new eigenmode amplitudes (in Price's propagator-matrix method) or the 2×2 interface reflection and transmission matrices (in the scattering-matrix method). In both methods each layer, of thickness δz, was 'traversed' by means of a phase matrix $\boldsymbol{\Delta}$ that multiplied the amplitude of each eigenmode α by a complex phase factor $\exp(-ik_0 q_\alpha \delta z)$, and the process was then repeated at each successive layer.

Let the eigenmode wave fields be represented, as in (1.127), by the 4-vector

$$\mathbf{g}_\alpha := (E_x\,,\, -E_y\,,\, \mathcal{H}_x\,,\, \mathcal{H}_y\,)_\alpha\,, \qquad \alpha = \pm 1, \pm 2$$

the sign of α indicating the direction of propagation with respect to the z-axis, normal to the stratification. The normalized eigenfields are given by

$$\hat{\mathbf{g}}_\alpha := (1\,,\, -E_y/E_x\,,\, \mathcal{H}_x\,/E_x\,,\, \mathcal{H}_y\,/E_x)_\alpha \tag{1.157}$$

Continuity of the tangential wave-field components across an interface separating layers $(\nu - 1)$ and ν at $z = z_{\nu-1}$ takes the form

$$\sum_\alpha a_\alpha^{(\nu-1)} \hat{\mathbf{g}}_\alpha^{\nu-1} = \sum_\alpha a_\alpha^{(\nu)} \hat{\mathbf{g}}_\alpha^{\nu} \quad \text{or} \quad \mathbf{G}^{\nu-1}\boldsymbol{a}^{\nu-1} = \mathbf{G}^{\nu}\boldsymbol{a}^{\nu} \tag{1.158}$$

where a_α is the amplitude of the eigenmode $\mathbf{g}_\alpha$, and

$$\mathbf{G} := [\hat{\mathbf{g}}_1\, \hat{\mathbf{g}}_2\, \hat{\mathbf{g}}_{-1}\, \hat{\mathbf{g}}_{-2}] =: [\mathbf{G}_+\,,\, \mathbf{G}_-]\,, \qquad \boldsymbol{a} := \begin{bmatrix} a_1 \\ a_2 \\ a_{-1} \\ a_{-2} \end{bmatrix} =: \begin{bmatrix} \boldsymbol{a}_+ \\ \boldsymbol{a}_- \end{bmatrix} \tag{1.159}$$

Consequently, we obtain from (1.158),

$$\boldsymbol{a}^\nu = [\mathbf{G}^\nu]^{-1}\, \mathbf{G}^{\nu-1}\boldsymbol{a}^{\nu-1} \tag{1.160}$$

Alternatively, we could calculate a scattering matrix $\mathbf{S} \equiv \mathbf{S}^{\nu,\nu-1}$ for the interface $(\nu - 1, \nu)$. $\mathbf{S}$ is defined by the relation

$$\begin{bmatrix} \boldsymbol{a}_-^{\nu-1} \\ \boldsymbol{a}_+^{\nu} \end{bmatrix} \equiv \boldsymbol{a}_{out} =: \mathbf{S}\boldsymbol{a}_{in} \equiv \begin{bmatrix} \mathbf{r}_+ & \mathbf{t}_- \\ \mathbf{t}_+ & \mathbf{r}_- \end{bmatrix} \begin{bmatrix} \boldsymbol{a}_+^{\nu-1} \\ \boldsymbol{a}_-^{\nu} \end{bmatrix} \tag{1.161}$$

in which $\mathbf{S}$ has been partitioned into 2×2 interface reflection and transmission matrices, $\mathbf{r}_\pm$ and $\mathbf{t}_\pm$, with the signed subscripts indicating the direction of incidence with respect to the z-axis. Rearranging the terms in (1.158) we obtain, with the aid of (1.159),

$$\mathbf{G}_{out}\boldsymbol{a}_{out} := \left[-\mathbf{G}_-^{\nu-1}, \mathbf{G}_+^{\nu}\right]\begin{bmatrix} \boldsymbol{a}_-^{\nu-1} \\ \boldsymbol{a}_+^{\nu} \end{bmatrix} = \left[\mathbf{G}_+^{\nu-1}, -\mathbf{G}_-^{\nu}\right]\begin{bmatrix} \boldsymbol{a}_+^{\nu-1} \\ \boldsymbol{a}_-^{\nu} \end{bmatrix} =: \mathbf{G}_{in}\boldsymbol{a}_{in} \tag{1.162}$$

Hence, recalling (1.161), we get

$$\boldsymbol{a}_{out} = \mathbf{G}_{out}^{-1}\mathbf{G}_{in}\boldsymbol{a}_{in} = \mathbf{S}\boldsymbol{a}_{in} \tag{1.163}$$

or

$$\mathbf{S} = \mathbf{G}_{out}^{-1}\mathbf{G}_{in} \tag{1.164}$$

so that the interface scattering matrix is determined by the relation between the modal 4-polarizations on both sides of the interface.

Coming back to the propagator formalism, we note that (1.160) gave the eigenmode amplitudes $\boldsymbol{a}^{\nu}(z_{\nu-1})$ just above the interface $(\nu-1,\nu)$ in terms of the amplitudes $\boldsymbol{a}^{\nu-1}(z_{\nu-1})$ just below it. The amplitudes $\boldsymbol{a}^{\nu}(z_{\nu})$ just below the following interface $(\nu,\nu+1)$, i.e. at the upper end of the homogeneous layer ν, whose thickness is δz, become

$$\boldsymbol{a}^{\nu}(z_{\nu}) = \boldsymbol{\Delta}^{\nu}\boldsymbol{a}^{\nu}(z_{\nu-1}) = \boldsymbol{\Delta}^{\nu}[\mathbf{G}^{\nu}]^{-1}\mathbf{G}^{\nu-1}\boldsymbol{a}^{\nu-1}(z_{\nu-1}) =: \mathbf{P}^{\nu}\boldsymbol{a}^{\nu-1}(z_{\nu-1}) \tag{1.165}$$

where the phase matrix $\boldsymbol{\Delta}^{\nu}$ is given by

$$\boldsymbol{\Delta}^{\nu} := \begin{bmatrix} \boldsymbol{\Delta}_+^{\nu} & 0 \\ 0 & \boldsymbol{\Delta}_-^{\nu} \end{bmatrix}, \qquad \boldsymbol{\Delta}_\pm^{\nu} := \begin{bmatrix} \exp(-ik_0q_{\pm 1}\delta z) & 0 \\ 0 & \exp(-ik_0q_{\pm 2}\delta z) \end{bmatrix} \tag{1.166}$$

and $\mathbf{P}^{\nu}$ is the propagator or transfer matrix for the layer ν.

Proceeding from the lowest interface at $z = z_0$ between the medium and free space to any layer $\nu = s$ we get, with the aid of (1.165),

$$\boldsymbol{a}^{s}(z_s) = \mathbf{P}^{s}\mathbf{P}^{s-1}\ldots\mathbf{P}^{1}\boldsymbol{a}^{0}(z_0) =: \mathbf{P}(z_s, z_0)\boldsymbol{a}^{0}(z_0) \tag{1.167}$$

(If the uppermost 'layer' s is a homogeneous infinite half-space, then $\mathbf{P}^{s}$ will of course not contain the phase matrix $\boldsymbol{\Delta}^{s}$.) The propagator $\mathbf{P}$ links four eigenmode amplitudes at one interface to four amplitudes at another, and from it one can extract in principle the four reflection and transmission matrices for the entire slab, $\mathbf{R}_\pm$ and $\mathbf{T}_\pm$, the signed subscripts indicating the direction of incidence, as in (1.161). Suppose that $\mathbf{P}$ is partitioned so that (1.167) can be written in the form

$$\boldsymbol{a}^{s} \equiv \begin{bmatrix} \boldsymbol{a}_+^{s} \\ \boldsymbol{a}_-^{s} \end{bmatrix} = \begin{bmatrix} \mathbf{P}_1 & \mathbf{P}_2 \\ \mathbf{P}_3 & \mathbf{P}_4 \end{bmatrix}\begin{bmatrix} \boldsymbol{a}_+^{0} \\ \boldsymbol{a}_-^{0} \end{bmatrix} \equiv \mathbf{P}\boldsymbol{a}^{0} \tag{1.168}$$

The amplitudes can be regrouped to yield

$$\begin{bmatrix} \boldsymbol{a}_-^0 \\ \boldsymbol{a}_+^s \end{bmatrix} =: \boldsymbol{a}_{out} = \begin{bmatrix} -\mathsf{P}_4^{-1}\mathsf{P}_3 & \mathsf{P}_4^{-1} \\ \mathsf{P}_1 - \mathsf{P}_2\mathsf{P}_4^{-1}\mathsf{P}_3 & \mathsf{P}_2\mathsf{P}_4^{-1} \end{bmatrix} \begin{bmatrix} \boldsymbol{a}_+^0 \\ \boldsymbol{a}_-^s \end{bmatrix}$$
$$=: \mathsf{S}\boldsymbol{a}_{in} \equiv \begin{bmatrix} \mathsf{R}_+ & \mathsf{T}_- \\ \mathsf{T}_+ & \mathsf{R}_- \end{bmatrix} \begin{bmatrix} \boldsymbol{a}_+^0 \\ \boldsymbol{a}_-^s \end{bmatrix} \tag{1.169}$$

An equivalent result has been given by Volland [129, eq. 49], and a somewhat different form is derived later in Sec. 3.3, eq. (3.112).

The drawback of the propagator method is that the physical processes (reflection, mode conversion, etc.) are not apparent from the intermediate computed results. Suppose however that we start the repeated matrix multiplication from the upper end to obtain, by analogy with (1.167),

$$\boldsymbol{a}^r(z_r) = \mathsf{P}^r\mathsf{P}^{r+1}\ldots\mathsf{P}^{s-1}\boldsymbol{a}^s(z_s) = \mathsf{P}\boldsymbol{a}^s(z_s)\,, \qquad r < s$$

and terminate with r=0. Then we could let $\boldsymbol{a}^s(z_s)$ represent two upgoing wave amplitudes at the top of the medium, and try to follow their development backwards. But if one of the waves is strongly evanescent, then at least one of the matrix elements will 'blow up' in the backward iteration. If, on the other hand, we set the upgoing evanescent amplitude to zero, then we get information on the 'penetrating mode' only.

These difficulties are overcome when the reflection and transmission matrices themselves are the dependent variables calculated in the iterative matrix multiplication, as we see in the following section.

The Altman-Cory thin-layer scattering-matrix method

Measurements made in satellites in the 1960's and early 1970's (Injun 3, Alouette 1, OGO-6) revealed many new propagation effects within the ionosphere associated, for instance, with ion-cyclotron whistlers or ion-cutoff whistlers at frequencies below 1kHz. At much higher frequencies, 100–150 kHz, the explanations proposed for phenomena such as the 'coupling echo' [79], [33, Sec. 16.14] and the Z-trace coupling and reflection process had never been confirmed by a detailed computer simulation. The thin-layer scattering-matrix technique developed by Altman and Cory [3,4] was aimed to investigate such problems. The method was later employed by Kennett [80], who derived the governing recursive equations independently by a somewhat different method, to analyse problems of elastic wave propagation in stratified media, and it has since become a widely used technique for solving propagation problems in seismology (see, for instance, Fryer and Frazer [55]).

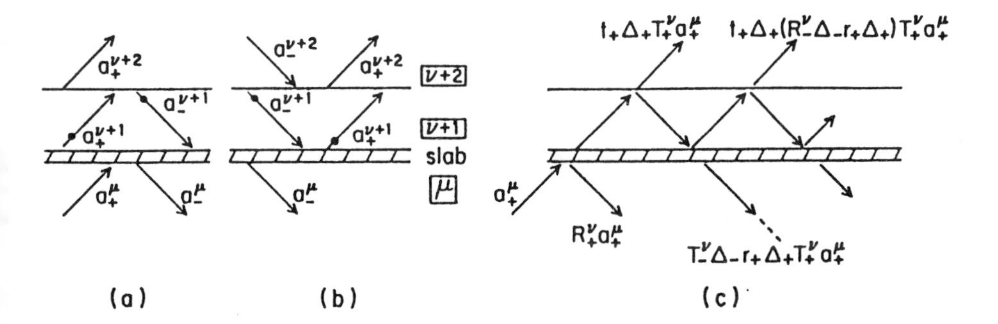

Figure 1.2: Derivation of the recursion relations in the thin-layer scattering-matrix method. (a) Incidence from below. (b) Incidence from above. (c) The multiple reflection derivation.

Consider a multilayer plane-stratified slab imbedded in a plane-stratified multilayer medium. The uppermost layer of the slab is labelled ν, and the following layers outside the slab, $(\nu+1)$ and $(\nu+2)$. The layer just below the slab is labelled μ. Let $\mathbf{R}_\pm^\nu$ and $\mathbf{T}_\pm^\nu$ denote the 2×2 reflection and transmission matrices for the slab, and $\mathbf{R}_\pm^{\nu+1}$ and $\mathbf{T}_\pm^{\nu+1}$ the matrices for the enlarged slab formed by adding the additional layer $(\nu+1)$. The corresponding matrices $\mathbf{r}_\pm$ and $\mathbf{t}_\pm$ for the interfaces between layers $(\nu+1)$ and $(\nu+2)$ can be determined with the aid of (1.164), (1.162) and (1.159), since the eigenvalues q_α and the eigenvectors $\mathbf{g}_\alpha$ are calculated for all layers, as mentioned in the previous section. In order to determine $\mathbf{R}_\pm^{\nu+1}$ and $\mathbf{T}_\pm^{\nu+1}$ when $\mathbf{R}_\pm^\nu$ and $\mathbf{T}_\pm^\nu$ are known, we consider a positive-going wave whose component eigenmodes have amplitudes $\boldsymbol{a}_+^\mu=(\boldsymbol{a}_1^\mu, \boldsymbol{a}_2^\mu)$, incident on the slab from below, see Fig. 1.2(a) in which the slab is represented as an equivalent interface between layers μ and $(\nu+1)$. The up- and downgoing amplitudes at the *bottom* and *top* of the layer $(\nu+1)$ are denoted $\boldsymbol{a}_+^{\nu+1}$ and $\boldsymbol{a}_-^{\nu+1}$ respectively. The phase matrices $\boldsymbol{\Delta}_\pm^{\nu+1}$ (1.166) give the complex phase change of each eigenmode in traversing the layer, but note that the negative-going modes $(\alpha=-1,-2)$ are now followed through the layer in the negative sense, so that δz in the exponent is negative for them. In Fig. 1.2(b) the incident wave is represented by $\boldsymbol{a}_-^{\nu+2}$, and the downgoing transmitted wave below the slab by $\boldsymbol{a}_-^\mu$. Application of of the scattering (reflection and transmission) matrices to each interface (with the slab considered as an equivalent interface) yields the following two sets of equations. For upgoing incidence, Fig. 1.2(a),

$$\begin{aligned} \boldsymbol{a}_+^{\nu+2} &= \mathsf{t}_+\boldsymbol{\Delta}_+\boldsymbol{a}_+^{\nu+1} =: \mathsf{T}_+^{\nu+1}\boldsymbol{a}_+^{\mu} && (a)\\ \boldsymbol{a}_-^{\nu+1} &= \mathsf{r}_+\boldsymbol{\Delta}_+\boldsymbol{a}_+^{\nu+1} && (b)\\ \boldsymbol{a}_+^{\nu+1} &= \mathsf{R}_-^{\nu}\boldsymbol{\Delta}_-\boldsymbol{a}_-^{\nu+1} + \mathsf{T}_+^{\nu}\boldsymbol{a}_+^{\mu} && (c)\\ \boldsymbol{a}_-^{\mu} &= \mathsf{T}_-^{\nu}\boldsymbol{\Delta}_-\boldsymbol{a}_-^{\nu+1} + \mathsf{R}_+^{\nu}\boldsymbol{a}_+^{\mu} =: \mathsf{R}_+^{\nu+1}\boldsymbol{a}_+^{\mu} && (d) \end{aligned} \qquad (1.170)$$

and for downgoing incidence, Fig. 1.2(b),

$$\begin{aligned} \boldsymbol{a}_-^{\mu} &= \mathsf{T}_-^{\nu}\boldsymbol{\Delta}_-\boldsymbol{a}_-^{\nu+1} =: \mathsf{T}_-^{\nu+1}\boldsymbol{a}_-^{\nu+2} && (a)\\ \boldsymbol{a}_+^{\nu+1} &= \mathsf{R}_-^{\nu}\boldsymbol{\Delta}_-\boldsymbol{a}_-^{\nu+1} && (b)\\ \boldsymbol{a}_-^{\nu+1} &= \mathsf{r}_+\boldsymbol{\Delta}_+\boldsymbol{a}_+^{\nu+1} + \mathsf{t}_-\boldsymbol{a}_-^{\nu+2} && (c)\\ \boldsymbol{a}_+^{\nu+2} &= \mathsf{t}_+\boldsymbol{\Delta}_+\boldsymbol{a}_+^{\nu+1} + \mathsf{r}_-\boldsymbol{a}_-^{\nu+2} =: \mathsf{R}_-^{\nu+1}\boldsymbol{a}_-^{\nu+2} && (d) \end{aligned} \qquad (1.171)$$

Elimination of $\boldsymbol{a}_+^{\nu+1}$ and $\boldsymbol{a}_-^{\nu+1}$ from (1.170*b*, *c* and *d*) and from (1.170*a*, *b* and *c*) respectively, yields expressions for $\mathsf{R}_+^{\nu+1}$ and $\mathsf{T}_+^{\nu+1}$:

$$\begin{aligned} \boldsymbol{a}_-^{\mu} &= \left\{\mathsf{R}_+^{\nu} + \mathsf{T}_-^{\nu}\boldsymbol{\Delta}_-\left[\mathsf{I} - \mathsf{r}_+\boldsymbol{\Delta}_+\mathsf{R}_-^{\nu}\boldsymbol{\Delta}_-\right]^{-1}\mathsf{r}_+\boldsymbol{\Delta}_+\mathsf{T}_+^{\nu}\right\}\boldsymbol{a}_+^{\mu} &&=: \mathsf{R}_+^{\nu+1}\boldsymbol{a}_+^{\mu}\\ \boldsymbol{a}_+^{\nu+2} &= \mathsf{t}_+\boldsymbol{\Delta}_+\left[\mathsf{I} - \mathsf{R}_-^{\nu}\boldsymbol{\Delta}_-\mathsf{r}_+\boldsymbol{\Delta}_+\right]^{-1}\mathsf{T}_+^{\nu}\boldsymbol{a}_+^{\mu} &&=: \mathsf{T}_+^{\nu+1}\boldsymbol{a}_+^{\mu} \end{aligned}$$

A corresponding pair of equations for $\mathsf{R}_-^{\nu+1}$ and $\mathsf{T}_-^{\nu+1}$ is obtained from (1.171 *a*–*d*) to give the following recursion equations:

$$\begin{aligned} \mathsf{R}_+^{\nu+1} &= \mathsf{R}_+^{\nu} + \mathsf{T}_-^{\nu}\boldsymbol{\Delta}_-\left[\mathsf{I} - \mathsf{r}_+\boldsymbol{\Delta}_+\mathsf{R}_-^{\nu}\boldsymbol{\Delta}_-\right]^{-1}\mathsf{r}_+\boldsymbol{\Delta}_+\mathsf{T}_+^{\nu}\\ \mathsf{T}_+^{\nu+1} &= \mathsf{t}_+\boldsymbol{\Delta}_+\left[\mathsf{I} - \mathsf{R}_-^{\nu}\boldsymbol{\Delta}_-\mathsf{r}_+\boldsymbol{\Delta}_+\right]^{-1}\mathsf{T}_+^{\nu}\\ \mathsf{R}_-^{\nu+1} &= \mathsf{r}_- + \mathsf{t}_+\boldsymbol{\Delta}_+\left[\mathsf{I} - \mathsf{R}_-^{\nu}\boldsymbol{\Delta}_-\mathsf{r}_+\boldsymbol{\Delta}_+\right]^{-1}\mathsf{R}_-^{\nu}\boldsymbol{\Delta}_-\mathsf{t}_-\\ \mathsf{T}_-^{\nu+1} &= \mathsf{T}_-^{\nu}\boldsymbol{\Delta}_-\left[\mathsf{I} - \mathsf{r}_+\boldsymbol{\Delta}_+\mathsf{R}_-^{\nu}\boldsymbol{\Delta}_-\right]^{-1}\mathsf{t}_- \end{aligned} \qquad (1.172)$$

These equations, in spite of their formidable appearance, are quite simply programmed since they involve only 2×2 matrices. The above derivation is perhaps the most direct, but the original derivation [4] relied on a multiple reflection approach which we shall now briefly describe since it provides physical insight into the constraints imposed on step size in methods based on homogeneous thin strata.

Consider an upgoing wave, represented by $\boldsymbol{a}_+^{\mu}$, incident on the slab from below. The wave is partially reflected ($\rightarrow \mathsf{R}_+^{\nu}\boldsymbol{a}_+^{\mu}$) and partially transmitted with subsequent phase change in traversing the layer ($\rightarrow \boldsymbol{\Delta}_+\mathsf{T}_+^{\nu}\boldsymbol{a}_+^{\mu}$). It is again partially reflected and partially transmitted at the interface (ν+1, ν+2).

This process gives an infinite geometric series of matrix products, representing multiply reflected and transmitted waves, which may be summed to yield the overall reflection and transmission matrices $\mathbf{R}_+^{\nu+1}$ and $\mathbf{T}_+^{\nu+1}$:

$$\begin{aligned}\mathbf{R}_+^{\nu+1} &= \mathbf{R}_+^{\nu} + \mathbf{T}_-^{\nu}\boldsymbol{\Delta}_-\left[\mathbf{I} + (\mathbf{r}_+\boldsymbol{\Delta}_+\mathbf{R}_-^{\nu}\boldsymbol{\Delta}_-) + (\mathbf{r}_+\boldsymbol{\Delta}_+\mathbf{R}_-^{\nu}\boldsymbol{\Delta}_-)^2 + \ldots\right]\mathbf{r}_+\boldsymbol{\Delta}_+\mathbf{T}_+^{\nu} \\ &= \mathbf{R}_+^{\nu} + \mathbf{T}_-^{\nu}\boldsymbol{\Delta}_-\left[\mathbf{I} - \mathbf{r}_+\boldsymbol{\Delta}_+\mathbf{R}_-^{\nu}\boldsymbol{\Delta}_-\right]^{-1}\mathbf{r}_+\boldsymbol{\Delta}_+\mathbf{T}_+^{\nu} \qquad (1.173)\end{aligned}$$

$$\begin{aligned}\mathbf{T}_+^{\nu+1} &= \mathbf{t}_+\boldsymbol{\Delta}_+\left[\mathbf{I} + (\mathbf{R}_-^{\nu}\boldsymbol{\Delta}_-\mathbf{r}_+\boldsymbol{\Delta}_+) + (\mathbf{R}_-^{\nu}\boldsymbol{\Delta}_-\mathbf{r}_+\boldsymbol{\Delta}_+)^2 \ldots\right]\mathbf{T}_+^{\nu} \\ &= \mathbf{t}_+\boldsymbol{\Delta}_+\left[\mathbf{I} - \mathbf{R}_-^{\nu}\boldsymbol{\Delta}_-\mathbf{r}_+\boldsymbol{\Delta}_+\right]^{-1}\mathbf{T}_+^{\nu} \qquad (1.174)\end{aligned}$$

which are just the first two recursion relations found in (1.172). We may similarly derive the other two by considering incidence from above.

The following features of the recursion relations (1.172) should be noted.

- If we are interested only in the overall reflection matrix $\mathbf{R}_+$ of the medium, a *single* recursion relation, the third in (1.172) is all that is needed. This yields ostensibly only $\mathbf{R}_-$ corresponding to incidence from above, but if we reverse the direction of iteration, i.e. start from above and add layers at the lower end, then this gives the reflection coefficient for incidence from below.

- If we are interested in a pair of matrices only, say $\mathbf{R}_+$ and $\mathbf{T}_+$ for incidence from below, then a *pair* of recursion relations—the third and fourth in (1.172)—will suffice, with the iteration starting again from above and the slab becoming progressively thicker from its lower end.

- If, however, we wish to follow the detailed propagation processes in the medium, to follow for instance the development of upgoing waves incident from below, we would require the computed values of $\mathbf{T}_+(z, z_0)$ and $\mathbf{R}_+(z, z_0)$ for increasing z. Then all four recursion relations are needed, with the slab thickness $(z - z_0)$ increasing progressively in the direction of propagation, the first 'slab' in the iteration being just the first interface at the required reference level z_0.

- If we require the *total field* at a given level z, with contributions both from the wave, incident from below, and from downgoing wave fields reflected from above, one should also know the reflection matrix $\mathbf{R}_+(z_H, z)$ for the overlying slab (z_H, z) so that all multiply-reflected waves in the infinitely thin layer separating the overlying and underlying slabs at z may be summed. This method is applied in Sec. 5.3.1 in our discussion of transfer matrices in the multilayer medium, and has been used by Cory et al. [45] to evaluate the total field that would be measured by a rocket launched into the ionosphere. The consecutive values of $\mathbf{R}_+(z_H, z)$ may be found by use of the third recursion relation in

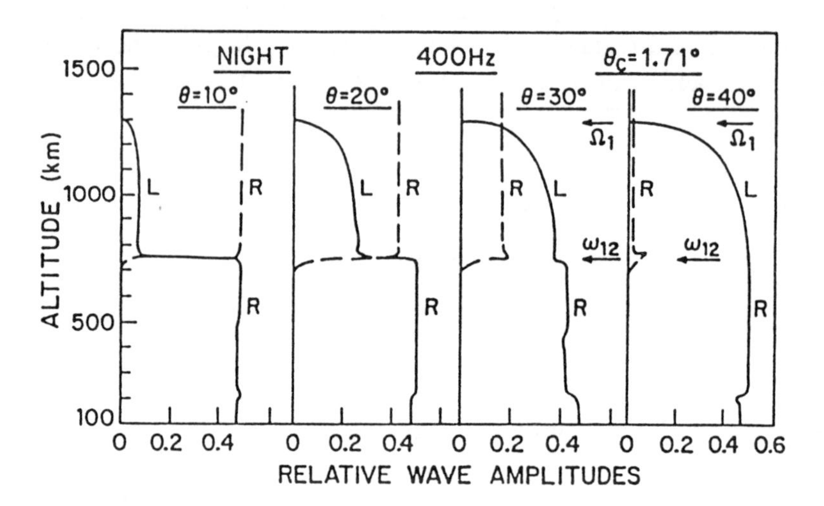

Figure 1.3: Relative wave amplitudes expressed by the transmission matrix elements, T^+_{LR} and T^+_{RR}, in the thin-layer scattering matrix method, illustrating proton whistler formation (*after Fijalkow et al.* [54]).

(1.172) in a single computer run that starts at a high altitude z_H and samples consecutive values of z down to the starting height.

An application of the thin-layer scattering-matrix method due to Fijalkow et al. [54] is shown in Fig. 1.3, in which the mechanism of formation of ion-cyclotron (proton) whistlers is illustrated. The relative wave amplitudes of the upgoing left-polarized (L) ion-cyclotron whistler and a right-polarized (R) electron whistler at a height z, given by the transmission matrix elements $T^+_{LR}(z, z_0)$ and $T^+_{RR}(z, z_0)$ respectively, are shown for a vertically incident upgoing R mode of unit amplitude at a height z_0 below the nighttime ionosphere. Wave amplitudes are normalized to be proportional to the z-component of the Poynting vector. The upgoing L-mode (proton whistler) is seen to be completely absorbed at the Ω_1 level, where the proton gyrofrequency equals the wave frequency. θ is the angle between the external magnetic field b and the (vertical) wave normal. At the *crossover level* [75], [33, Sec. 13.9] marked ω_{12}, the refractive indices of the two modes are equal ($n_L = n_R$), or nearly so, for an appreciable angular range ($0 \leq \theta \leq 20°$ in the model used here). The wave polarizations become equal and real at a critical angle $\theta = \theta_c$ ($\approx 1.7°$ here) and remain real, with polarization reversal as the ω_{12} level is crossed, for all $\theta > \theta_c$. Appreciable intermode coupling may occur here [74], the modal behaviour being closely analogous to that of the magnetoionic modes at the $X = 1$ level (1.66) at low frequencies [6].

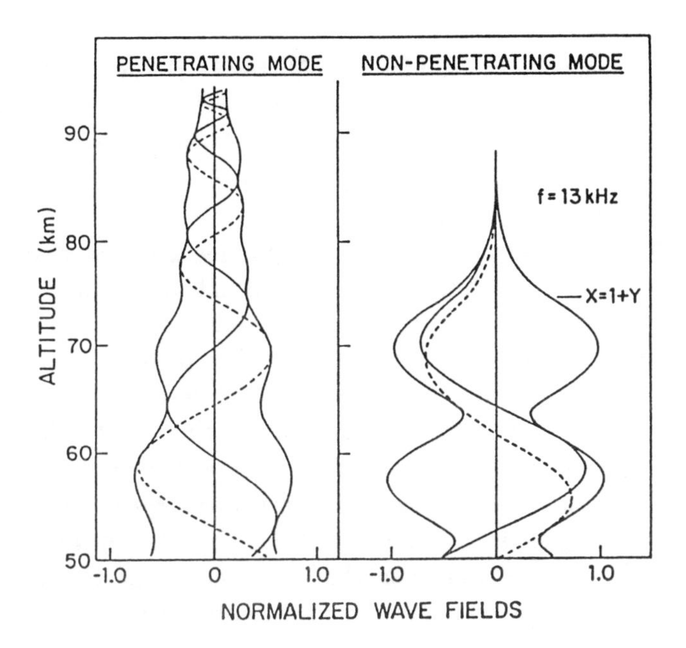

Figure 1.4: The wave-field envelopes of the penetrating and non-penetrating modes in Pitteway's full-wave method, illustrating standing-wave and evanescent structure. The inner curves give instantaneous values of the field at a quarter-period time difference (*after Cory* [44]).

For large values of θ ($\gtrsim 30°$) the electron whistler mode R is seen to pass smoothly, with polarization reversal, into a proton whistler mode L. For small angles, $\theta \lesssim 20°$, 'limiting polarization' conditions exist [6], [33, Secs. 17.10 and 17.11] and the composite, physical wave passes through it with no polarization change, in spite of the modal reshuffling at this level. This computer simulation shows that the modal behaviour at crossover is determined not by the proximity to critical coupling, as is commonly assumed, but by the angular range over which limiting polarization conditions exist.

In Fig. 1.4 a different type of computer output is shown, viz. the wave-field pattern in the ionosphere as given by Pitteway's full-wave method for an upgoing whistler-type wave at a frequency of 13kHz. The computed penetrating and non-penetrating wave outputs [44] show standing wave structure due to the reflected waves, and evanescent wave structure in the non-penetrating upgoing wave due to total reflection at the $X = 1 + Y$ level (1.66).

The problem of step size

In the full-wave methods of Budden and Pitteway the numerical integration must follow the detailed wave-field structure in the ionosphere, so that the number of integration steps required (typically 50 steps per free-space wavelength) is proportional to the frequency, and becomes inconveniently large at medium frequencies($\gtrsim$ 300kHz).

In the homogeneous-strata methods, the influence of the step size is manifested by two effects. The first is the approximation in the complex phase, which could perhaps be remedied if we assumed a linear, or other, variation of q_α within a layer, so that the complex phase would be given by $\exp(-ik_0 \int q_\alpha \, dz)$ in that layer. The other effect is more subtle.

We saw in the multiple-reflection analysis leading to (1.173) and (1.174), that the overall reflection coefficient in a layer depended on the first, third and higher odd-order partial reflections. The transmission coefficients are modified by second and higher even-order reflections. If reflections beyond the first order can be neglected in a thin layer, then the overall transmission coefficient, given by the product of the transmission coefficients of successive interfaces, reduces to a WKB approximation [27], [5, Sec. 2], which is unaffected by further reduction in size of the interval. The same is true of the first-order reflected waves. But if second- and third-order reflections are not negligible, then their phases relative to the zero-order (transmitted) and first-order (reflected) waves will be different from that in two 'half layers'. The error will be largest when the eigenvalues q_α are changing rapidly.

In the scattering-matrix and propagator methods, this may be handled automatically. The sorting and proper ordering of the eigenmodes in the computer relies on the criterion of continuity. If the jump in the complex value of one of the q_α from one layer to the next is too large, then the interval is automatically divided into 5 (or 10) sub-intervals, and the process is repeated. The errors due to large changes in q_α are thereby appreciably reduced, but the upper usable frequency in these mehods has nevertheless been found to be limited to about 500 kHz. The *matrizant methods* described in the following sections are designed to overcome this limitation.

1.5.4 Matrizant methods

The matrizant methods, as developed by Keller and Keller [78] and Volland [128, 129, 130] are based on matrix multiplication as in the propagator method described earlier, but the variation of the parameters of the medium within each elementary layer or interval is now taken into account. In one approach developed by Inoue and Horowitz [71], Rawer and Suchy [105, Sec. 13$\iota, \kappa, \lambda, \mu$]

and employed by Bossy [23], the wave fields at one height are related to those at another height by a 4×4 matrix called a *matrizant*. The matrizant may be expressed as a power series within each layer, if one assumes that the propagation matrix $\mathbf{T}$ in the Clemmow-Heading equations (1.127) varies linearly within the layer [71, eq. (15)], or that it can be represented by a n-th order polynomial [105, Sec. 13]. Insofar as this method uses the wave fields as the independent variables, it is subject to the same problem of numerical swamping encountered previously. Inoue and Horowitz [71] have described a method, differing from that of Pitteway, for solving this problem.

A second approach relies on the development of the matrizant as an infinite series of integrals of increasing multiplicity [57, Vol. 2, Chap. 14, Secs. 5,6,7,8], [78], [129]. In a variant due to Keller and Keller [78], developed with the aim of obtaining a rapidly converging form of the series, the matrizant relates eigenmode amplitudes rather than wave fields, in which the rapid phase variations have been 'transformed out' of the modal amplitudes, but introduced into the integrands of the multiple integrals in each step. This method has been successfully applied by Pitteway and Horowitz [99] to analyze propagation problems at high frequencies (up to 10 MHz), with the same small number of steps required in principle at any frequency. Bossy [25,24] has modified the matrizant method by means of a 'hybrid' technique that overcomes the problem of swamping and imparts greater flexibility to the matrizant method. The matrizants are first calculated for successive intervals using either of the two methods descibed above, and these are then cast into the form of transfer matrices (propagators) that relate the complex eigenmode amplitudes at the two bounding surfaces of each layer. From these the scattering matrices for each interval are derived, as in (1.169), and thence the scattering matrix for a slab that becomes progressively thicker from either end, as in the Altman-Cory method, so that the propagation processes within the medium become immediately apparent from the computer output.

We shall now discuss these methods in detail.

The Inoue-Horowitz power-series development

Inoue and Horowitz [71], as also Volland [129], introduce the matrizant $\mathbf{M}(z,z_\nu)$ within an interval $(z_\nu \leq z \leq z_{\nu+1})$ at the level of the Clemmow-Heading equations

$$\mathbf{e}'(z) = -ik_0\mathbf{T}(z)\mathbf{e}(z) \equiv \boldsymbol{\tau}(z)\mathbf{e}(z)\,, \qquad \mathbf{e} := (E_x\,,\,E_y\,,\,H_x\,,\,H_z) \tag{1.175}$$

The transverse components of $\mathbf{H}$ rather than of $\mathcal{H}$ are used (Z_0 is absorbed into $\mathbf{T}$), but this is of no consequence; $\mathbf{M}(z,z_\nu)$ relates the wave fields $\mathbf{e}(z)$ and

$\mathbf{e}(z_\nu)$ at any two levels, z and z_ν,

$$\mathbf{e}(z) =: \mathbf{M}(z, z_\nu)\mathbf{e}(z_\nu) \tag{1.176}$$

When inserted into (1.175), with z as the independent variable and z_ν fixed, this yields (with the prime denoting d/dz),

$$\mathbf{M}'(z, z_\nu) = -ik_0\mathbf{T}(z)\mathbf{M}(z, z_\nu) \equiv \tau(z)\mathbf{M}(z, z_\nu) \tag{1.177}$$

We now suppose, with Rawer and Suchy [105], that $\tau(z)$ may be expanded as an nth order polynomial within the interval $(z_\nu, z_{\nu+1})$. (This is feasible if the parameters of the medium are known analytic functions of z. Inoue and Horowitz assume linear variation of τ within the interval, using the prescribed values of the parameters to calculate τ at the end points, z_ν and $z_{\nu+1}$.) We suppose also that $\mathbf{M}(z,z_\nu)$ may be expressed as an infinite power series within the interval:

$$\tau(z) = \tau_0 + (z - z_\nu)\tau_1 + (z - z_\nu)^2\tau_2 + \ldots + (z - z_\nu)^n\tau_n \tag{1.178}$$

$$\mathbf{M}(z, z_\nu) = \mathbf{I}^{(4)} + (z - z_\nu)\mathbf{M}_1 + (z - z_\nu)^2\mathbf{M}_2 + \ldots \tag{1.179}$$

with $\mathbf{I}^{(4)}$ denoting the 4×4 unit matrix. Inserting (1.178) and (1.179) into (1.177) we find the following recursion relations for the coefficient matrices in the power-series expansion for $\mathbf{M}(z,z_\nu)$ [71, eq. (15)], [105, eq. (13.36a)]:

$$\mathbf{M}_{r+1} = \frac{1}{r+1}\sum_{s=0}^{n}\tau_s\mathbf{M}_{r-s}, \quad \text{with} \quad \mathbf{M}_0 = \mathbf{I}^{(4)} \quad \text{and} \quad \mathbf{M}_{r-s} = 0 \quad \text{for} \quad s > r \tag{1.180}$$

where n, it will be recalled, is the order of the polynomial approximation for τ (1.178). In practical applications, cf. Bossy [23], the power series (1.179) is truncated when the the largest matrix element in the term $(z_{\nu+1} - z_\nu)^r\mathbf{M}_r$ is less than some predetermined value (say 10^{-8}).

The matrizant $\mathbf{M}(z_\nu,z_0)$, relating fields $\mathbf{e}(z_\nu)$ at an arbitrary height z_ν to those at a fixed reference height z_0, is given finally by the product of the (sub-) matrizants for the elementary intervals,

$$\mathbf{M}(z_\nu, z_0) = \mathbf{M}(z_\nu, z_{\nu-1})\mathbf{M}(z_{\nu-1}, z_{\nu-2})\ldots\mathbf{M}(z_1, z_0) \tag{1.181}$$

Bossy [23], employing analytic ionospheric models and a fifth-order polynomial for τ in each interval, has obtained 7-figure accuracy by this method at a frequency of 100 kHz, using 1 km intervals which are appreciably larger than those that may be used in the homogeneous strata methods. The method is subject to numerical swamping, as pointed out earlier, and one of the motivating factors in its further development was to overcome this difficulty.

We note in conclusion an interesting variant of the above method employed by Inoue and Horowitz [71]. The formal solution of (1.177) is

$$\mathsf{M}(z,z_\nu) = \exp\left(-ik_0 \int_{z_\nu}^{z} \mathsf{T}(z)\, dz\right) \tag{1.182}$$

which may be written in closed form (rather than as an infinite power series) in terms of exponential functions of the eigenvalues q_α of T, times 4×4 projectors. Expressions for 3×3 projectors are given in eq.(A1) of Appendix A.1. Their generalization to 4×4 projectors is straightforward [57, Vol. 1, Chap. 5, Sec. 2.1]. This method simplifies, and is then a useful alternative to the power-series method, when linear variation of T within each interval may be assumed [71, eqs. (18) and (19)].

The Keller and Keller method

A solution of (1.177)

$$\mathsf{M}'(z, z_\nu) = \tau(z)\mathsf{M}(z, z_\nu)$$

may be derived by direct integration of the equation and use of successive approximations for M on the right-hand side. With $\mathsf{M}(z_\nu,z_\nu)=\mathsf{I}^{(4)}$, we get

$$\begin{aligned}
\mathsf{M}(z, z_\nu) &= \mathsf{I}^{(4)} + \int_{z_\nu}^{z} \tau(\varsigma_1)\mathsf{M}(\varsigma_1, z_\nu)\, d\varsigma_1 \\
&= \mathsf{I}^{(4)} + \int_{z_\nu}^{z} \tau(\varsigma_1)\, d\varsigma_1 + \int_{z_\nu}^{z} \tau(\varsigma_1)\, d\varsigma_1 \int_{z_\nu}^{\varsigma_1} \tau(\varsigma_2)\mathsf{M}(\varsigma_2, z_\nu)\, d\varsigma_2 \\
&= \mathsf{I}^{(4)} + \int_{z_\nu}^{z} \tau(\varsigma_1)\, d\varsigma_1 + \int_{z_\nu}^{z} \tau(\varsigma_1)\, d\varsigma_1 \int_{z_\nu}^{\varsigma_1} \tau(\varsigma_2)\, d\varsigma_2 \\
&\quad + \int_{z_\nu}^{z} \tau(\varsigma_1)\, d\varsigma_1 \int_{z_\nu}^{\varsigma_1} \tau(\varsigma_2)\, d\varsigma_2 \int_{z_\nu}^{\varsigma_2} \tau(\varsigma_3)\, d\varsigma_3 + \ldots
\end{aligned} \tag{1.183}$$

The resultant series, given by Gantmacher [57, Vol. 2, Chap. 14, Sec. 5] and employed by Volland [128, 129, 130], is uniformly and absolutely convergent, but its rate of convergence is rather slow. Keller and Keller [78] have modified this form of solution in order to obtain a more rapidly convergent series, better suited to numerical analysis.

We start again with the Clemmow-Heading equations (1.175), but decompose the fields $\mathrm{e}=(E_x, E_y, H_x, H_y)$ into the eigenmodes of the medium, i.e. into the normalized eigenvectors $\hat{\mathrm{e}}_\alpha$ ($\alpha=\pm1, \pm2$) of the matrix T multiplied by the corresponding eigenmode amplitudes a_α:

$$\begin{aligned} \mathbf{e}' &= -ik_0\mathbf{T}\mathbf{e}\,, \qquad \mathbf{e} = \mathbf{E}\,\boldsymbol{a} \\ \mathbf{E} &:= [\hat{\mathbf{e}}_1\,, \hat{\mathbf{e}}_2\,, \hat{\mathbf{e}}_{-1}\,, \hat{\mathbf{e}}_{-2}]\,, \qquad \boldsymbol{a} := \begin{bmatrix} a_1 \\ a_2 \\ a_{-1} \\ a_{-2} \end{bmatrix} \end{aligned} \tag{1.184}$$

Hence

$$\boldsymbol{a}' = -ik_0\mathbf{E}^{-1}\mathbf{T}\mathbf{E}\boldsymbol{a} - \mathbf{E}^{-1}\mathbf{E}'\boldsymbol{a} \tag{1.185}$$

The eigenmodes are assumed to be linearly independent (if collisions are present the singularities of $\mathbf{T}$, i.e. the coupling points, usually appear at complex heights, cf. [33, Sec. 16.1], and are not encountered in integrations along the real height axis); hence $\mathbf{E}$ is non-singular and $\mathbf{T}$ in (1.185) is diagonalized by $\mathbf{E}$. Eq. (1.185) becomes

$$\boldsymbol{a}' = \Big[-ik_0\boldsymbol{\Lambda} - (\boldsymbol{\Gamma}_D + \boldsymbol{\Gamma})\Big]\boldsymbol{a}\,, \qquad \boldsymbol{\Gamma}_D + \boldsymbol{\Gamma} := \mathbf{E}^{-1}\mathbf{E}'\,, \qquad \boldsymbol{\Lambda} := \mathbf{E}^{-1}\mathbf{T}\mathbf{E} \tag{1.186}$$

with $\boldsymbol{\Lambda}$ diagonal; $\boldsymbol{\Gamma}_D$ contains the diagonal terms of $\mathbf{E}^{-1}\mathbf{E}'$ and $\boldsymbol{\Gamma}$ the off-diagonal terms. Combining the diagonal matrices, we get

$$\boldsymbol{a}' = -ik_0\mathbf{Q}\boldsymbol{a} - \boldsymbol{\Gamma}\boldsymbol{a}\,, \qquad \mathbf{Q} := \boldsymbol{\Lambda} - \frac{i}{k_0}\boldsymbol{\Gamma}_D \tag{1.187}$$

Now the amplitudes a_α in (1.187) carry the rapidly varying phases of the eigenmodes, which would have to be followed by means of small step sizes in the numerical integration. These phases are therefore 'transformed out' by the replacement of a_α by another set of slowly varying amplitudes f_α,

$$\boldsymbol{a}(z) = \exp\left(-ik_0\int_{z_0}^{z}\mathbf{Q}\,dz\right)\mathbf{f}(z)\,, \qquad \boldsymbol{a}(z_0) = \mathbf{f}(z_0)\,, \qquad \mathbf{f} := \begin{bmatrix} f_1 \\ f_2 \\ f_{-1} \\ f_{-2} \end{bmatrix} \tag{1.188}$$

and (1.187) becomes

$$\begin{aligned} \boldsymbol{a}' + ik_0\mathbf{Q}\boldsymbol{a} &= \exp\left(-ik_0\int_{z_0}^{z}\mathbf{Q}\,dz\right)\mathbf{f}' \\ &= -\boldsymbol{\Gamma}\exp\left(-ik_0\int_{z_0}^{z}\mathbf{Q}\,dz\right)\mathbf{f} \end{aligned} \tag{1.189}$$

The final result is

$$\begin{aligned}\mathbf{f}'(z) &= \left[-\exp\left(ik_0\int_{z_0}^{z}\mathbf{Q}\,dz\right)\mathit{\Gamma}\exp\left(-ik_0\int_{z_0}^{z}\mathbf{Q}\,dz\right)\right]\mathbf{f}(z) \\ &=: \mathcal{T}(z,z_0)\,\mathbf{f}(z)\end{aligned} \tag{1.190}$$

thereby defining the matrix $\mathcal{T}$ which replaces $\mathbf{T}$ or $\boldsymbol{\tau}$ in the Clemmow-Heading formulation (1.175), while the slowly varying amplitudes $\mathbf{f}$ replace the wave fields $\mathbf{e}$.

We now introduce the matrizant in the form of a transfer matrix to relate amplitudes $\mathbf{f}(z)$ and $\mathbf{f}(z_0)$ at a height z and a fixed reference height z_0 respectively,

$$\mathbf{f}(z) = \mathcal{M}(z,z_0)\mathbf{f}(z_0) \tag{1.191}$$

$$\mathbf{f}'(z) = \mathcal{M}'(z,z_0)\mathbf{f}(z_0) \tag{1.192}$$

so that (1.191) in (1.190) gives

$$\mathcal{M}'(z,z_0) = \mathcal{T}(z,z_0)\mathcal{M}(z,z_0) \tag{1.193}$$

The solution, as in (1.183), is

$$\mathcal{M}(z,z_0) = \mathsf{I}^{(4)} + \int_{z_0}^{z}\mathcal{T}(\varsigma_1)\,d\varsigma_1 + \int_{z_0}^{z}\mathcal{T}(\varsigma_1)\,d\varsigma_1\int_{z_0}^{\varsigma_1}\mathcal{T}(\varsigma_2)\,d\varsigma_2 + \ldots \tag{1.194}$$

Due to rapid convergence of the series, truncation after the first integral usually yields satisfactory results, while inclusion of the double integral permits large step sizes to be employed. Since eigenmode amplitudes rather than fields are computed, the problem of numerical swamping is less important (see the discussion in Sec. 1.5.3). The eigenmode amplitudes a_α at any level may now be recovered from the f_α by means of (1.188), (the integrals in (1.189) have already been determined in the matrizant calculation), to give with the aid of (1.191) and (1.188),

$$\begin{aligned}\boldsymbol{a}(z) &= \exp\left(-ik_0\int_{z_0}^{z}\mathbf{Q}\,dz\right)\mathbf{f}(z) = \exp\left(-ik_0\int_{z_0}^{z}\mathbf{Q}\,dz\right)\mathcal{M}(z,z_0)\boldsymbol{a}(z_0) \\ &=: \mathbf{P}(z,z_0)\boldsymbol{a}(z_0)\end{aligned} \tag{1.195}$$

The matrix

$$\mathbf{P}(z,z_0) = \exp\left(-ik_0\int_{z_0}^{z}\mathbf{Q}\,dz\right)\mathcal{M}(z,z_0) \tag{1.196}$$

is seen to be equivalent to the propagator matrix (1.167) discussed in Sec. 1.5.3. The wave fields can be recovered, if needed, from the amplitudes $\boldsymbol{a}(z)$ by means of

$$\mathbf{e} = \mathbf{E}\,\boldsymbol{a}, \qquad \boldsymbol{a} = \mathbf{E}^{-1}\mathbf{e} \tag{1.197}$$

as in (1.184), the inverse relation yielding amplitudes from fields in the power-series method discussed previously.

This method of Keller and Keller has been successfully applied by Pitteway and Horowitz [99], who have carried the series in (1.194) to the double integral. They showed that for all frequencies between 20 kHz with an integration range of 20 km, to 10 Mhz with a range of 100 km, the entire integration could be performed in 30 steps, irrespective of frequency, to give three figure accuracy. In calculating the matrizant (1.194), one must determine $\mathcal{T}$ in the integrand from oscillatory phase integrals (1.190), and these were calculated by them on the assumption of linear variation of $\mathbf{Q}$ in each interval. In evaluating the integrals it was assumed that

$$\int e^{ax+bx^2}\, dx \approx \int (1+bx^2)e^{ax}\, dx \tag{1.198}$$

It would seem to us that this method is the most suited for numerical calculations at high frequencies ($\gtrsim$1 Mhz), but we are unaware of any numerical work, besides that of Pitteway and Horowitz, done by this technique. It is therefore not clear whether numerical difficulties are encountered when downgoing evanescent waves, for instance, are integrated upwards so that one or two of the elements in the matrizant then blow up exponentially.

Bossy's hybrid method

We now describe the technique introduced by Bossy [25,24] primarily as a means of overcoming numerical swamping in the power-series method in which wave fields at two levels, z_ν and z_0, are related by matrizants (1.176) formed by repeated multiplication of of sub-matrizants for consecutive elementary steps or intervals (1.181):

$$\mathbf{e}(z_\nu) = \mathsf{M}(z_\nu, z_0)\mathbf{e}(z_0)\,, \qquad \mathsf{M}(z, z_0) = \mathsf{M}(z_\nu, z_{\nu-1})\mathsf{M}(z_{\nu-1}, z_{\nu-2})\ldots\mathsf{M}(z_1, z_0) \tag{1.199}$$

The (sub-)matrizant $\mathsf{M}(z_\nu,z_{\nu-1})$ for any elementary step or layer is readily transformed with the aid of (1.197) into a propagator (transfer matrix) P relating up- and downgoing eigenmode amplitudes at the two bounding surfaces of the layer,

$$\begin{aligned} \boldsymbol{a}(z_\nu) &= \mathsf{E}(z_\nu)^{-1}\mathsf{M}(z_\nu, z_{\nu-1})\mathsf{E}(z_{\nu-1})\boldsymbol{a}(z_{\nu-1}) \\ &=: \mathsf{P}(z_\nu, z_{\nu-1})\boldsymbol{a}(z_{\nu-1}) \end{aligned} \tag{1.200}$$

In terms of the two upgoing and two downgoing eigenmode amplitudes, $\boldsymbol{a}_+$

and $\boldsymbol{a}_-$, this becomes

$$\begin{bmatrix} \boldsymbol{a}_+(z_\nu) \\ \boldsymbol{a}_-(z_\nu) \end{bmatrix} = \begin{bmatrix} \mathbf{P}_1 & \mathbf{P}_2 \\ \mathbf{P}_3 & \mathbf{P}_4 \end{bmatrix} \begin{bmatrix} \boldsymbol{a}_+(z_{\nu-1}) \\ \boldsymbol{a}_-(z_{\nu-1}) \end{bmatrix} \tag{1.201}$$

in which the propagator $\mathbf{P}$ has been split into four 2×2 matrices as in (1.168). The four 2×2 reflection and transmission matrices, $\mathbf{r}_\pm$ and $\mathbf{t}_\pm$, are now computed for this, and for all, elementary layers as in (1.169),

$$\begin{aligned} \boldsymbol{a}_{out} := \begin{bmatrix} \boldsymbol{a}_-(z_{\nu-1}) \\ \boldsymbol{a}_+(z_\nu) \end{bmatrix} &= \begin{bmatrix} -\mathbf{P}_4^{-1}\mathbf{P}_3 & \mathbf{P}_4^{-1} \\ \mathbf{P}_1 - \mathbf{P}_2\mathbf{P}_4^{-1}\mathbf{P}_3 & \mathbf{P}_2\mathbf{P}_4^{-1} \end{bmatrix} \begin{bmatrix} \boldsymbol{a}_+(z_{\nu-1}) \\ \boldsymbol{a}_-(z_\nu) \end{bmatrix} \\ &= \begin{bmatrix} \mathbf{r}_+ & \mathbf{t}_- \\ \mathbf{t}_+ & \mathbf{r}_- \end{bmatrix} \begin{bmatrix} \boldsymbol{a}_+(z_{\nu-1}) \\ \boldsymbol{a}_-(z_\nu) \end{bmatrix} =: \mathbf{S}\,\boldsymbol{a}_{in} \end{aligned} \tag{1.202}$$

Suppose we have computed the reflection and transmission matrices, $\mathbf{R}_\pm^\nu$ and $\mathbf{T}_\pm^\nu$, for the slab bounded by z_0 and z_ν, and we now add to it the elementary layer $(z_\nu, z_{\nu+1})$ whose reflection and transmission matrices are $\mathbf{r}_\pm$ and $\mathbf{t}_\pm$, as in (1.202). Let $\mathbf{R}_\pm^{\nu+1}$ and $\mathbf{T}_\pm^{\nu+1}$ denote the corresponding matrices for the composite slab $(z_0, z_{\nu+1})$. The problem is identical to that discussed in Sec. 1.5.4 and illustrated in Fig. 1.2, except that now $\mathbf{r}_\pm$ and $\mathbf{t}_\pm$ refer to the overlying elementary layer rather than the interface $(\nu+1, \nu+2)$. Furthermore, the phase matrices $\mathbf{\Delta}_\pm$, (1.166) and (1.170), now reduce to unit matrices $\mathbf{I}^{(4)}$, since the overlying layer is contiguous with the underlying slab, viz. $\delta z=0$. The reflection and transmission matrices, $\mathbf{R}_\pm^{\nu+1}$ and $\mathbf{T}_\pm^{\nu+1}$, are now determined by the same recursion relations as in (1.172), with $\mathbf{\Delta}_\pm \to \mathbf{I}^{(4)}$:

$$\begin{aligned} \mathbf{R}_+^{\nu+1} &= \mathbf{R}_+^\nu + \mathbf{T}_-^\nu \left[\mathbf{I} - \mathbf{r}_+\mathbf{R}_-^\nu\right]^{-1} \mathbf{r}_+\mathbf{T}_+^\nu \\ \mathbf{T}_+^{\nu+1} &= \mathbf{t}_+ \left[\mathbf{I} - \mathbf{R}_-^\nu\mathbf{r}_+\right]^{-1} \mathbf{T}_+^\nu \\ \mathbf{R}_-^{\nu+1} &= \mathbf{r}_- + \mathbf{t}_+ \left[\mathbf{I} - \mathbf{R}_-^\nu\mathbf{r}_+\right]^{-1} \mathbf{R}_-^\nu\mathbf{t}_- \\ \mathbf{T}_-^{\nu+1} &= \mathbf{T}_-^\nu \left[\mathbf{I} - \mathbf{r}_+\mathbf{R}_-^\nu\right]^{-1} \mathbf{t}_- \end{aligned} \tag{1.203}$$

These relations, given by Bossy and Claes [25] and Bossy [24], impart to the matrizant method the flexibility of the thin-layer scattering-matrix technique, but with the advantage of being able to work with large step sizes, and thereby to extend considerably the upper frequency limit.

This technique is applicable also to the Keller and Keller method, in which eigenmodes and amplitudes are recovered as in (1.195). If the problem of numerical swamping does not arise, then it is unnecessary to employ the recursion relations (1.203). Instead, the propagator matrix $\mathbf{P}(z,z_0)$ (1.196) for all slab thicknesses, derived from the repeated product of sub-matrizants for each elementary layer, may be transformed directly into scattering matrices by means of (1.169). It seems to us that computer output in terms of scattering matrix elements is the most convenient form for extracting information on the physical propagation processes.

Chapter 2

Eigenmode reciprocity in k-space

2.1 Reciprocity in physical space and in k-space

2.1.1 Overview

Until the mid 1960's the problem of reciprocity in electromagnetics had been developing in two separate, and seemingly unrelated directions. As early as 1896 Lorentz [91] had demonstrated that if two independent current distributions, $\mathbf{J}_1(\mathbf{r})$ and $\mathbf{J}_2(\mathbf{r})$, generated electromagnetic fields, $\mathbf{E}_1(\mathbf{r})$, $\mathbf{H}_1(\mathbf{r})$ and $\mathbf{E}_2(\mathbf{r})$, $\mathbf{H}_2(\mathbf{r})$ respectively, in free space, then

$$\int \mathbf{E}_1(\mathbf{r}) \cdot \mathbf{J}_2(\mathbf{r})\, d^3r = \int \mathbf{E}_2(\mathbf{r}) \cdot \mathbf{J}_1(\mathbf{r})\, d^3r \tag{2.1}$$

and this was recognized as an expression of the 'interchangeability' of transmitting and receiving antennas. This, or an equivalent formulation,

$$\nabla \cdot (\mathbf{E}_1 \times \mathbf{H}_2 - \mathbf{E}_2 \times \mathbf{H}_1) = 0 \tag{2.2}$$

became to be known as the Lorentz reciprocity theorem, and will be discussed in some detail in Chap. 4. The theorem was used, inter alia, to deduce the properties of transmitting antennas if their properties as receiving antennas were known. Eckersley [51], for instance, used the theorem to deduce the radiation pattern of a transmitting antenna as modified by an imperfectly conducting ground below it, by solving the simpler problem of its response as a receiving antenna.

Sommerfeld [110] and Dällenbach [47] pointed out that the theorem would hold in anisotropic media provided that the electric permittivity $\boldsymbol{\varepsilon}$, the magnetic permeability $\boldsymbol{\mu}$ and the conductivity $\boldsymbol{\sigma}$ were symmetric tensors. Rumsey [106] and Cohen [40] noted that a modified form of Lorentz reciprocity would hold also for non-symmetric tensors provided that the second (reciprocal) system of currents and fields were taken in a 'transposed medium', characterized by the transposed tensors $\boldsymbol{\varepsilon}^T$, $\boldsymbol{\mu}^T$ and $\boldsymbol{\sigma}^T$. Harrington and Villeneuve [63] applied the theorem to gyrotropic media, such as magnetoplasmas or ferrites, in which the 'transposed medium' is just the original medium with the direction of the external magnetic field reversed. Kong and Cheng [84] and Kerns [81] extended the result to bianisotropic media (see Sec. 2.2.2) and introduced the concept of a 'complementary' or 'adjoint' medium, which generalizes the earlier concept of the transposed medium.

A parallel, and seemingly unrelated line of development treated what we shall call 'reciprocity in (transverse-) **k**-space', which in its early form dealt with the symmetry properties of the scattering matrices in a plane-stratified ionospheric magnetoplasma. Budden [29] and Barron and Budden [21] found that the 2×2 reflection matrix for plane-wave incidence on a plane-stratified magnetoplasma was the transpose of the reflection matrix for another symmetrically disposed direction of incidence, which we shall subsequently call the 'conjugate direction'. (Because of Snell's law, the component $\mathbf{k}_t$ of the propagation vector in the stratification plane—the 'transverse' component—is the same for the incoming plane wave and for the outgoing, scattered waves.) Pitteway and Jespersen [100] and Heading [66] found similar results relating the transmission coefficients for upgoing waves incident on the ionosphere in a given direction, and downgoing waves incident in a symmetrically disposed, conjugate direction. These results were later generalized by Suchy and Altman [118,119,12,13] who showed that the 4×4 scattering matrices could be expressed in terms of suitably defined eigenmode amplitudes within the gyrotropic medium, and not only in terms of linearly polarized base modes in free space outside of the scattering medium. This result was further extended by Altman et al. [10] to include bianisotropic media, and it was shown that a wide range of 'adjoint' or 'complementary' reciprocal media could be generated by means of orthogonal transformations (rotation, reflection or inversion) of the transposed medium.

2.1.2 From physical space to k-space

The two lines of development just described converged from both directions. A passive antenna is a scattering object, and any dielectric scattering object will re-radiate by virtue of the currents induced by the external fields incident on it. Lorentz reciprocity will apply to such scattering objects (see, for instance, Rumsey [106]). Harrington and Villeneuve [63] showed that if a scattering object, characterized by constitutive tensors, $\boldsymbol{\varepsilon}$, $\boldsymbol{\mu}$ and $\boldsymbol{\sigma}$, be considered as

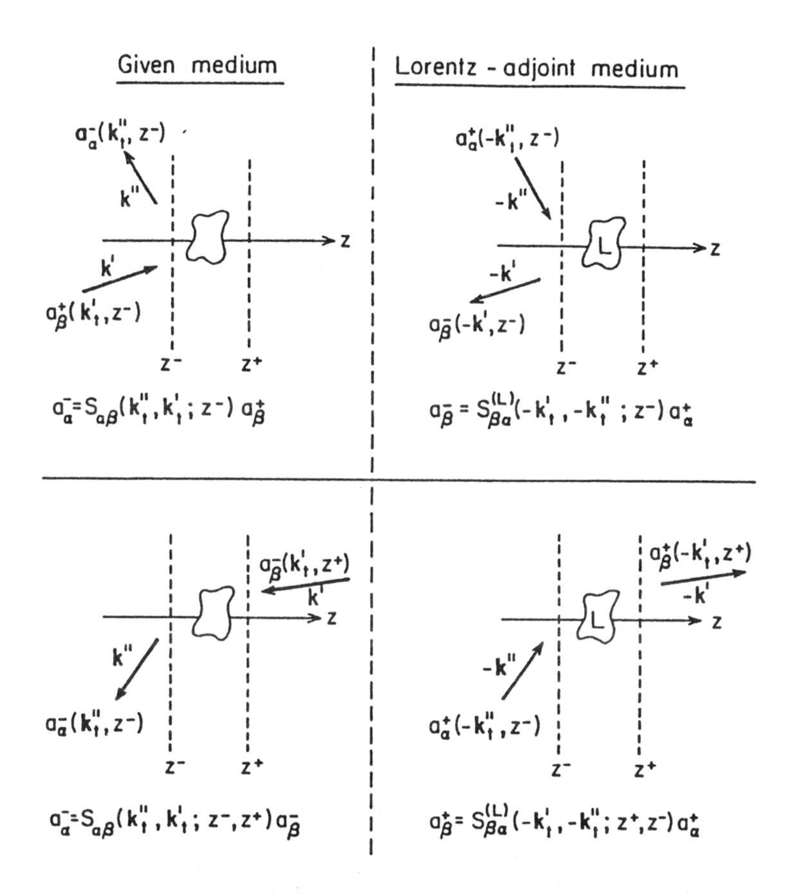

Figure 2.1: Scattering relations illustrated schematically for object with given or Lorentz-adjoint medium. In all cases $S_{\alpha\beta}(\mathbf{k}_t'', \mathbf{k}_t') = S_{\beta\alpha}^{(L)}(-\mathbf{k}_t', -\mathbf{k}_t'')$.

a generalized N terminal-pair network, with $\mathbf{V}$ and $\mathbf{I}$ representing column matrices of 'terminal' voltages and currents at the surface of the scatterer, one may define a scattering matrix $\mathbf{S}$ through the relation

$$\mathbf{V} =: \mathbf{S}\,\mathbf{I} \tag{2.3}$$

They showed that if the medium of the object had transposed constitutive tensors $\boldsymbol{\varepsilon}^T$, $\boldsymbol{\mu}^T$ and $\boldsymbol{\sigma}^T$, the scattering matrix would be transposed to $\mathbf{S}^T$.

It was the work of Kerns [81], however, that bridged the gap from reciprocity in real (physical) space to reciprocity in **k**-space. Let us suppose, with Kerns, that a scattering object in free space, Fig. 2.1, is contained between two imaginary planes, z^- and z^+. We consider an incoming wave, with an electric wave field $\mathbf{E}^{in}(z^-)$ or $\mathbf{E}^{in}(z^+)$ incident on the object from the left or right respectively. We shall adapt Kerns' notation to that used by us. The transverse component (transverse to the z-axis) of the electric field, $\mathbf{E}_t^{in}$, may be Fourier analysed in the $z = z^-$ or $z = z^+$ planes. Any Fourier component having a transverse wave vector

$$\mathbf{k}_t \equiv (k_x, k_y) \qquad \text{with} \qquad |\mathbf{k}_t| = ({k_0}^2 - {k_z}^2)^{1/2}, \qquad k_0 := \omega(\varepsilon_0 \, \mu_0)^{1/2}$$

may be decomposed into two 'modes', in which the electric fields, $\mathbf{E}_{t1}$ and $\mathbf{E}_{t2}$, are respectively parallel and perpendicular to the plane of incidence. The basis vectors along these fields will be

$$\hat{\epsilon}_{\parallel} := \mathbf{k}_t / \mid \mathbf{k}_t \mid \equiv \hat{\epsilon}_{\pm 1} \qquad \text{and} \qquad \hat{\epsilon}_{\perp} := \hat{\mathbf{z}} \times \hat{\epsilon}_{\pm 1} \equiv \hat{\epsilon}_{\pm 2}$$

Fourier analysis of $\mathbf{E}_t^{in}(z^{\mp})$ yields the spectral amplitude densities, $\underset{\sim}{A}_{\alpha}^{\pm}(\mathbf{k}_t, z^{\mp})$, in transverse-$\mathbf{k}$ space, with $\alpha = 1, 2$ or $\alpha = -1, -2$ for positive- or negative-going waves respectively:

$$\mathbf{E}_t^{in}(z^{\mp}) = \frac{1}{2\pi} \iint \underset{\sim}{A}_{\alpha}^{\pm}(\mathbf{k}_t, z^{\mp}) \, \hat{\epsilon}_{\alpha} \, \exp\left[-i(k_x x + k_y y)\right] dk_x \, dk_y \tag{2.4}$$

integrated over the entire transverse-$\mathbf{k}$ plane, with assumed summation over the characteristic polarizations $\alpha = 1, 2$ for $z = z^-$, or $\alpha = -1, -2$ for $z = z^+$. Phase factors $\exp(\mp i k_z z^{\mp})$ have been included in the spectral amplitudes $\underset{\sim}{A}_{\alpha}^{\pm}$. Underlying tildes ($\sim$) are used in this section to denote quantities that represent densities in transverse-$\mathbf{k}$ space.

The outgoing scattered wave fields $\mathbf{E}_t^{out}(z^{\pm})$ may similarly be Fourier analyzed to yield outgoing amplitude densities, $\underset{\sim}{A}_{\alpha}^{\pm}(\mathbf{k}_t, z^{\pm})$. It is convenient to define normalized amplitude densities, $\underset{\sim}{a}_{\alpha}^{\pm}$:

$$\underset{\sim}{a}_{\alpha}^{\pm} := {\eta_{\alpha}}^{1/2} \underset{\sim}{A}_{\alpha}^{\pm}, \qquad \alpha = \pm 1, \pm 2$$

where

$$\eta_{\pm 1} := \omega \varepsilon_0 / \mid k_z \mid = \frac{k}{\mid k_z \mid} \sqrt{\frac{\varepsilon_0}{\mu_0}} \qquad \text{and} \qquad \eta_{\pm 2} := \mid k_z \mid / \omega \mu_0 = \frac{\mid k_z \mid}{k} \sqrt{\frac{\varepsilon_0}{\mu_0}}$$

are the characteristic wave admittances [81, eqs. (1.2-5) and (1.2-6)]. Then $|\underset{\sim}{a}_{\alpha}^{\pm}|^2$ will represent the spectral densities (in transverse-$\mathbf{k}$ space) of the z-component of the time-averaged energy fluxes across the surfaces $z = z^+$ or $z = z^-$:

$$\langle P_{z^{\mp}, \alpha} \rangle = -\frac{1}{2} \iint |\underset{\sim}{a}_{\alpha}^{\pm}(\mathbf{k}_t, z^{\mp})|^2 \, dk_x \, dk_y, \qquad \alpha = \pm 1, \pm 2 \quad \text{for} \quad z = z^{\mp} \tag{2.5}$$

for incoming waves, and

$$\langle P_{z^\pm,\alpha}\rangle = +\frac{1}{2}\iint |\underset{\sim}{a}_\alpha^\pm(\mathbf{k}_t, z^\pm)|^2\, dk_x\, dk_y, \qquad \alpha = \pm 1, \pm 2 \quad \text{for} \quad z = z^\pm \quad (2.6)$$

for outgoing waves, in which, for simplicity, we have ignored the contributions from evanescent modes [81, eqs. (1.4–2) and (1.4–3)].

Outgoing and incoming modal amplitude densities will be related by elements of a scattering-density matrix $\underset{\sim}{\mathbf{S}}$. Symbolically, we may write [81, eq. (1.3–1)]

$$\underset{\sim}{a}_\alpha^{out}(\mathbf{k}_t'') = \iint \underset{\sim}{S}_{\alpha\beta}(\mathbf{k}_t'', \mathbf{k}_t')\, \underset{\sim}{a}_\beta^{in}(\mathbf{k}_t')\, dk_x'\, dk_y' \tag{2.7}$$

which, for back- and forward-scattered waves respectively, becomes

$$\underset{\sim}{a}_\alpha^\mp(\mathbf{k}_t'', z^\mp) = \iint \underset{\sim}{S}_{\alpha\beta}(\mathbf{k}_t'', \mathbf{k}_t'; z^\mp)\, \underset{\sim}{a}_\beta^\pm(\mathbf{k}_t', z^\mp)\, dk_x'\, dk_y'$$

and

$$\underset{\sim}{a}_\alpha^\pm(\mathbf{k}_t'', z^\pm) = \iint \underset{\sim}{S}_{\alpha\beta}(\mathbf{k}_t'', \mathbf{k}_t'; z^\pm, z^\mp)\, \underset{\sim}{a}_\beta^\pm(\mathbf{k}_t', z^\mp)\, dk_x'\, dk_y'$$

Now let the medium of the scattering object be replaced by a '(Lorentz-) adjoint' medium. (Just what is meant by this is explained in Sec. 3.4. In the case of a magnetoplasma, it means the given medium in which the external magnetic field has been reversed in direction). Suppose also that all outgoing wave vectors $\mathbf{k}''$ are reversed in direction ($\mathbf{k}'' \to -\mathbf{k}''$) so that they become *incoming* wave fields. Kerns' (Lorentz-) adjoint scattering theorem [81, eq. (1.5–5)] states that in this case the outgoing wave fields will be just the incoming wave fields in the original problem with their wave vectors reversed ($\mathbf{k}' \to -\mathbf{k}'$). In the case of back-scattering this means that

$$\underset{\sim}{S}_{\alpha\beta}(\mathbf{k}_t'', \mathbf{k}_t'; z^\mp) = \underset{\sim}{S}_{\beta\alpha}^{(L)}(-\mathbf{k}_t', -\mathbf{k}_t''; z^\mp) \tag{2.8}$$

and for forward scattering

$$\underset{\sim}{S}_{\alpha\beta}(\mathbf{k}_t'', \mathbf{k}_t'; z^\pm, z^\mp) = \underset{\sim}{S}_{\beta\alpha}^{(L)}(-\mathbf{k}_t', -\mathbf{k}_t''; z^\mp, z^\pm) \tag{2.9}$$

where $\underset{\sim}{\mathbf{S}}^{(L)}$ is the scattering-density matrix for the Lorentz-adjoint medium. These relations are illustrated schematically in Fig. 2.1.

Suppose now that the incoming wave field in (2.7) is that of a single plane wave with a transverse propagation vector $\mathbf{k}_t$. Then $\underset{\sim}{a}_\beta^{in}$ becomes a Dirac delta function (aside from a multiplying factor) in transverse-$\mathbf{k}$ space,

$$\underset{\sim}{a}_\beta^{in}(\mathbf{k}_t') = a_\beta^{in}(\mathbf{k}_t)\delta(\mathbf{k}_t' - \mathbf{k}_t) = a_\beta^\pm(z^\mp)\delta(\mathbf{k}_t' - \mathbf{k}_t), \qquad a_\beta^{in}(\mathbf{k}_t) = \iint \underset{\sim}{a}_\beta^{in}(\mathbf{k}_t')\, dk_x'\, dk_y'$$

and (2.7) becomes

$$\underset{\approx}{a}_{\alpha}^{out}(\mathbf{k}_t'') = \underset{\approx}{S}_{\alpha,\beta}(\mathbf{k}_t'', \mathbf{k}_t)\, a_{\beta}^{in}(\mathbf{k}_t) \tag{2.10}$$

Finally we let the scattering object be a plane-stratified slab, situated between the planes $z = z^-$ and $z = z^+$, i.e. all constitutive parameters of the medium are functions of the z-coordinate only. Because of Snell's law, the scattering-density matrix $\underset{\approx}{S}_{\alpha\beta}$ becomes a delta function in $\mathbf{k}_t$-space for plane-wave incidence, and the amplitude densities of both the incident and scattered waves will also be delta functions in $\mathbf{k}_t$-space:

$$\begin{aligned} \underset{\approx}{S}_{\alpha\beta}(\mathbf{k}_t'', \mathbf{k}_t) &= S_{\alpha\beta}(\mathbf{k}_t)\, \delta(\mathbf{k}_t'' - \mathbf{k}_t) \\ \underset{\approx}{a}_{\alpha}^{out}(\mathbf{k}_t'') &= a_{\alpha}^{out}(\mathbf{k}_t)\, \delta(\mathbf{k}_t'' - \mathbf{k}_t) = a_{\alpha}^{\mp}(z^{\mp})\, \delta(\mathbf{k}_t'' - \mathbf{k}_t) \end{aligned} \tag{2.11}$$

If $\underset{\approx}{S}_{\alpha\beta}$ and $\underset{\approx}{a}_{\alpha}^{out}(\mathbf{k}_t'')$ are now substituted into (2.10), and the equation then integrated over all $\mathbf{k}_t''$, we obtain

$$a_{\alpha}^{out}(\mathbf{k}_t) = S_{\alpha\beta}(\mathbf{k}_t)\, a_{\beta}^{in}(\mathbf{k}_t) \tag{2.12}$$

The scattering relation (2.7) then reduces to a straightforward matrix relation, with each incident mode, $\alpha = \pm 1, \pm 2$, generating two reflected (back-scattered) and two transmitted (forward-scattered) modes. With

$$\boldsymbol{a}_+ := \begin{bmatrix} a_1 \\ a_2 \end{bmatrix}, \qquad \boldsymbol{a}_- := \begin{bmatrix} a_{-1} \\ a_{-2} \end{bmatrix}$$

we have

$$\begin{aligned} \boldsymbol{a}_{out} := \begin{bmatrix} \boldsymbol{a}_-(z^-) \\ \boldsymbol{a}_+(z^+) \end{bmatrix} &= \begin{bmatrix} \mathsf{R}_+(\mathbf{k}_t; z^-) & \mathsf{T}_-(\mathbf{k}_t; z^-, z^+) \\ \mathsf{T}_+(\mathbf{k}_t; z^+, z^-) & \mathsf{R}_-(\mathbf{k}_t; z^+) \end{bmatrix} \begin{bmatrix} \boldsymbol{a}_+(z^-) \\ \boldsymbol{a}_-(z^+) \end{bmatrix} \\ &= \mathsf{S}(\mathbf{k}_t)\, \boldsymbol{a}_{in} \end{aligned} \tag{2.13}$$

defining thereby the 4-element modal-amplitude column matrices, $\boldsymbol{a}_{in}$ and $\boldsymbol{a}_{out}$, and the 2×2 reflection and transmission matrices, $\mathsf{R}_\pm$ and $\mathsf{T}_\pm$, which constitute the scattering matrix S. (The signed subscripts indicate the direction of the incident mode with respect to the z-axis.)

Kerns' theorem, (2.8) and (2.9), reduces in this case to

$$\mathsf{S}(\mathbf{k}_t) = \tilde{\mathsf{S}}^{(L)}(-\mathbf{k}_t) \tag{2.14}$$

$$\mathsf{R}_\pm(\mathbf{k}_t; z^\mp) = \tilde{\mathsf{R}}_\pm^{(L)}(-\mathbf{k}_t; z^\mp), \qquad \mathsf{T}_\pm(\mathbf{k}_t; z^+, z^-) = \tilde{\mathsf{T}}_\mp^{(L)}(-\mathbf{k}_t; z^-, z^+)$$

with, typically

$$R_{12}^+(\mathbf{k}_t; z^-) = R_{21}^{(L)+}(-\mathbf{k}_t; z^-), \qquad T_{12}^+(\mathbf{k}_t; z^+, z^-) = T_{21}^{(L)-}(-\mathbf{k}_t; z^-, z^+)$$

where $T^{\pm}_{\alpha\beta}$ and $R^{\pm}_{\alpha\beta}$, with $\alpha, \beta = 1$ or 2, denote elements of the 2×2 matrices $\mathbf{T}_{\pm}$ and $\mathbf{R}_{\pm}$.

This restricted form of Kerns' scattering theorem will be discussed in Sec. 3.4 in the general context of scattering theorems in plane-stratified media. In Sec. 7.3 Kerns' scattering theorem will be generalized to the case in which an anisotropic scattering object is immersed in a homogeneous or plane-stratified anisotropic medium.

2.1.3 Reciprocity in transverse-k space: a review of the earlier scattering theorems

The interest in the ionosphere, until the mid-fifties, lay primarily in its ability to reflect radio waves. Vertical ionospheric sounding had been employed since the mid-thirties to determine ionospheric structure and maximum usable frequencies for radio communication between fixed ground stations. Point-to-point long-wave and very-long-wave radio links had been tested experimentally to determine diurnal and seasonal variations, as well as the directional dependence of the ionospheric reflection coefficients. The first heroic efforts in the early fifties, especially by Budden and his coworkers [30,31], to produce full-wave computer programs to solve the differential equations governing the propagation of radio waves in a plane-stratified magnetoplasma, were aimed at producing, as their primary output, a set of reflection coefficients for arbitrary directions of incidence. Various equalities were then discovered in the numerically computed reflection coefficients for certain symmetrically disposed directions of incidence. The analytical proof of these 'reciprocity theorems' was found only later, after the theorems were already known from the computer output [29,21].

In 1953 Storey [114] showed both experimentally and theoretically that very-low-frequency (whistler) waves, guided by the earth's magnetic field, could penetrate through the ionospheric $X = 1$ level (where waves of similar polarization but higher frequency would normally have been reflected — cf. Sec. 1.2) to reach a magnetically conjugate point in the opposite hemisphere. (The frequency dispersion of these waves—the higher frequencies arriving before the lower—generated a whistling sound of falling frequency when the audio-frequency electromagnetic waves were received by an antenna connected to an audio amplifying system. Hence the name 'whistler'.) Storey's findings were one of the motivating factors in developing computer programs, such as that due to Pitteway [98], to calculate very-low-frequency transmission coefficients for propagation through the ionosphere for plane wave incidence from both below and above the ionosphere. Equalities between the transmission coefficients of downgoing whistler waves and upgoing 'penetrating modes' were again found in the computer output, for certain symmetrically disposed planes of incidence, and the analytical proof then followed [100].

Heading [66] undertook a systematic analysis of reciprocity (scattering) relations in plane-stratified magnetoplasmas, by considering certain general symmetry properties of Maxwell's second-order differential equations in such media. Equalities were again found between elements of the reflection and transmission matrices for certain pairs of symmetrically related directions of plane-wave incidence. The scattering matrix elements were defined, as in Budden's treatment [29], in terms of linearly polarized base modes in the free space bounding the medium.

At this stage there was still no obvious connection between the results of Kerns previously discussed, as applied to plane-stratified media, and those of Barron and Budden, Pitteway and Jespersen, and Heading. Kerns' scattering theorem involved an 'adjoint medium', which in the case of a magnetoplasma meant a magnetic-field reversed medium, with wave vectors reversed in direction too. The work of Budden and others, on the other hand, compared scattering matrix elements in the same medium, but with different directions of incidence.

The thin-layer scattering-matrix numerical technique developed by Altman and Cory [3,4] (see Sec. 1.5.3) led, fortuitously, to a generalized form of the scattering theorem in plane-stratified media. In this method the elements of the reflection and transmission matrices were just the quantities which were recursively summed in the numerical procedure in which thin elementary layers were added stepwise to the plane-stratified slab. The scattering matrix elements related amplitudes of eigenmodes emerging from slabs of varying thicknesses, imbedded in the given statified medium, to the amplitudes of eigenmodes incident on the slab. The 'amplitude' of an eigenmode was taken initially to be the (square root of the) z-component, normal to the stratification, of the time-averaged Poynting flux of the eigenmode. The computed output yielded the elements of the scattering matrix $\mathbf{S}(\mathbf{k}_t, \phi)$, (2.13), for given values of the transverse wave vector $\mathbf{k}_t$, and for given azimuthal angles, ϕ, between the plane of incidence and the magnetic meridian plane (the plane containing the external magnetic field, $\mathbf{b}$, and the normal to the stratification, $\hat{\mathbf{z}}$). The scattering matrix $\mathbf{S}(\mathbf{k}_t, \phi)$, when the plane of incidence was at an azimuthal angle ϕ, was found to be the transpose of that for a *conjugate* orientation in which the azimuthal angle was $(\pi - \phi)$,

$$\mathbf{S}(\mathbf{k}_t, \phi) = \widetilde{\mathbf{S}}(\mathbf{k}_t, \pi - \phi) \tag{2.15}$$

as long as the medium was lossless. The exact equality broke down as soon as collisional losses were introduced. On the basis of a procedure due to Budden and Jull (1964) in their treatment of reciprocity of magnetoionic rays [35], the complex conjugate wave fields, $\mathbf{E}^*$ and $\mathbf{H}^*$, appearing in the expression for the mean Poynting flux (see Sec. 2.3.1), were replaced by the computed adjoint wave fields, $\overline{\mathbf{E}}$ and $\overline{\mathbf{H}}$. This meant that the complex-conjugate transverse wave polarizations, ρ^*, appearing in the expression for the mean Poynting flux (see eq. 2.65 in Sec. 2.3.1) were replaced by the adjoint wave polarizations, $\bar{\rho} = -\rho$ (2.66). (The equality $\rho^* = \bar{\rho} = -\rho$ holds only for loss-free media). The eigenmode scattering theorem (2.15), reported by Altman in 1971 [2], was found to be exact, but the analytic proof was only found much later by Suchy and Altman [119, 12].

This chapter deals with some of the properties of the adjoint Maxwell equations, and their use in the derivation of the eigenmode scattering theorem. In Chap. 3 we consider the generalization of the theorem to base modes which are not eigenmodes of the medium, and discuss some of the earlier reciprocity (scattering) theorems in the light of the generalized theorem.

2.1.4 From transverse-k space back to physical space

The link back from reciprocity in transverse-**k** space to reciprocity in physical space was found by Schatzberg and Altman [8, 108]. Their procedure was to Fourier-analyse currents and fields in transverse-**k** space, and then to set up the angular spectrum of plane-wave eigenmodes associated with an element of current, $\mathbf{J}_1(\mathbf{k}_t, z')\,dz'$, flowing in an elementary layer of thickness dz' in the medium. With the aid of the scattering matrix $\mathbf{S}(\mathbf{k}_t; z, z')$ a dyadic Green's function, $\mathbf{G}(\mathbf{k}_t; z, z')$, was determined, so that the overall field $\mathbf{e}_1(\mathbf{k}_t, z)$ at a level z in the medium was given by

$$\mathbf{e}_1(\mathbf{k}_t, z) = \int \mathbf{G}(\mathbf{k}_t; z, z')\,\mathbf{J}_1(\mathbf{k}_t, z')\,dz' \tag{2.16}$$

In a similar fashion a second, independent current distribution, $\mathbf{J}_2(\mathbf{k}_t, z')$, generated a field $\mathbf{e}_2(\mathbf{k}_t, z)$.

A mirroring (reflection) transformation of the currents and fields with respect to the magnetic meridian $(\mathbf{b}, \hat{\mathbf{z}})$ plane, yielded the 'conjugate' currents and fields, $\mathbf{J}^c(\mathbf{k}_t^c, z')$ and $\mathbf{e}^c(\mathbf{k}_t^c, z)$. Here, $\mathbf{k}_t^c$ has been formed by reversing the sign of $\mathbf{k}_t$ (this will later be seen to be an expression of time reversal, which is inherent in the reciprocity process), and then the y-component, normal to the $(\mathbf{b}, \hat{\mathbf{z}})$ plane, is again sign-reversed by the reflection transformation to give

$$\mathbf{k}_t = k_0(s_x, s_y), \qquad \mathbf{k}_t^c = k_0(-s_x, s_y) \tag{2.17}$$

With the aid of the scattering theorem (2.15), which may be written in the form

$$\mathbf{S}(s_x, s_y) = \mathbf{S}^T(-s_x, s_y) \tag{2.18}$$

a simple relation was found between the given and conjugate dyadic Green's functions, $\mathbf{G}(\mathbf{k}_t; z, z')$ and $\mathbf{G}^c(\mathbf{k}_t^c; z, z')$, in $\mathbf{k}_t$ space. An inverse Fourier transfomation in $\mathbf{k}_t$-space led finally to a Lorentz-type reciprocity relation in real space [8, 108]

$$\int \mathbf{e}_1(\mathbf{r}) \cdot \mathbf{J}_2(\mathbf{r})\, d^3r = \int \mathbf{e}_2^c(\mathbf{r}) \cdot \mathbf{J}_1^c(\mathbf{r})\, d^3r \tag{2.19}$$

Eq. (2.19) is derived in Chap. 5. It will be noted that this result does not contain any feature that would indicate that its validity is restricted to plane-stratified media. In fact it is shown in Chap. 6 that in any medium that has 'conjugation symmetry', the reciprocity relation (2.19) will apply. A medium will be said to possess 'conjugation symmetry' if, after being 'time reversed', it can be mapped into itself by means of an orthogonal transformation. A 'time-reversed' magnetoplasma, for example, is one in which the external magnetic field has been reversed.

2.2 The adjoint wave fields

2.2.1 The need for an auxiliary set of equations adjoint to Maxwell's equations

The field equations of physics may generally be written as a system of first-order partial differential equations or, on elimination of some of the field variables, as higher-order equations. The Maxwell field, containing both electric and magnetic components, may be described by six first-order differential equations with six possible source terms, components of the electric and equivalent magnetic currents. The coefficients of the field components in these equations will be determined by the constitutive relations of the medium considered, as expressed by the constitutive tensor which relates the field vectors **D** and **B** to **E** and **H**. Examples of such tensors and their characteristic symmetries are discussed in Sec. 4.1.

To reveal the basic symmetries of the fields it is useful to make use of an auxiliary or *adjoint* set of equations, which will be satisfied by ***adjoint field variables***. These hypothetical adjoint fields will then exhibit a ***reciprocity relation*** with respect to the fields in the original problem. The adjoint fields will in general be non-physical, insofar as they satisfy the non-physical adjoint equations, but frequently they can be related in a simple and direct way to the physical fields in another *conjugate* problem, derived from the original by some sort of mapping transfomation (such as reflection). The reciprocity relation between the given and adjoint problems then leads to a reciprocity relation between fields (or between currents and fields, if sources are present) in the two physical configurations of the given and conjugate problems.

In the case of plane-stratified media both the given Maxwell field and the adjoint field can in principle be decomposed into characteristic wave fields or eigenmodes. It will be shown that the given and adjoint eigenmodes are biorthogonal, a property which provides a simple procedure for decomposing a wave field into its constituent eigenmodes, and for determining their amplitudes in a manner suitable for application in a general scattering theorem which will be derived in Sec. 2.5.

2.2.2 Maxwell's equations in anisotropic, plane-stratified media

The electric and magnetic wave fields in an anisotropic or bianisotropic medium will be related in general by a 6×6 constitutive tensor K:

$$\begin{bmatrix} \mathbf{D} \\ \mathbf{B} \end{bmatrix} = \begin{bmatrix} \boldsymbol{\varepsilon} & \boldsymbol{\xi} \\ \boldsymbol{\eta} & \boldsymbol{\mu} \end{bmatrix} \begin{bmatrix} \mathbf{E} \\ \mathbf{H} \end{bmatrix} \equiv \mathsf{K}\mathbf{e} \tag{2.20}$$

We note that the fundamental fields, defined by the Lorentz force on an electric charge q

$$\mathbf{F} = q(\mathbf{E} + \mathbf{v} \times \mathbf{B})$$

are $\mathbf{E}$ and $\mathbf{B}$, whereas $\mathbf{D}$ and $\mathbf{H}$ are derived fields which contain the additional contributions of electric polarization and magnetization currents. Nevertheless it is convenient, for the sake of symmetry of Maxwell's equations, to represent the constitutive tensor K in this form. $\boldsymbol{\varepsilon}$ is the 3×3 electric permittivity tensor, and $\boldsymbol{\mu}$ is the magnetic permeability tensor which, for media having no magnetic activity (such as plasmas, with or without an ambient magnetic field) is just the scalar permeability μ_0 of free space, $\boldsymbol{\mu} = \mu_0 \mathsf{I}^{(3)}$, where $\mathsf{I}^{(n)}$ represents the $n \times n$ unit matrix. The 3×3 coupling matrices $\boldsymbol{\xi}$ and $\boldsymbol{\eta}$ in (2.20) are usually zero except for a small class of *bianisotropic* media in which a magnetic field produces electric polarization, and an electric field magnetizes the medium. (Moving media are bianisotropic since the electric and magnetic fields are coupled by the Lorentz transformation, and so too are the magneto-electric or so-called Tellegen media [124] in which the elementary electric dipoles also have magnetic moment. Bianisotropic media have been discussed by Post [101], Kong and Cheng [37,82,83], van Bladel [127] and others, and are considered in some detail in Sec. 4.1.).

With time-harmonic $\exp(i\omega t)$ variation of all field quantities, Maxwell's equations take the form

$$[i\omega \mathbf{K} + \mathbf{D}]\,\mathbf{e}(\mathbf{r}) = -\mathbf{j}(\mathbf{r}) \tag{2.21}$$

with $\mathbf{D}$, the differential operator, given by

$$\mathbf{D} := \begin{bmatrix} 0 & -\nabla \times \mathbf{I}^{(3)} \\ \nabla \times \mathbf{I}^{(3)} & 0 \end{bmatrix} = \mathbf{D}^T \tag{2.22}$$

The generalized wave-field and current vectors, $\mathbf{e}$ and $\mathbf{j}$, are given by

$$\mathbf{e} := \begin{bmatrix} \mathbf{E} \\ \mathbf{H} \end{bmatrix}, \qquad \mathbf{j} := \begin{bmatrix} \mathbf{J}_{\mathrm{e}} \\ \mathbf{J}_{\mathrm{m}} \end{bmatrix} \tag{2.23}$$

where $\mathbf{J}_{\mathrm{e}}$ and $\mathbf{J}_{\mathrm{m}}$ are the electric and equivalent magnetic current densities.

If we split the differential operator $\mathbf{D}$ into three cartesian differential operators, (2.21) becomes

$$\left[i\omega \mathbf{K} + \mathbf{U}_x \frac{\partial}{\partial x} + \mathbf{U}_y \frac{\partial}{\partial y} + \mathbf{U}_z \frac{\partial}{\partial z}\right] \mathbf{e}(\mathbf{r}) = -\mathbf{j}(\mathbf{r}) \tag{2.24}$$

where

$$\mathbf{U}_x := \begin{bmatrix} \mathbf{0} & -\hat{\mathbf{x}} \times \mathbf{I}^{(3)} \\ \hat{\mathbf{x}} \times \mathbf{I}^{(3)} & \mathbf{0} \end{bmatrix} = \mathbf{U}_x{}^T, \quad \mathbf{U}_y := \begin{bmatrix} \mathbf{0} & -\hat{\mathbf{y}} \times \mathbf{I}^{(3)} \\ \hat{\mathbf{y}} \times \mathbf{I}^{(3)} & \mathbf{0} \end{bmatrix} = \mathbf{U}_y{}^T$$

$$\mathbf{U}_z := \begin{bmatrix} \mathbf{0} & -\hat{\mathbf{z}} \times \mathbf{I}^{(3)} \\ \hat{\mathbf{z}} \times \mathbf{I}^{(3)} & \mathbf{0} \end{bmatrix} = \mathbf{U}_z{}^T = \left[\begin{array}{ccc|ccc} & & & 0 & 1 & 0 \\ & \mathbf{0} & & -1 & 0 & 0 \\ & & & 0 & 0 & 0 \\ \hline 0 & -1 & 0 & & & \\ 1 & 0 & 0 & & \mathbf{0} & \\ 0 & 0 & 0 & & & \end{array}\right] \tag{2.25}$$

(Note that $\mathbf{U}_x$, $\mathbf{U}_y$, $\mathbf{U}_z$ and $\mathbf{D}$ are all symmetric). Now assume the medium to be plane stratified with the z-axis normal to the stratification. We denote the projection of $\mathbf{k}$ on the stratification plane by $\mathbf{k}_t$,

$$\mathbf{k}_t := (k_x, k_y) = k_0(s_x, s_y) \tag{2.26}$$

where s_x and s_y are propagation constants (Snell's law). Fourier-transforming $\mathbf{e}(\mathbf{r})$ and $\mathbf{j}(\mathbf{r})$ in (2.24) in the transverse (stratification) plane, we have typically

$$\mathbf{e}(\mathbf{r}) = \frac{k_0^2}{4\pi^2} \iint \mathbf{e}(\mathbf{k}_t, z) \exp[-ik_0(s_x x + s_y y)]\, ds_x\, ds_y \tag{2.27}$$

$$\mathbf{e}(\mathbf{k}_t, z) = \iint \mathbf{e}(\mathbf{r}) \exp[ik_0(s_x x + s_y y)]\, dx\, dy \tag{2.28}$$

Substitution in (2.24), (with K independent of x and y), yields

$$ik_0 \left[c\mathsf{K}(z) - s_x \mathsf{U}_x - s_y \mathsf{U}_y - \frac{i}{k_0} \mathsf{U}_z \frac{d}{dz}\right] \mathbf{e}(\mathbf{k}_t, z) = -\mathbf{j}(\mathbf{k}_t, z)$$

or, more concisely

$$ik_0 \left[\mathsf{C} - \frac{i}{k_0} \mathsf{U}_z \frac{d}{dz}\right] \mathbf{e}(\mathbf{k}_t, z) \equiv \mathsf{L}\, \mathbf{e}(\mathbf{k}_t, z) = -\mathbf{j}(\mathbf{k}_t, z) \tag{2.29}$$

where

$$\mathsf{C} := [c\mathsf{K} - s_x \mathsf{U}_x - s_y \mathsf{U}_y]$$

and L is the Maxwell operator:

$$\mathsf{L} := ik_0 \left[\mathsf{C} - \frac{i}{k_0} \mathsf{U}_z \frac{d}{dz}\right]$$

2.2.3 Eigenmodes in the plane-stratified medium

In order to find the eigenmodes of the plane-stratified medium we set the source term in (2.29) to zero

$$\mathsf{L}\,\mathbf{e} := ik_0 \left[\mathsf{C} - \frac{i}{k_0} \mathsf{U}_z \frac{d}{dz}\right] \mathbf{e}(\mathbf{k}_t, z) = 0 \tag{2.30}$$

and assume local plane-wave solutions

$$\mathbf{e}_\alpha(\mathbf{k}_t, z) = \mathbf{e}_\alpha(\mathbf{k}_t) \exp(-ik_0\, q_\alpha z) \tag{2.31}$$

to obtain the eigenmode equation

$$[\mathsf{C} - q_\alpha\, \mathsf{U}_z]\, \mathbf{e}_\alpha(\mathbf{k}_t, z) = 0 \tag{2.32}$$

Since there are two null rows and columns in U_z, the equation

$$\det [\mathsf{C} - q_\alpha\, \mathsf{U}_z] = 0 \tag{2.33}$$

gives a quartic equation in q_α (the Booker quartic (1.109), discussed in Sec. 1.3), yielding two positive- and two negative-going waves with respect to the z-axis, corresponding to $\alpha = \pm 1, \pm 2$. The eigenvectors $\mathbf{e}_\alpha$ give the characteristic wave polarizations (the the ratios of the various wave-field components) corresponding to each eigenvalue q_α.

2.2.4 The Lagrange identity and the bilinear concomitant

We now construct the equation adjoint to (2.30) by changing the sign of the differential operator d/dz and replacing C by its transpose C^T (i.e. replacing K by K^T):

$$\bar{\mathsf{L}}\,\bar{\mathrm{e}} := ik_0 \left[\mathsf{C}^T + \frac{i}{k_0}\mathsf{U}_z \frac{d}{dz}\right] \bar{\mathrm{e}}(\mathbf{k}_t, z) = 0 \tag{2.34}$$

where $\bar{\mathsf{L}}$ is the adjoint Maxwell operator, and $\bar{\mathrm{e}}(\mathbf{k}_t,\mathrm{z})$ now denotes an *adjoint wave field*, satisfying the adjoint Maxwell equations.

We should note at this point that in the case of a cold magnetoplasma permeated by an external magnetic field $\mathbf{b}$, the constitutive tensor $\mathsf{K}\equiv\mathsf{K}(\mathbf{b})$ has the general form

$$\mathsf{K} = \left[\begin{array}{cc} \boldsymbol{\varepsilon} & \mathbf{0} \\ \mathbf{0} & \mu_0\,\mathsf{I}^{(3)} \end{array}\right] =: \left[\begin{array}{cc} \varepsilon_0(\mathsf{I}^{(3)} + \chi) & \mathbf{0} \\ \mathbf{0} & \mu_0\,\mathsf{I}^{(3)} \end{array}\right] \tag{2.35}$$

defining, for later use, the susceptibility matrix χ; $\boldsymbol{\varepsilon}\equiv\boldsymbol{\varepsilon}(\mathbf{b})$(1.38) is given by

$$\frac{\boldsymbol{\varepsilon}}{\varepsilon_0} = S\,(\mathsf{I} - \hat{\mathbf{b}}\hat{\mathbf{b}}^T) - iD\,\hat{\mathbf{b}} \times \mathsf{I} + P\,\hat{\mathbf{b}}\hat{\mathbf{b}}^T\,, \qquad \hat{\mathbf{b}} := \mathbf{b}/|\mathbf{b}|$$

S, D and P being parameters of the medium, cf. (1.37), (1.40) and (1.39); $\boldsymbol{\varepsilon}(\hat{\mathbf{b}})$ is clearly gyrotropic, i.e. $\boldsymbol{\varepsilon}(-\mathbf{b})=\boldsymbol{\varepsilon}^T(\mathbf{b})$, by virtue of the antisymmetric term $\hat{\mathbf{b}}\times\mathsf{I}$, and so too are $\mathsf{K}(\mathbf{b})$ (2.35) and $\mathsf{C}(\mathbf{b})$ (2.29),

$$\mathsf{K}(-\mathbf{b}) = \mathsf{K}^T(\mathbf{b})\,, \qquad \mathsf{C}(-\mathbf{b}) = \mathsf{C}^T(\mathbf{b}) \tag{2.36}$$

The given and adjoint operators L and $\bar{\mathsf{L}}$ will obey the *Lagrange identity*

$$\bar{\mathrm{e}}^T\mathsf{L}\,\mathrm{e} - \mathrm{e}^T\bar{\mathsf{L}}\,\bar{\mathrm{e}} = \nabla\cdot\mathbf{P} \tag{2.37}$$

where the vector $\mathbf{P}$ is called the *bilinear concomitant* [94, Sec. 7.5]. In our case, (2.34), the differential operator is just the z-component of ∇, and remembering that $\bar{\mathrm{e}}^T\mathsf{C}\,\mathrm{e}$ is a scalar which is equal to its transpose, and therefore eliminated on subtraction in (2.37), we find

$$\left[\bar{\mathrm{e}}^T\mathsf{U}_z\frac{d\mathrm{e}}{dz} + \mathrm{e}^T\mathsf{U}_z\frac{d\bar{\mathrm{e}}}{dz}\right] = \frac{d}{dz}\left[\bar{\mathrm{e}}^T\mathsf{U}_z\,\mathrm{e}\right]$$

which, with $\mathsf{L}\,\mathrm{e} = \bar{\mathsf{L}}\,\bar{\mathrm{e}} = 0$, (2.30) and (2.34), gives

$$\frac{d}{dz}\left[\bar{\mathrm{e}}^T\mathsf{U}_z\,\mathrm{e}\right] = 0 \tag{2.38}$$

Hence the (z-component of the) bilinear concomitant vector is a constant

$$\bar{\mathrm{e}}^T\mathsf{U}_z\,\mathrm{e} = P_z = \mathrm{const} \tag{2.39}$$

an important result that we shall require later.

If we were to consider an arbitrary source-free medium, i.e. not necessarily plane-stratified, governed by (2.21) with $\mathbf{j}(\mathbf{r}) = 0$, we would write the formally adjoint equation as before [46, p. 234–236] by replacing K by its transpose K^T, and the differential operator D (2.22) by its negative transpose $-\mathsf{D}^T$. If therefore the Maxwell system, (2.21) or (2.24) with $\mathbf{j}(\mathbf{r}) = 0$, is given by

$$\mathsf{L}\,\mathsf{e} := [i\omega\mathsf{K} + \mathsf{D}]\,\mathsf{e}(\mathbf{r}) = \left[i\omega\mathsf{K} + \mathsf{U}_x\frac{\partial}{\partial x} + \mathsf{U}_y\frac{\partial}{\partial y} + \mathsf{U}_z\frac{\partial}{\partial z}\right]\mathsf{e}(\mathbf{r}) = 0 \tag{2.40}$$

the formally adjoint equation, with $\mathsf{D}^T = \mathsf{D}$, (2.22), will be

$$\bar{\mathsf{L}}\,\bar{\mathsf{e}} := \left[i\omega\mathsf{K}^T - \mathsf{D}^T\right]\bar{\mathsf{e}}(\mathbf{r}) = \left[i\omega\mathsf{K}^T - \mathsf{U}_x\frac{\partial}{\partial x} - \mathsf{U}_y\frac{\partial}{\partial y} - \mathsf{U}_z\frac{\partial}{\partial z}\right]\bar{\mathsf{e}}(\mathbf{r}) = 0 \tag{2.41}$$

Application of the Lagrange identity, (2.37), then yields the result

$$\nabla\cdot\mathbf{P} = 0, \qquad \mathbf{P} := \mathbf{E}\times\overline{\mathbf{H}} + \overline{\mathbf{E}}\times\mathbf{H} \tag{2.42}$$

and the expression $\bar{\mathsf{e}}^T\mathsf{U}_z\,\mathsf{e}$ appearing in (2.39) is seen to be the z-component of the Poynting-like product in (2.42). This bilinear concomitant vector $\mathbf{P}$ was introduced by Budden and Jull [35] in their study of reciprocity of ray paths in magnetoionic media, and a variant of it was used by Pitteway and Jespersen [100] to derive their reciprocity theorem discussed in Sec. 3.2.4.

It should be remarked that the above prescription ($\mathsf{K} \to \mathsf{K}^T$, $\mathsf{D} \to -\mathsf{D}^T$) for forming the adjoint system is not unique, and any other prescription that will satisfy a Lagrange identity like (2.37) is equally valid. The particular form chosen by us, yields a bilinear concomitant vector $\mathbf{P}$ (2.42) which reduces to the time-averaged Poynting vector in loss-free media (see Sec. 2.3.1), and is particularly useful in the applications discussed in this and the next chapters. Other prescriptions may be formulated by certain orthogonal transformations of the adjoint Maxwell system. They have been used by Kong and Cheng [84] and by Kerns [81], and are useful in analysing Lorentz-type reciprocity when the waves which are compared travel in opposite directions (i.e. when the wave vectors are reversed in **k**-space, or the roles of receiving and transmitting antennas are interchanged in real space). Such transformed adjoint systems are introduced in Sec. 3.4, and discussed in some detail in Chaps. 4 and 6.

2.2.5 Biorthogonality of the given and adjoint eigenmodes

We now derive another important result that links the given and adjoint eigenvectors. Assuming local plane-wave solutions to the adjoint equation (2.34) of the form

$$\bar{\mathbf{e}}_\beta(\mathbf{k}_t, z) = \bar{\mathbf{e}}_\beta(\mathbf{k}_t)\exp(ik_0\,\bar{q}_\beta\, z) \tag{2.43}$$

we obtain the adjoint eigenmode equation

$$\left[\mathbf{C}^T - \bar{q}_\beta \mathbf{U}_z\right]\bar{\mathbf{e}}_\beta(\mathbf{k}_t, z) = 0 \tag{2.44}$$

The eigenvalues are determined by

$$\det\left[\mathbf{C}^T - \bar{q}_\beta \mathbf{U}_z\right] = 0 \tag{2.45}$$

which is seen to give the same quartic equation in q as (2.33). Hence the given and adjoint eigenvalues are identical

$$\bar{q}_\beta = q_\beta \tag{2.46}$$

This implies that the given and adjoint modal refractive indices are also equal

$$\bar{n}_\beta(s_x, s_y, \bar{q}_\beta) = n_\beta(s_x, s_y, q_\beta) \tag{2.47}$$

Note however that q_β and $\bar{q}_\beta$ appear in the plane-wave representations (2.31) and (2.43) with opposite signs, but since both representations have the same $\exp(i\omega t)$ time dependence, this means that the given and adjoint waves propagate in opposite directions with respect to the z-axis (but of course in the same transverse direction, since $\bar{\mathbf{k}}_t = \mathbf{k}_t$).

Again applying the Lagrange identity (2.37) to the eigenmode equations (2.32) and (2.44), and remembering that $\bar{q}_\beta = q_\beta$, we find that

$$(q_\beta - q_\alpha)\bar{\mathbf{e}}_\beta^T \mathbf{U}_z\, \mathbf{e}_\alpha = 0 \tag{2.48}$$

which gives the well known biorthogonality relation [105,18,53] between the given and adjoint eigenmodes

$$\bar{\mathbf{e}}_\beta^T \mathbf{U}_z\, \mathbf{e}_\alpha = \text{const}\,\delta_{\alpha\beta} = \delta_{\alpha\beta}\, P_{z,\alpha} \tag{2.49}$$

with the aid of (2.39). If the eigenmodes are suitably normalized this relation may be written

$$\hat{\bar{\mathbf{e}}}_\beta^T\, \mathbf{U}_z\, \hat{\mathbf{e}}_\alpha = \delta_{\alpha\beta}\,\mathrm{sgn}(\alpha), \qquad \alpha,\,\beta = \pm 1, \pm 2 \tag{2.50}$$

To discuss the nature of the normalization, it will be necessary to define the *amplitude* of an eigenmode, and this will be crucial to the scattering theorems which will be derived later, in which ingoing and outgoing eigenmode amplitudes will be related.

2.3 The amplitude of an eigenmode

2.3.1 Amplitude in a loss-free medium

In discussing the propagation of a characteristic (eigen-) mode in a plane-stratified medium in which there are no collisional losses, it is useful to define the modal amplitude as the square root of the z-component (normal to the stratification) of the time-averaged Poynting vector (see, for instance, [100, 126]). If the medium varies slowly, so that there are no losses due to reflection or to mode coupling, it will be shown that the amplitude is conserved, i.e. it will remain constant even though the parameters of the medium vary in the direction normal to the stratification.

The time-averaged Poynting vector $\langle \mathbf{S} \rangle$ is given by

$$\langle \mathbf{S} \rangle = \mathbf{E} \times \mathbf{H}^\star + \mathbf{E}^\star \times \mathbf{H} \tag{2.51}$$

aside from a factor $1/4$ which we have absorbed into $\langle \mathbf{S} \rangle$, and its z-component is given by

$$\langle S_z \rangle = \tilde{\mathrm{e}}^\star \mathsf{U}_z \, \mathrm{e} \tag{2.52}$$

Now the complex-conjugate wave field $\mathrm{e}^\star$ obeys an equation given by the complex conjugate of (2.30)

$$\left[\mathsf{C}^\star + \frac{i}{k_0} \mathsf{U}_z \frac{d}{dz}\right] \mathrm{e}^\star(\mathbf{k}_t, z) = \left[\mathsf{C}^T + \frac{i}{k_0} \mathsf{U}_z \frac{d}{dz}\right] \mathrm{e}^\star(\mathbf{k}_t, z) = 0 \tag{2.53}$$

since the dielectric tensor $\boldsymbol{\varepsilon}$ in C is hermitian. [The hermiticity of $\boldsymbol{\varepsilon}$, or of K, can be shown to stem from the requirement of energy conservation (see for instance [1, p. 9] or [113, p. 65]) and conversely, as in the present discussion, will be shown to lead to energy conservation].

Similarly, the complex-conjugate eigenwave field for a progressive plane wave (q_α real) becomes, from (2.32)

$$[\mathsf{C}^\star - q_\alpha \, \mathsf{U}_z] \, \mathrm{e}^\star = \left[\mathsf{C}^T - q_\alpha \, \mathsf{U}_z\right] \, \mathrm{e}^\star = 0 \tag{2.54}$$

Thus the complex-conjugate wave fields in loss-free media obey the adjoint Maxwell equations, and eqs. (2.38) and (2.39) will apply here too, with

$$\tilde{\mathrm{e}}^\star \mathsf{U}_z \, \mathrm{e} = P_z = \mathrm{const} \tag{2.55}$$

Comparison with (2.52) gives

$$\langle S_z \rangle = P_z = \mathrm{const} \tag{2.56}$$

so that in loss-free media the (z-component of the) bilinear concomitant is seen to be just the (z- component of the) mean Poynting vector, as already noted by Budden and Jull [35] and others [118, 119, 12], which expresses conservation of mean energy flux. Analogy with (2.49) also gives the biorthogonality of the given and complex-conjugate eigenmodes in loss-free media

$$\tilde{\mathbf{e}}_\alpha^\star \mathbf{U}_z \, \mathbf{e}_\beta = \delta_{\alpha\beta} \, P_{z,\alpha} \tag{2.57}$$

We could now define a modal amplitude (or at least its modulus) by equating its square to the modal energy flux

$$\tilde{\mathbf{e}}_\alpha^\star \mathbf{U}_z \, \mathbf{e}_\beta = \mathrm{sgn}(\alpha) \, \delta_{\alpha\beta} \, |a_\alpha|^2 \tag{2.58}$$

and then define normalized modal wave fields, $\hat{\mathbf{e}}_\alpha$ or $\hat{\mathbf{e}}_\alpha^\star$, by dividing the given fields by the modulus of the amplitudes:

$$\mathbf{e}_\alpha = |a_\alpha| \; \hat{\mathbf{e}}_\alpha \qquad\qquad \mathbf{e}_\alpha^\star = |a_\alpha| \; \hat{\mathbf{e}}_\alpha^\star$$

thereby letting the normalized wave fields carry the phase information of the given fields. Such a procedure is manifestly unsatisfactory, in that a normalized wave field would not be uniquely defined at a given level, and it is preferable to let the complex amplitude carry the phase information by letting it have the same phase as one of the components of $\mathbf{e}_\alpha$, say $\mathbf{e}_x$ or $\mathbf{e}_\xi$ (depending on the coordinate system in which the components of $\mathbf{e}_\alpha$ are expressed). We then have

$$\mathbf{e}_\alpha = a_\alpha \hat{\mathbf{e}}_\alpha \qquad\qquad \mathbf{e}_\alpha^\star = a_\alpha^\star \, \hat{\mathbf{e}}_\alpha^\star \tag{2.59}$$

In either case a normalized modal wave field is that which generates unit energy flux normal to the stratification:

$$(\hat{\mathbf{e}}_\alpha^\star)^T \mathbf{U}_z \, \hat{\mathbf{e}}_\beta = \mathrm{sgn}(\alpha) \, \delta_{\alpha\beta} \tag{2.60}$$

Now an arbitrary wave field $\mathbf{e}(\mathbf{k}_t)$ can be expressed as a linear superposition of the eigenmodes of the medium

$$\mathbf{e}(\mathbf{k}_t) = \sum_{\alpha=\pm 1, \pm 2} a_\alpha \hat{\mathbf{e}}_\alpha(\mathbf{k}_t) \tag{2.61}$$

where, by virtue of (2.60),

$$a_\alpha = (\hat{\mathbf{e}}_\alpha^\star)^T \mathbf{U}_z \, \mathbf{e} \, \mathrm{sgn}(\alpha), \qquad\qquad a_\alpha^\star = (\hat{\mathbf{e}}_\alpha)^T \mathbf{U}_z \, \mathbf{e}^\star \mathrm{sgn}(\alpha) \tag{2.62}$$

and hence, with $\mathbf{e} = \mathbf{e}(\mathbf{k}_t)$,

$$\begin{aligned} \langle S_z \rangle = \tilde{\mathbf{e}}^\star \mathbf{U}_z \, \mathbf{e} &= a_1^\star a_1 + a_2^\star a_2 - a_{-1}^\star a_{-1} - a_{-2}^\star a_{-2} \\ &= \sum_\alpha |a_\alpha|^2 \; \mathrm{sgn}(\alpha) \end{aligned} \tag{2.63}$$

We have thus expressed the energy flux normal to the stratification of an arbitrary wave field as the algebraic sum of the energy fluxes of the component eigenmodes.

Results analogous to those derived in this section (modal orthogonality in loss-free media and separation of overall energy flux into contributions of the component eigenmodes) have been given by Marcuse [92, Sec. 8.5] in his discussion of optical fibres and dielectric waveguides having cylindrical symmetry. There the form of the modes is dictated by the geometry of the problem (i.e. by the boundary conditions) and by a radiation condition at infinity, but the formalism is somewhat similar. In Sec. 2.6 we discuss the problem of curved stratified media in some detail.

Suppose we wish to determine the z-component of the energy flux associated with an eigenmode in a loss-free magnetoplasma. One method (not necessarily the simplest) would be to determine the eigenmode components (the wave polarizations) in the (ξ, η, ς) coordinate system (1.77), in which the ς-axis is along the wave-normal direction and the ξ-axis is in the plane spanned by the wave normal and the external magnetic field, cf. (1.82)–(1.85) in Sec. 1.2,

$$\begin{aligned} \mathbf{E} &:= (E_\xi,\, E_\eta,\, E_\varsigma) = (1,\, \rho,\, \sigma)\, E_\xi \\ \mathbf{H} &:= (H_\xi,\, H_\eta,\, H_\varsigma) = Y_0(-\rho,\, 1,\, 0)\, nE_\xi \end{aligned} \tag{2.64}$$

where ρ is purely imaginary and σ purely real, as may be seen from (1.81) in Sec. 1.2 with S, P, D and n^2 all real. $Y_0 \equiv 1/Z_0 := (\varepsilon_0/\mu_0)^{1/2}$ is the free-space admittance. The mean Poynting vector becomes

$$\begin{aligned} \langle \mathbf{S} \rangle &= Y_0\{-(\sigma + \sigma^\star),\ -(\rho^\star\sigma + \rho\sigma^\star),\ 2(1 + \rho\rho^\star)\}\ nE_\xi^\star E_\xi \\ &= 2Y_0\{-\sigma,\ 0,\ 1 - \rho^2\}\, nE_\xi^\star E_\xi \end{aligned} \tag{2.65}$$

and the z-component of $\langle \mathbf{S} \rangle$, as well as the components of $\mathbf{e}_\alpha$ if required, are then determined by a coordinate transformation from the $(\xi,\ \eta,\ \varsigma)$ to the (x, y, z) system (1.80).

2.3.2 Amplitude of an eigenmode in the general case

Normalization of the wave fields

For lossy media the constitutive tensors are no longer hermitian, and the orthogonality of eigenmodes with respect to the complex-conjugate modes is thereby lost, together with the manifest advantage of being able to express modal amplitudes via the complex-conjugate wave fields.

The *adjoint* wave fields, however, retain their biorthogonality with respect to the given fields (2.49), and we may use this property in the definition of modal amplitudes which will be valid for lossy media too. The constant bilinear concomitant $P_z = \bar{\mathbf{e}}^T \mathbf{U}_z \mathbf{e}$ (2.39) evidently no longer represents the z-component of the mean Poynting vector if absorption is present, since the energy flux would attenuate in the direction of propagation of the wave. The point is that the amplitude of an eigenmode, a_α, is no longer equal in magnitude to the amplitude, $\bar{a}_\alpha$, of the adjoint eigenmode since the constancy of the Poynting cross product (2.49) implies that as $\mathbf{e}_\alpha$ attenuates, $\bar{\mathbf{e}}_\alpha$ will grow correspondingly.

To obtain the adjoint eigenmode components it is convenient to express field quantities in the (ξ, η, ς) system, as in (2.64). In a magnetoplasma, as pointed out in Sec. 2.2.4, the adjoint medium is obtained by reversing the direction of the external magnetic field b, so that the transverse wave polarization $\rho{:=}E_\eta/E_\xi$ (1.84) in this system is reversed in sign, while the longitudinal polarization $\sigma{:=}E_\varsigma/E_\xi$ (1.85) is unchanged, cf. (1.78),

$$\bar{\rho}_\alpha = -\rho_\alpha, \qquad \bar{\sigma}_\alpha = \sigma_\alpha \tag{2.66}$$

(the corresponding relations for the complex-conjugate polarizations

$$\rho^\star_\alpha = -\rho_\alpha, \qquad \sigma^\star_\alpha = \sigma_\alpha$$

are valid only in loss-free media). Hence, if the modal field

$$\mathbf{e}_\alpha \equiv (E_\xi,\ E_\eta,\ E_\varsigma;\ H_\xi,\ H_\eta,\ H_\varsigma)_\alpha$$

has the form

$$\mathbf{E}_\alpha = (1,\ \rho_\alpha,\ \sigma_\alpha)\, E_{\alpha\xi}, \qquad \mathbf{H}_\alpha = Y_0(-\rho_\alpha,\ 1,\ 0)\, n_\alpha\, E_{\alpha\xi} \tag{2.67}$$

the adjoint field will be

$$\overline{\mathbf{E}}_\alpha = (1,\ -\rho_\alpha,\ \sigma_\alpha)\overline{E}_{\alpha\xi}, \qquad \overline{\mathbf{H}}_\alpha = Y_0(\rho_\alpha,\ 1,\ 0)\, n_\alpha \overline{E}_{\alpha\xi} \tag{2.68}$$

and we may form the bilinear concomitant vector $\mathbf{P}$ from the Poynting-like product

$$\begin{aligned} \mathbf{P}_\alpha &= \mathbf{E}_\alpha \times \overline{\mathbf{H}}_\alpha + \overline{\mathbf{E}}_\alpha \times \mathbf{H}_\alpha \\ &= 2Y_0(-\sigma_\alpha,\ 0,\ 1 - {\rho_\alpha}^2)\, n_\alpha \overline{E}_{\alpha\xi} E_{\alpha\xi} \end{aligned} \tag{2.69}$$

which is formally identical to (2.65), except that both σ and ρ^2 may now be complex. The z-component may now be obtained by a coordinate transformation (1.80) to the (x, y, z) system

$$P_{z,\alpha} = \bar{\mathbf{e}}^T_\alpha \mathbf{U}_z \mathbf{e}_\alpha = 2Y_0(-\sigma_\alpha,\ 0,\ 1 - {\rho_\alpha}^2)_z\, n_\alpha \overline{E}_{\alpha\xi} E_{\alpha\xi} \tag{2.70}$$

This is a convenient representation to use for normalizing eigenmodes. If we choose

$$\widehat{E}_{\alpha\xi} = \widehat{\widehat{E}}_{\alpha\xi} = \left\{2Y_0 n_\alpha \operatorname{sgn}(\alpha)\,(-\sigma_\alpha,\ 0,\ 1-\rho_\alpha{}^2)_z\right\}^{-\frac{1}{2}} \tag{2.71}$$

we can define the normalized eigenfields through (2.67) and (2.68)

$$\begin{aligned} \hat{\mathbf{e}}_\alpha &= (1,\ \rho_\alpha,\ \sigma_\alpha\ ;\ \ -Y_0\, n_\alpha \rho_\alpha,\ Y_0\, n_\alpha,\ 0)\,\widehat{E}_{\alpha\xi} \\ \hat{\bar{\mathbf{e}}}_\alpha &= (1,\ -\rho_\alpha,\ \sigma_\alpha\ ;\ \ Y_0\, n_\alpha \rho_\alpha,\ Y_0\, n_\alpha,\ 0)\,\widehat{\widehat{E}}_{\alpha\xi} \end{aligned} \tag{2.72}$$

which yield immediately the required biorthogonality normalization

$$(\hat{\bar{\mathbf{e}}}_\alpha)^T \mathbf{U}_z\, \hat{\mathbf{e}}_\beta = \delta_{\alpha\beta} \operatorname{sgn}(\alpha) \tag{2.73}$$

Eigenmode amplitudes

We now relate a modal wave field to a normalized field via the modal amplitude, as in the previous section,

$$\mathbf{e}_\alpha = a_\alpha\, \hat{\mathbf{e}}_\alpha, \qquad\qquad \bar{\mathbf{e}}_\alpha = \bar{a}_\alpha\, \hat{\bar{\mathbf{e}}}_\alpha \tag{2.74}$$

so that (2.49) becomes

$$\bar{\mathbf{e}}_\alpha^T \mathbf{U}_z\, \mathbf{e}_\beta = \operatorname{sgn}(\alpha)\, \delta_{\alpha\beta}\ \bar{a}_\alpha\, a_\alpha \tag{2.75}$$

Now an arbitrary wave field $\mathbf{e}(\mathbf{k}_t)$, as well as its adjoint, can be expressed in terms of the eigenmodes

$$\mathbf{e}(\mathbf{k}_t) = \sum_{\alpha=\pm1,\pm2} a_\alpha\, \hat{\mathbf{e}}_\alpha(\mathbf{k}_t), \qquad\qquad \bar{\mathbf{e}}(\mathbf{k}_t) = \sum_{\alpha=\pm1,\pm2} \bar{a}_\alpha\, \hat{\bar{\mathbf{e}}}_\alpha(\mathbf{k}_t) \tag{2.76}$$

whence

$$a_\alpha = (\hat{\bar{\mathbf{e}}}_\alpha)^T \mathbf{U}_z\, \mathbf{e}\ \operatorname{sgn}(\alpha), \qquad \bar{a}_\alpha = \hat{\mathbf{e}}_\alpha^T \mathbf{U}_z\, \bar{\mathbf{e}}\ \operatorname{sgn}(\alpha) \tag{2.77}$$

by virtue of (2.73) and (2.76).

Finally we have the generalized Poynting flux density (2.39)

$$\begin{aligned} P_z &= [\mathbf{E}\times\overline{\mathbf{H}} + \overline{\mathbf{E}}\times\mathbf{H}]_z = \bar{\mathbf{e}}^T \mathbf{U}_z\, \mathbf{e} \\ &= \bar{a}_1 a_1 + \bar{a}_2 a_2 - \bar{a}_{-1} a_{-1} - \bar{a}_{-2} a_{-2} \\ &= \sum_\alpha \bar{a}_\alpha\, a_\alpha \operatorname{sgn}(\alpha) = \text{const} \end{aligned} \tag{2.78}$$

expressed as the sum of the generalized flux densities of the eigenmodes, by analogy with (2.63).

We remark in conclusion that the procedure adopted here for determining an eigenmode amplitude in an absorbing medium may seem somewhat cumbersome, but it is straightforward and easily incorporated into a computer program for calculating wave fields in plane-stratified media. In most cases of practical interest the aim of such calculations is to determine fields or scattering coefficients outside the absorbing regions, where the squares of the modal amplitudes reduce simply to the z-components (normal to the stratification) of the Poynting flux of each mode. For our purposes, however, the important result is that modal amplitudes can *in principle* be defined in absorbing (and hence in all) media which, in conjunction with modal biorthogonality, permits the decomposition of generalized energy flux into the sum of the contributions of each of the eigenmodes.

2.4 The conjugate wave fields

2.4.1 The physical content of the conjugate problem

In our review in Sec. 2.1 we noted that earlier scattering (reciprocity) theorems for plane-stratified magnetoplasmas related the ingoing and outgoing amplitudes of waves incident from two different directions — the given and *conjugate* directions. If the transverse components (i.e. in the plane of the statification) of the incident wave vector are $\mathbf{k}_t = k_0(s_x, s_y)$, those of the conjugate wave vector are defined to be $\mathbf{k}_t^c = k_0(-s_x, s_y)$ see Fig. 2.2. To characterize the relation between the given and conjugate wave vectors geometrically, Barron and Budden [21], Pitteway and Jespersen [100] and others, when discussing incidence on the earth's ionosphere from below, have pointed out that the planes of incidence in the two cases are 'symmetrically disposed about the vertical East–West plane, at right angles to the magnetic meridian plane', i.e. if the plane of incidence in the one case is at an azimuthal angle ϕ with respect to the meridian plane, the conjugate plane of incidence is at an angle $(\pi - \phi)$.

This characterization, although perfectly true, concealed the physical nature of the symmetry. With the hindsight provided by a number of later papers, [108, 9, 11], we note that the given and conjugate planes of incidence are reflections with respect to the *magnetic meridian plane*. But this is only part of the story.

If we take the original (given) problem and perform a reflection mapping, $\mathcal{R}$, with respect to some arbitrary plane, then all proper (polar) vectors, such as the position vector $\mathbf{r}$, the electric field $\mathbf{E}(\mathbf{r})$ or the electric current density $\mathbf{J}_e(\mathbf{r})$, undergo 'geometric mirroring', in the sense that physical arrows would

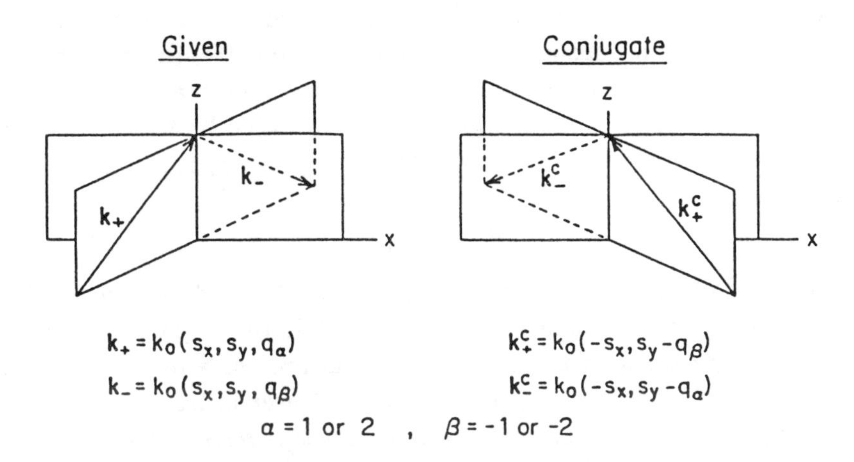

Figure 2.2: Given and conjugate eigenmodes.
The z-axis is normal to the stratification, and the external magnetic field lies in the (x, z) plane (the magnetic meridian plane).

be imaged by a mirror. All axial (pseudo-) vectors, on the other hand, such as the wave field $\mathbf{H}(\mathbf{r})$, the external magnetic field $\mathbf{b}$ or the equivalent magnetic current density $\mathbf{J}_\mathrm{m}(\mathbf{r})$ undergo mirroring too, but in addition, are ***reversed in direction***. Such mappings will be considered in some detail in Chap. 6. It is well known (and will be demonstrated in Sec. 6.2) that Maxwell's equations are invariant under such orthogonal mappings.

We now perform a time-reversal transformation, $\mathcal{T}$, on the reflected problem. The operation $\mathcal{T}$ can be visualized by imagining the original process to have been recorded on a movie film, and then observed when the film is run backwards. Maxwell's equations are invariant under time reversal, as will be demonstrated in Chap. 7, and it will be shown in particular that quantities such as as $\mathbf{H}$, $\mathbf{B}$, $\mathbf{J}_\mathrm{e}$ and $\mathbf{S}$ (the Poynting vector) are odd (i.e. change sign) under time reversal, whereas $\mathbf{E}$, $\mathbf{D}$ and $\mathbf{J}_\mathrm{m}$ are even. For our purposes this means that the combined action of $\mathcal{R}$ and $\mathcal{T}$ leaves the original external magnetic field $\mathbf{b}$ unchanged, i.e.

$$\mathcal{T}\mathcal{R}\,\mathbf{b} = \mathcal{R}\mathcal{T}\,\mathbf{b} = \mathbf{b}$$

and the mapped eigenmodes (i.e. reflected and time-reversed) will remain eigenmodes of the (unchanged) mapped medium.

Absorption losses in the medium require special attention. These will be expressed in the constitutive tensor K through an imaginary term $i\nu$, where ν is the effective collision frequency. Time reversal, as will be shown in Sec. 7.2, has the effect of changing the sign of the collision term, or to be more precise, converts the constitutive tensor into its complex conjugate, thereby changing the sign not only of ν, but of $\mathbf{b}$ which appears also as an imaginary term, $i\mathbf{b}$, in gyrotropic media. The effect of time reversal will then be to transform the eigenvalue q_α in the plane-wave representation, $\exp(-ik_0\, q_\alpha\, z)$ into its negative complex conjugate,

$$\mathcal{T} q_\alpha = -q_\alpha^\star$$

so that a damped plane wave, propagating in the positive z-direction, would be transformed into a growing plane wave propagating in the negative z-direction. However, insofar as we wish to describe physical processes in a physical absorbing medium after applying our reflection-time-reversal transformation, we shall not transform the collision frequency. Under this ***restricted time reversal*** the wave eigenvectors $\mathbf{k}_\alpha$will reverse their directions

$$\mathbf{k}_\alpha(s_x, s_y, q_\alpha) \rightarrow \mathbf{k}_{-\alpha}(-s_x, -s_y, -q_\alpha)$$

(but not $q_\alpha \rightarrow -q_\alpha^\star$, which would yield growing waves), and the signs of the magnetic wave-field components will also be changed, leading to a reversal in direction of the Poynting vector.

This, then, is the rationale of the mathematical procedure (in itself quite rigorous) which will now be used to generate the 'conjugate eigenmodes' by a reflection-time-reversal transformation.

2.4.2 The conjugating transformation

The restricted time-reversal procedure

We start off by exhibiting explicitly the components, s_x and s_y, of $\mathbf{k}_t$ (2.26) in the eigenmode equation (2.32), as well as the dependence of K, and consequently of $\mathbf{e}_\alpha$, on the external magnetic field $\mathbf{b}$,

$$[c\,\mathsf{K}(\mathbf{b}) - s_x\,\mathsf{U}_x - s_y\,\mathsf{U}_y - q_\alpha\,\mathsf{U}_z]\;\mathbf{e}_\alpha(\mathbf{b}; s_x,\, s_y) = 0 \qquad (2.79)$$

We reverse the direction of $\mathbf{b}$, so that the adjoint eigenmode equation, (2.44) in conjunction with (2.36), is satisfied by the field $\bar{\mathbf{e}}_\alpha(s_x,s_y)$, adjoint to $\mathbf{e}_\alpha(\mathbf{b};\, s_x, s_y)$

$$\mathsf{L}(-\mathbf{b})\,\bar{\mathbf{e}}_\alpha \equiv \bar{\mathsf{L}}\,\bar{\mathbf{e}}_\alpha := ik_0\,[c\mathsf{K}(-\mathbf{b}) - s_x\mathsf{U}_x - s_y\mathsf{U}_y - q_\alpha\mathsf{U}_z]\;\bar{\mathbf{e}}_\alpha(s_x, s_y) = 0 \qquad (2.80)$$

where we have used the result (2.46), $\bar{q}_\alpha = q_\alpha$, and it will be remembered that $\bar{\mathbf{e}}_\alpha$ has the plane-wave ansatz $\exp(ik_0 q_\alpha z)$ (2.43).

We note that (2.80) is also satisfied by the Maxwell field $\mathbf{e}_\alpha(-\mathbf{b}; s_x, s_y)$, i.e. by a physical wave field in the magnetic-field reversed medium, that has exactly the same wave polarization as the adjoint mode, but of course a different z-dependence. This polarization, and specifically the relation between the $\mathbf{E}$ and $\mathbf{H}$ fields, prescribes the direction of the Poynting flux, which will be consistent with the direction imposed by the sign of $\Im m(q_\alpha)$ in the Maxwell eigenmode $\mathbf{e}_\alpha(-\mathbf{b}; s_x, s_y)$, but inconsistent with the direction of propagation of the (unphysical) adjoint eigenmode.

Next we apply the *Poynting-vector reversing operator* $\bar{\mathsf{I}}$

$$\bar{\mathsf{I}} \equiv \bar{\mathsf{I}}^{(6)} := \begin{bmatrix} \mathsf{I}^{(3)} & 0 \\ 0 & -\mathsf{I}^{(3)} \end{bmatrix} = \bar{\mathsf{I}}^{-1} = \bar{\mathsf{I}}^{T} \tag{2.81}$$

[the direction of the Poynting vector of the wave field $\bar{\mathsf{I}}\mathbf{e}$ is opposite to that of the field $\mathbf{e}$] to (2.80):

$$\bar{\mathsf{I}}\left[c\mathsf{K}(-\mathrm{b}) - s_x\mathsf{U}_x - s_y\mathsf{U}_y - q_\alpha\mathsf{U}_z\right]\bar{\mathsf{I}}\,\bar{\mathsf{I}}\,\mathbf{e}_\alpha(-\mathrm{b}; s_x, s_y) = 0 \tag{2.82}$$

in which, for clarity, $\bar{\mathbf{e}}_\alpha(s_x, s_y)$ has been replaced by $\mathbf{e}_\alpha(-\mathbf{b}; s_x, s_y)$. Noting that

$$\bar{\mathsf{I}}\,\mathsf{U}_i\,\bar{\mathsf{I}} = -\mathsf{U}_i \qquad (i = x, y, z),$$

[see (2.25)], and

$$\bar{\mathsf{I}}\,\mathsf{K}\,\bar{\mathsf{I}} = \mathsf{K}$$

when K is of the form given by (2.35), we get

$$\left[c\mathsf{K}(-\mathrm{b}) + s_x\mathsf{U}_x + s_y\mathsf{U}_y + q_\alpha\mathsf{U}_z\right]\bar{\mathsf{I}}\,\mathbf{e}_\alpha(-\mathrm{b}; s_x, s_y) = 0 \tag{2.83}$$

This completes the (restricted) time-reversal transformation of the Maxwell system (2.79), and we now proceed to reflect the system with respect to the magnetic-meridian plane.

Reflection of wave fields

In general a (polar-) vector field, such as $\mathbf{E}(\mathbf{r})$, will be mapped by reflection with respect to the magnetic meridian plane, $y = 0$, into $\mathbf{E}'(\mathbf{r}') = \mathcal{R}\mathbf{E}(\mathbf{r})$, where

$$\mathbf{E}'(\mathbf{r}') = \mathsf{q}_y\mathbf{E}(\mathbf{r}), \qquad \mathbf{r}' = \mathsf{q}_y\mathbf{r}, \qquad \mathsf{q}_y := \begin{bmatrix} 1 & 0 & 0 \\ 0 & -1 & 0 \\ 0 & 0 & 1 \end{bmatrix} \tag{2.84}$$

On the other hand an axial-vector field such as $\mathbf{H(r)}$ is, in addition, reversed in sign on reflection, so that the overall reflected electromagnetic field $\mathcal{R}\mathbf{e(r)} \equiv \mathbf{e}'(\mathbf{r}')$ is given by

$$\mathbf{e}'(\mathbf{r}') = \mathbf{Q}_y\mathbf{e}(\mathbf{r}) \equiv \begin{bmatrix} \mathbf{q}_y & 0 \\ 0 & -\mathbf{q}_y \end{bmatrix} \begin{bmatrix} \mathbf{E(r)} \\ \mathbf{H(r)} \end{bmatrix}; \qquad \mathbf{r}' = \mathbf{q}_y\mathbf{r} \tag{2.85}$$

We apply the reflection matrix $\mathbf{Q}_y = \mathbf{Q}_y{}^{-1}$ to (2.83)

$$\mathbf{Q}_y \left[c\mathsf{K}(-\mathbf{b}) - s_x\mathbf{U}_x - s_y\mathbf{U}_y - q_\alpha\mathbf{U}_z\right] \mathbf{Q}_y \left\{\mathbf{Q}_y\, \bar{\mathsf{I}}\mathbf{e}_\alpha(-\mathbf{b}; s_x, s_y)\right\} = 0 \tag{2.86}$$

and note that

$$\mathbf{Q}_y\mathbf{U}_x\mathbf{Q}_y = \mathbf{U}_x, \qquad \mathbf{Q}_y\mathbf{U}_y\mathbf{Q}_y = -\mathbf{U}_y, \qquad \mathbf{Q}_y\mathbf{U}_z\mathbf{Q}_y = \mathbf{U}_z \tag{2.87}$$

Furthermore, if the magnetic field $\mathbf{b}$ is parallel to the $y = 0$ plane, then K, given by (2.35), with $\boldsymbol{\varepsilon}$ given by (1.45),

$$\mathsf{K} = \begin{bmatrix} \boldsymbol{\varepsilon} & \mathbf{0} \\ \mathbf{0} & \mu_0\,\mathsf{I}^{(3)} \end{bmatrix}, \qquad \boldsymbol{\varepsilon} = \varepsilon_0 \begin{bmatrix} S - C\hat{b}_x^2 & iD\hat{b}_z & -C\hat{b}_x\hat{b}_z \\ -iD\hat{b}_z & S & iD\hat{b}_x \\ -C\hat{b}_x\hat{b}_z & -iD\hat{b}_x & S - C\hat{b}_z^2 \end{bmatrix} \tag{2.88}$$

is magnetic-field reversed by $\mathbf{Q}_y$:

$$\mathbf{Q}_y\mathsf{K}(\mathbf{b})\mathbf{Q}_y = \mathsf{K}(-\mathbf{b}) \tag{2.89}$$

Hence (2.86) becomes

$$\begin{aligned} &\left[c\mathsf{K}(\mathbf{b}) + s_x\mathbf{U}_x - s_y\mathbf{U}_y + q_\alpha\mathbf{U}_z\right] \mathbf{Q}_y\, \bar{\mathsf{I}}\,\mathbf{e}_\alpha(-\mathbf{b}; s_x, s_y) \\ &\qquad = \left[\mathsf{C}(\mathbf{b}; -s_x, s_y) - q^c_{-\alpha}\mathbf{U}_z\right] \mathbf{e}^c_{-\alpha} = 0 \end{aligned} \tag{2.90}$$

with the notation of (2.29), and we have thereby formally identified the transformed (time-reversed, reflected) wave field as the *conjugate eigenmode*:

$$-\,q^c_{-\alpha} = q_\alpha = \bar{q}_\alpha, \qquad \mathbf{e}^c_{-\alpha}(\mathbf{b}; -s_x, s_y) = \mathbf{Q}_y\, \bar{\mathsf{I}}\,\mathbf{e}_\alpha(-\mathbf{b}; s_x, s_y) \tag{2.91}$$

with $q_\alpha = \bar{q}_\alpha$ taken from (2.46). In terms of the adjoint eigenmode this gives

$$\mathbf{e}^c_{-\alpha}(-s_x, s_y) = \mathbf{Q}_y\, \bar{\mathsf{I}}\,\bar{\mathbf{e}}_\alpha(s_x, s_y) \equiv \mathbf{Q}^c_y\bar{\mathbf{e}}_\alpha(s_x, s_y) \tag{2.92}$$

where the diagonal matrix $\mathbf{Q}^c_y$ is given by

$$\mathbf{Q}^c_y \equiv \mathbf{Q}_y\, \bar{\mathsf{I}} = \begin{bmatrix} \mathbf{q}_y & 0 \\ 0 & \mathbf{q}_y \end{bmatrix}; \qquad \mathbf{q}_y := \begin{bmatrix} 1 & 0 & 0 \\ 0 & -1 & 0 \\ 0 & 0 & 1 \end{bmatrix} \tag{2.93}$$

with

$$\mathbf{Q}_y^c = \tilde{\mathbf{Q}}_y^c = \left[\mathbf{Q}_y^c\right]^{-1}$$

Since the adjoint operation is involutary, i.e. $\bar{\bar{\mathbf{e}}} = \mathbf{e}$, (2.92) may be written as

$$\bar{\mathbf{e}}_{-\alpha}^c = \mathbf{Q}_y^c \mathbf{e}_\alpha \tag{2.94}$$

determining the adjoint of a mode in the conjugate system. We note that the *conjugating matrix* $\mathbf{Q}_y^c$ imposes 'geometrical mirroring' on both polar and axial vectors, i.e. it does not reverse the sign of the reflected (axial-vector) wave fields. This leads to a reversal of the direction of the z-component of the Poynting vector, so that upgoing waves are transformed into downgoing.

The conjugate modal amplitudes

We now apply (2.92) and (2.94) to relate the normalized eigenvectors and their adjoints in the given and conjugate problems:

$$\hat{\mathbf{e}}_\alpha^c = \mathbf{Q}_y^c \hat{\bar{\mathbf{e}}}_{-\alpha}, \qquad \hat{\bar{\mathbf{e}}}_\alpha = \mathbf{Q}_y^c \hat{\mathbf{e}}_{-\alpha} \tag{2.95}$$

and use them, with the aid of (2.77), to determine the amplitudes of a conjugate eigenmode a_α^c and its adjoint $\bar{a}_\alpha^c$:

$$\begin{aligned} a_\alpha^c &= \left(\hat{\bar{\mathbf{e}}}_\alpha^c\right)^T \mathbf{U}_z \mathbf{e}^c \operatorname{sgn}(\alpha) = \left[\mathbf{Q}_y^c \hat{\mathbf{e}}_{-\alpha}\right]^{\mathrm{T}} \mathbf{U}_z \left[\mathbf{Q}_y^c \bar{\mathbf{e}}\right] \operatorname{sgn}(\alpha) \\ &= -\hat{\mathbf{e}}_{-\alpha}^T \mathbf{U}_z \bar{\mathbf{e}} \operatorname{sgn}(\alpha) \end{aligned} \tag{2.96}$$

since

$$\mathbf{Q}_y^c \mathbf{U}_z \mathbf{Q}_y^c = -\mathbf{U}_z$$

and hence, with $-\operatorname{sgn}(\alpha) = \operatorname{sgn}(-\alpha)$, we find

$$a_\alpha^c = \bar{a}_{-\alpha}, \qquad \bar{a}_\alpha^c = a_{-\alpha} \tag{2.97}$$

2.4.3 Resumé

Before proceeding let us retrace some of the relevant steps we have taken till now in this chapter. We considered a solution to Maxwell's equations in a plane-stratified medium, consisting of a set of eigenmodes having a common value of $\mathbf{k}_t$, the projection of the propagation vector $\mathbf{k}$ on the stratification plane, which is transverse to the z-axis. We constructed mathematically a set of adjoint eigenmodes, biorthogonal to the original set, and used the biorthogonality condition to define amplitudes of the given and adjoint eigenmodes at any level, z. Next, we performed a conjugating transformation (reflection and time reversal) of these eigenmodes to obtain a set of conjugate eigenmodes which was shown to be a solution of Maxwell's equations (or the adjoint equations) in the conjugate problem, in which the plane of incidence is a mirror image with respect to the magnetic meridian plane of the original plane of incidence. Finally, a simple relation was found between the eigenmode amplitudes in the given and conjugate problems, which will be required in the next section to derive the scattering theorem.

The reader may well ask why we are using this somewhat elaborate conjugating transformation, when we could have reached the same end result by a more direct transformation which maps $\mathbf{k}_t(s_x, s_y)$ into $\mathbf{k}_t^c(-s_x, s_y)$, as will indeed be demonstrated in Sec. 3.2.5. The reason is that the method described is much more general in its scope than that used in the special case of planar stratification, and will be applied in later chapters to problems possessing quite general spatial symmetries.

2.5 The eigenmode scattering theorem

2.5.1 The scattering matrix

The motivation for most numerical or analytic calculations of wave propagation through a plane-stratified medium, is to derive eventually the reflection, transmission and intermode coupling coefficients for plane-wave incidence from either end. These coefficients are conveniently grouped into the scattering matrix S.

Let $\hat{\mathbf{e}}_\alpha(z)$, $\alpha = \pm 1, \pm 2$, represent one of the 6-component normalized eigenvectors, defined in Sec. 2.3.2, at a level z, for positive- or negative-going characteristic waves propagating in a plane-stratified medium, with equal prescribed values of the transverse wave vector $\mathbf{k}_t = k_0(s_x, s_y)$; $\hat{\bar{\mathbf{e}}}_\alpha$ is the corresponding normalized adjoint eigenmode. The overall wave fields, $\mathbf{e}(z)$ and $\bar{\mathbf{e}}(z)$, at any level may be decomposed into the respective eigenvectors $\mathbf{e}_\alpha$, or their adjoints $\bar{\mathbf{e}}_\alpha$, as in (2.76) and (2.78)

$$\mathbf{e} = \sum_\alpha a_\alpha \hat{\mathbf{e}}_\alpha, \qquad \bar{\mathbf{e}}_\alpha = \sum_\alpha \bar{a}_\alpha \hat{\bar{\mathbf{e}}}_\alpha \tag{2.98}$$

where

$$a_\alpha = \hat{\bar{\mathbf{e}}}_\alpha^T \mathsf{U}_z \mathbf{e} \operatorname{sgn}(\alpha) \qquad \bar{a}_\alpha = \hat{\mathbf{e}}_\alpha^T \mathsf{U}_z \bar{\mathbf{e}} \operatorname{sgn}(\alpha) \tag{2.99}$$

It will be convenient to replace the summation representation in (2.98) by matrix notation:

$$\mathbf{e} = \mathsf{E}_+ \boldsymbol{a}_+ + \mathsf{E}_- \boldsymbol{a}_- = \mathsf{E}\,\boldsymbol{a}, \qquad \bar{\mathbf{e}} = \overline{\mathsf{E}}\,\bar{\boldsymbol{a}} \tag{2.100}$$

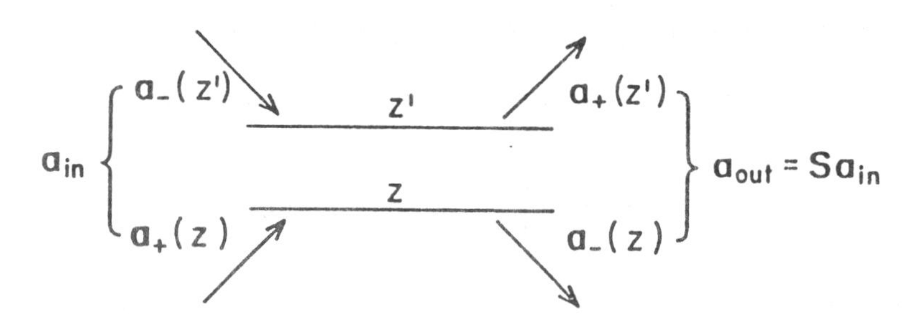

Figure 2.3: Incoming and outgoing eigenmodes related by the scattering matrix $\mathsf{S}(z, z')$.

where

$$\mathsf{E}_{\pm} = [\hat{\mathbf{e}}_{\pm 1}\ \hat{\mathbf{e}}_{\pm 2}], \qquad \mathsf{E} = [\mathsf{E}_{+}\ \mathsf{E}_{-}] = [\hat{\mathbf{e}}_{1}\ \hat{\mathbf{e}}_{2}\ \hat{\mathbf{e}}_{-1}\ \hat{\mathbf{e}}_{-2}] \tag{2.101}$$

$$\boldsymbol{a}_{\pm} = \begin{bmatrix} a_{\pm 1} \\ a_{\pm 2} \end{bmatrix}, \qquad \boldsymbol{a} = \begin{bmatrix} \boldsymbol{a}_{+} \\ \boldsymbol{a}_{-} \end{bmatrix} = [a_1\ a_2\ a_{-1}\ a_{-2}]^T$$

with adjoint quantities similarly defined.

Now consider the wave amplitudes $\boldsymbol{a}(z)$ and $\boldsymbol{a}(z')$ at two levels, z and z', with $z' > z$. In terms of the wave amplitudes $\boldsymbol{a}_{\pm}$ at z and z', we write in condensed notation

$$\boldsymbol{a}_{in} = \begin{bmatrix} \boldsymbol{a}_{+}(z) \\ \boldsymbol{a}_{-}(z') \end{bmatrix}, \qquad \boldsymbol{a}_{out} = \begin{bmatrix} \boldsymbol{a}_{-}(z) \\ \boldsymbol{a}_{+}(z') \end{bmatrix} \tag{2.102}$$

(see Fig. 2.3), and define the scattering matrix $\mathsf{S} = \mathsf{S}(s_x, s_y; z, z')$, and its adjoint $\overline{\mathsf{S}} = \overline{\mathsf{S}}(s_x, s_y; z, z')$, by means of

$$\boldsymbol{a}_{out} = \mathsf{S}\, \boldsymbol{a}_{in}, \qquad \bar{\boldsymbol{a}}_{out} = \overline{\mathsf{S}}\, \bar{\boldsymbol{a}}_{in} \tag{2.103}$$

Written out in full, in terms of the 2×2 reflection and transmission matrices, $\mathsf{R}_{\pm}$ and $\mathsf{T}_{\pm}$, this becomes

$$\begin{bmatrix} \boldsymbol{a}_{-}(z) \\ \boldsymbol{a}_{+}(z') \end{bmatrix} = \begin{bmatrix} \mathsf{R}_{+}(z) & \mathsf{T}_{-}(z, z') \\ \mathsf{T}_{+}(z', z) & \mathsf{R}_{-}(z') \end{bmatrix} \begin{bmatrix} \boldsymbol{a}_{+}(z) \\ \boldsymbol{a}_{-}(z') \end{bmatrix} \tag{2.104}$$

2.5.2 Derivation of the eigenmode scattering theorem

Relation between given and adjoint scattering matrices

Our derivation is based on the constancy of the bilinear concomitant vector, (2.39) and (2.78),

$$P_z = \bar{\mathbf{e}}^T \mathbf{U}_z \, \mathbf{e} = \sum_{\alpha} \bar{a}_\alpha a_\alpha \, \mathrm{sgn}(\alpha) = \mathrm{const} \tag{2.105}$$

Applying this result to the modal amplitudes at z' and z, we have

$$\bar{\boldsymbol{a}}_+^T(z')\,\boldsymbol{a}_+(z') - \bar{\boldsymbol{a}}_-^T(z')\,\boldsymbol{a}_-(z') = \bar{\boldsymbol{a}}_+^T(z)\,\boldsymbol{a}_+(z) - \bar{\boldsymbol{a}}_-^T(z)\,\boldsymbol{a}_-(z)$$

and, regrouping

$$\bar{\boldsymbol{a}}_+^T(z')\,\boldsymbol{a}_+(z') + \bar{\boldsymbol{a}}_-^T(z)\,\boldsymbol{a}_-(z) = \bar{\boldsymbol{a}}_+^T(z)\,\boldsymbol{a}_+(z) + \bar{\boldsymbol{a}}_-^T(z')\,\boldsymbol{a}_-(z') \tag{2.106}$$

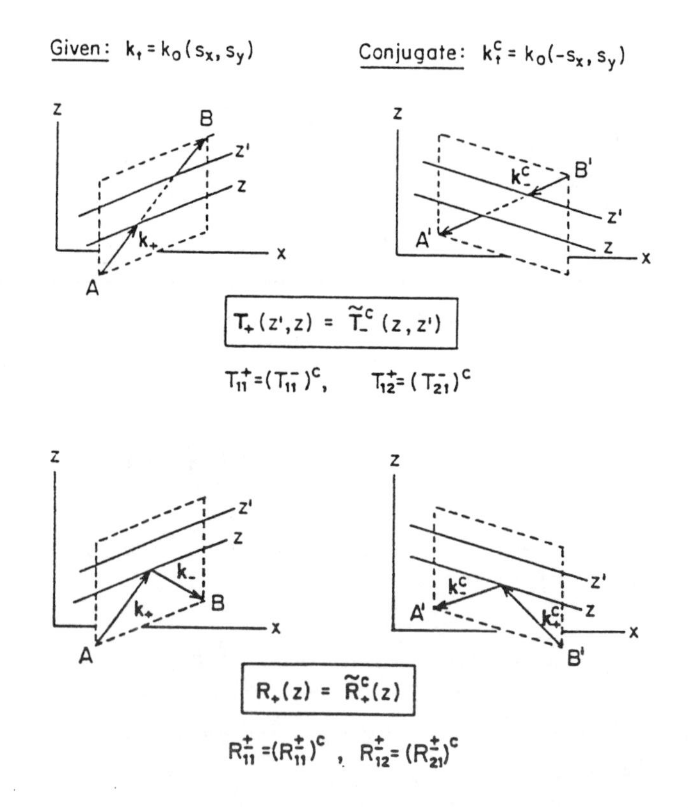

Figure 2.4: Reciprocity in k-space in a plane-stratified magnetoplasma. The z-axis is normal to the stratification, and the external magnetic field lies in the magnetic-meridian (x-z) plane.

so that, with (2.102)

$$\bar{a}_{out}^T\, a_{out} = \bar{a}_{in}^T\, a_{in} \tag{2.107}$$

Application of (2.103) yields

$$\bar{a}_{in}^T\, \overline{\mathbf{S}}^T \mathbf{S}\, a_{in} = \bar{a}_{in}^T\, a_{in}$$

and finally

$$\overline{\mathbf{S}}^T \mathbf{S} = \mathbf{I}^{(4)} = \mathbf{S}\,\overline{\mathbf{S}}^T \tag{2.108}$$

since $\overline{\mathbf{S}}^T = \mathbf{S}^{-1}$.

Relation between adjoint, conjugate and given scattering matrices

When the eigenmodes in Fig. 2.3 undergo a conjugating transformation, the incoming and outgoing amplitudes are correspondingly transformed, (2.97), to yield

$$a_{\pm}^c = \begin{bmatrix} a_{\pm 1}^c \\ a_{\pm 2}^c \end{bmatrix} = \begin{bmatrix} \bar{a}_{\mp 1} \\ \bar{a}_{\mp 2} \end{bmatrix} = \bar{a}_{\mp} \tag{2.109}$$

$$a_{in}^c = \begin{bmatrix} a_+^c(z) \\ a_-^c(z') \end{bmatrix} = \begin{bmatrix} \bar{a}_-(z) \\ \bar{a}_+(z') \end{bmatrix} = \bar{a}_{out}, \qquad a_{out}^c = \bar{a}_{in} \tag{2.110}$$

and since, by (2.103)

$$\bar{a}_{out} = \overline{\mathbf{S}}\, \bar{a}_{in}$$

this transforms to

$$a_{in}^c = \overline{\mathbf{S}}\, a_{out}^c = (\mathbf{S}^c)^{-1}\, a_{out}^c \tag{2.111}$$

by definition of $\mathbf{S}^c$. Hence, with $\overline{\mathbf{S}}^{-1} = \mathbf{S}^T$ from (2.108), we get

$$\mathbf{S}^c \equiv \begin{bmatrix} \mathbf{R}_+^c & \mathbf{T}_-^c \\ \mathbf{T}_+^c & \mathbf{R}_-^c \end{bmatrix} = \begin{bmatrix} \tilde{\mathbf{R}}_+ & \tilde{\mathbf{T}}_+ \\ \tilde{\mathbf{T}}_- & \tilde{\mathbf{R}}_- \end{bmatrix} \equiv \tilde{\mathbf{S}} \tag{2.112}$$

This is the eigenmode scattering theorem [119, 12], expressing 'reciprocity in k-space', that we set out to prove. The reciprocity relations

$$\mathbf{R}_{\pm}^c = \tilde{\mathbf{R}}_{\pm}, \qquad \mathbf{T}_{\pm}^c = \tilde{\mathbf{T}}_{\mp} \tag{2.113}$$

are illustrated in Fig. 2.4.

A word as to notation. The elements of the matrix $\mathbf{R}_{\pm}$ (or analogously $\mathbf{T}_{\pm}$) will be written as $R_{11}^{\pm}$, $R_{12}^{\pm}$, $R_{21}^{\pm}$ and $R_{22}^{\pm}$, the $\pm$ sign indicating the direction of incidence with respect to the z-axis. It will sometimes be convenient, however, when the modal species or polarization is specifically characterized, e.g. parallel ($\|$) or perpendicular ($\perp$) to the plane of incidence, right- or left-circular (r or ℓ), to adopt and extend Budden's [32] notation, so that ${}_{\|}R_{\perp}^{+}$ represents the conversion coefficient from a positive-going perpendicularly polarized incident mode to a reflected (converted) negative-going parallel-polarized mode. Similarly, ${}_{\ell}T_{r}^{-}$ means negative-going right circular to negative-going left circular.

2.6 Curved stratified media

2.6.1 Curvilinear coordinates

In all previous sections the spatial variation of the media under consideration was taken to be in one cartesian coordinate only, i.e. the media were assumed to be plane stratified. This allowed the full use of Fourier transformations in the two coordinates, x and y, transverse to the normal z-direction of the stratification. This restriction to plane stratification is usually sufficient when a curved stratified medium can be approximated by a plane stratified one in the region of propagation. But the question remains whether the scattering theorems, $\overline{\mathbf{S}}^{-1} = \mathbf{S}^T$ (2.108) and $\mathbf{S}^c = \mathbf{S}^T$ (2.112), hold also in curved stratified media. This problem has been addressed by Suchy and Altman [120]. Replacing the generalized Poynting flux densities, $P_{z,\alpha}$ (2.70) and P_z (2.78), by the corresponding *Poynting fluxes*, which are the integrals of $P_{z,\alpha}$ and P_z over (parts of) the curved stratification surfaces, the scattering theorems (2.108), and consequently (2.112), can be generalized to curved statified media.

To prove this statement we have to apply the Lagrange identity twice, first to a system of partial differential equations for the two coordinates u and v in the stratification surfaces, and then to a system of partial differential equations for u, v and w, where the w-coordinate is directed along the normal to these surfaces, thus generalizing (2.30).

Since we cannot employ the transverse Fourier transforms, (2.27) and (2.28), we decompose the differential operator

$$\nabla := \mathbf{g}^u \frac{\partial}{\partial u} + \mathbf{g}^v \frac{\partial}{\partial v} + \mathbf{g}^w \frac{\partial}{\partial w} \tag{2.114}$$

into a tangential part

$$\nabla_t := \mathbf{g}^u \frac{\partial}{\partial u} + \mathbf{g}^v \frac{\partial}{\partial v} \tag{2.115}$$

and a normal part

$$\nabla_w := \mathbf{g}^w \frac{\partial}{\partial w} \tag{2.116}$$

with the reciprocal set $\mathbf{g}^u$, $\mathbf{g}^v$, $\mathbf{g}^w$ of base vectors, obeying

$$\mathbf{g}^i \mathbf{g}_j = \delta_{ij} \, .$$

The base vectors $\mathbf{g}_u$, $\mathbf{g}_v$, $\mathbf{g}_w$ span the arc length element

$$d\mathbf{r} = \mathbf{g}_u du + \mathbf{g}_v dv + \mathbf{g}_w dw$$

[115, Sec. 1.14]. With the corresponding decomposition of the symmetric differential operator $\mathsf{D} = \mathsf{D}^T$ (2.22), viz

$$\mathsf{D} = \mathsf{D}_t + \mathsf{D}_w \tag{2.117}$$

into a tangential and a normal part

$$\mathsf{D}_t := \begin{bmatrix} 0 & -\nabla_t \times \mathsf{I} \\ \nabla_t \times \mathsf{I} & 0 \end{bmatrix} = \mathsf{D}_t^T, \qquad \mathsf{D}_w := \begin{bmatrix} 0 & -\nabla_w \times \mathsf{I} \\ \nabla_w \times \mathsf{I} & 0 \end{bmatrix} = \mathsf{D}_w^T \tag{2.118}$$

with $\mathsf{I} \equiv \mathsf{I}^{(3)}$, Maxwell's equations (2.21) and (2.22) become

$$[i\omega\mathsf{K} + \mathsf{D}_w + \mathsf{D}_t]\,\mathbf{e} = -\mathbf{j} \tag{2.119}$$

2.6.2 The biorthogonality relation

To establish a set of eigenmodes in the curved stratified medium, we proceed in a manner analogous to that in Sec. 2.2.3, equating the source term $\mathbf{j}$ to zero and keeping the normal coordinate w constant [53, Sec.8.2a]. Then all six (covariant) components $E_u \ldots H_w$ of the generalized wave-field vector $\mathbf{e} := [\mathbf{E}, \mathbf{H}]^T$ (2.23) have the same harmonic factor $\exp(-i\kappa w)$, where κ is the separation constant . Application of $\nabla_w := \mathbf{g}^w \, \partial/\partial w$ (2.116) leads to

$$\mathsf{D}_w \mathbf{e} = -i\kappa \mathsf{U}_w \mathbf{e} \qquad \text{with} \qquad \mathsf{U}_w := \begin{bmatrix} 0 & -\mathbf{g}^w \times \mathsf{I} \\ \mathbf{g}^w \times \mathsf{I} & 0 \end{bmatrix} = \mathsf{U}_w^T \tag{2.120}$$

and to the eigenvalue equation

$$\mathsf{L}_\alpha \mathbf{e}_\alpha := [i\omega\mathsf{K} - i\kappa_\alpha \mathsf{U}_w + \mathsf{D}_t]\,\mathbf{e}_\alpha = 0 \tag{2.121}$$

instead of (2.32), which applied to plane-stratified media.

The corresponding adjoint eigenvalue equation reads, with $\mathsf{U}_w = \mathsf{U}_w^T$ (2.120) and $\mathsf{D}_t = \mathsf{D}_t^T$ (2.118),

$$\bar{\mathsf{L}}_\alpha \bar{\mathbf{e}}_\alpha := \left[i\omega\mathsf{K}^T - i\,\bar{\kappa}_\alpha \mathsf{U}_w - \mathsf{D}_t\right] \bar{\mathbf{e}}_\alpha = 0 \tag{2.122}$$

The operators L_α and $\bar{\mathsf{L}}_\alpha$ may now be combined to form the Lagrange identity

$$\bar{\mathbf{e}}_\beta^T \, \mathsf{L}_\alpha \mathbf{e}_\alpha - \mathbf{e}_\alpha^T \, \bar{\mathsf{L}}_\beta \, \bar{\mathbf{e}}_\beta = \mathsf{D}_t : \mathbf{e}_\alpha \bar{\mathbf{e}}_\beta^T - i(\kappa_\alpha - \overline{\kappa}_\beta) \bar{\mathbf{e}}_\beta^T \mathsf{U}_w \, \mathbf{e}_\alpha \tag{2.123}$$

in which, for compactness, we have used the scalar 'double-dot product' introduced by Gibbs [58, Sec. 117]; see also [94, p.57]:

$$\mathsf{D} : \mathbf{e}\bar{\mathbf{e}}^T := \sum_{i,j} D_{ij}(e_j \bar{e}_i) = \sum_{i,j} (\bar{e}_i \, D_{ij} \, e_j + e_j \, D_{ji} \, \bar{e}_i)$$

with $D_{ij} = D_{ji}$ representing any symmetric differential operator. In (2.123), with $\mathsf{D} \rightarrow \mathsf{D}_t$ (2.118), $\mathbf{e}_\alpha := (\mathbf{E}_\alpha, \mathbf{H}_\alpha)$ (2.23) and $\bar{\mathbf{e}}_\beta := (\overline{\mathbf{E}}_\beta, \overline{\mathbf{H}}_\beta)$, the first term on the right-hand side is just the tangential divergence of the bilinear concomitant vector $\mathbf{P}_{\alpha\beta}$, viz.

$$\begin{aligned}
\mathsf{D}_t : \mathbf{e}_\alpha \bar{\mathbf{e}}_\beta^T &= \bar{\mathbf{e}}_\beta^T \mathsf{D}_t \, \mathbf{e}_\alpha + \mathbf{e}_\alpha^T \mathsf{D}_t \, \bar{\mathbf{e}}_\beta \\
&= \overline{\mathbf{E}}_\beta \cdot \nabla_t \times \mathbf{H}_\alpha + \overline{\mathbf{H}}_\beta \cdot \nabla_t \times \mathbf{E}_\alpha + \mathbf{E}_\alpha \cdot \nabla_t \times \overline{\mathbf{H}}_\beta + \mathbf{H}_\alpha \cdot \nabla_t \times \overline{\mathbf{E}}_\beta \\
&= \nabla_t \cdot (\mathbf{E}_\alpha \times \overline{\mathbf{H}}_\beta + \overline{\mathbf{E}}_\beta \times \mathbf{H}_\alpha) =: \nabla_t \cdot \mathbf{P}_{\alpha\beta}
\end{aligned} \tag{2.124}$$

The second term on the right-hand side of (2.123) contains its (contravariant) normal component:

$$\bar{\mathbf{e}}_\beta^T \mathsf{U}_w \, \mathbf{e}_\alpha = \mathbf{g}^w \cdot \left(\mathbf{E}_\alpha \times \overline{\mathbf{H}}_\beta + \overline{\mathbf{E}}_\beta \times \mathbf{H}_\alpha \right) = P_{\alpha\beta}^w \tag{2.125}$$

With $\mathsf{L}_\alpha \mathbf{e}_\alpha = 0$ (2.121) and $\bar{\mathsf{L}}_\beta \bar{\mathbf{e}}_\beta = 0$ (2.122), the Lagrange identity (2.123) gives

$$\nabla_t \cdot \mathbf{P}_{\alpha\beta} = i(\kappa_\alpha - \overline{\kappa}_\beta) P_{\alpha\beta}^w \tag{2.126}$$

To derive a biorthogonality relation for the eigenvectors $\mathbf{e}_\alpha$ and the adjoint eigenvectors $\bar{\mathbf{e}}_\beta$, we apply Gauss' divergence theorem in two dimensions

$$\int \nabla_t \cdot \mathbf{P}_{\alpha\beta} \, dS = \oint \hat{\boldsymbol{\nu}} \cdot \mathbf{P}_{\alpha\beta} \, ds \tag{2.127}$$

to a (finite part of a) stratification surface. (In the integral on the right, the boundary curve on the surface is encompassed in a right-hand sense about the normal $\mathbf{g}^w$. The unit normal vector $\hat{\boldsymbol{\nu}}$ lies on the surface and points in an outward direction with respect to the boundary curve.)

With the boundary conditions [53, eqs.8.2.4c and 1.1.23b], (see also Secs. 4.3 and 6.3 in this book),

$$\hat{\boldsymbol{\nu}} \times \mathbf{E}_\alpha = \mathbf{Z} \, \mathbf{H}_\alpha \qquad \hat{\boldsymbol{\nu}} \times \overline{\mathbf{E}}_\beta = -\mathbf{Z}^T \, \overline{\mathbf{H}}_\beta \tag{2.128}$$

the contour integral vanishes, and the integrated Lagrange identity (2.126), which becomes a Green's theorem, yields

$$(\kappa_\alpha - \overline{\kappa}_\beta) \int \mathbf{g}^w \cdot \mathbf{P}_{\alpha\beta}\, dS = 0 \tag{2.129}$$

As a further requirement for the derivation of a biorthogonality relation, we exclude modal concomitant vectors $\mathbf{P}_\alpha := \mathbf{E}_\alpha \times \overline{\mathbf{H}}_\alpha + \overline{\mathbf{E}}_\alpha \times \mathbf{H}_\alpha$ (2.124) that lie on a stratification surface, i.e. we require that

$$P_\alpha^w := \mathbf{g}^w \cdot \mathbf{P}_\alpha \neq 0 \tag{2.130}$$

For loss-free media with $\bar{\mathbf{e}} = \mathbf{e}^*$ (2.124), and therefore $P_\alpha^w = \langle S_\alpha^w \rangle$ (2.56), this condition excludes surface-wave modes whose (time-averaged) Poynting vectors $\langle \mathbf{S}_\alpha \rangle$ are tangential to the stratification surfaces.

Under the condition (2.130) we can derive from Green's theorem (2.129) first, the identity

$$\overline{\kappa}_\alpha = \kappa_\alpha \tag{2.131}$$

of the adjoint and the given eigenmodes $\overline{\kappa}_\alpha$ and κ_α, and second, the biorthogonality relation

$$\int \bar{\mathbf{e}}_\beta^T\, \mathsf{U}_w\, \mathbf{e}_\alpha\, dS = \delta_{\alpha\beta} \int P_\alpha^w\, dS \tag{2.132}$$

(Similar reasoning has been employed by Felsen and Marcuvitz [53, p.53] with the time t in place of the normal coordinate w.)

2.6.3 The generalized Poynting flux

We have obtained the simple harmonic w-dependence $\mathbf{e}_\alpha \sim \exp(-i\kappa_\alpha w)$ of the modal eigenvectors by keeping the normal coordinate w constant in the Maxwell system (2.119). An analogous dependence, $\exp\{-i(k_u u + k_v v)\}$, on the surface coordinates u and v is only possible if all coefficients of the (covariant) components $E_u \ldots H_w$ in the Maxwell system (2.119) do not depend on u and v. Since

$$\nabla_t \times \mathbf{E} = \frac{1}{\sqrt{g}} \left[\mathbf{g}_u \frac{\partial E_w}{\partial v} - \mathbf{g}_v \frac{\partial E_w}{\partial u} + \mathbf{g}_w \left(\frac{\partial E_v}{\partial u} - \frac{\partial E_u}{\partial v} \right) \right]$$

and

$$\mathbf{g}^w \times \mathbf{H} = \frac{1}{\sqrt{g}} (\mathbf{g}_v H_u - \mathbf{g}_u H_v)$$

with the Jacobian

$$\sqrt{g} := \mathbf{g}_u \times \mathbf{g}_v \cdot \mathbf{g}_w = (\mathbf{g}^u \times \mathbf{g}^v \cdot \mathbf{g}^w)^{-1}$$

[115, Secs. 1.15 and 1.41], this requires that the Jacobian be independent of u and v. The only coordinate systems satisfying this requirement are those with cartesian coordinates in which $g = 1$, and (circular) cylindrical coordinates ρ, φ, z with

$$u = \phi \qquad v = z \qquad w = \rho \qquad \sqrt{g} = \rho$$

[56, eqs. 19 and 21]. For these two cases the application of $\nabla_t = \mathbf{g}^u \, \partial/\partial u + \mathbf{g}^v \, \partial/\partial v$ (2.115) leads to

$$\mathsf{D}_t \mathbf{e} = (ik_u \mathsf{U}_u + ik_v \mathsf{U}_v)\mathbf{e} \tag{2.133}$$

with

$$\mathsf{U}_u := \begin{bmatrix} \mathbf{0} & -\mathbf{g}^u \times \mathsf{I} \\ \mathbf{g}^u \times \mathsf{I} & \mathbf{0} \end{bmatrix} \qquad \mathsf{U}_v := \begin{bmatrix} \mathbf{0} & -\mathbf{g}^v \times \mathsf{I} \\ \mathbf{g}^v \times \mathsf{I} & \mathbf{0} \end{bmatrix} \tag{2.134}$$

and to an algebraic eigenvalue problem

$$[i\omega \mathsf{K} - ik_u \mathsf{U}_u - ik_v \mathsf{U}_v - i\kappa_\alpha \mathsf{U}_w]\mathbf{e}_\alpha = 0 \tag{2.135}$$

With $u = x$, $v = y$, $w = z$, $k_u = k_0 s_x$, $k_v = k_0 s_y$, $\kappa_\alpha = k_0 q_\alpha$, we recover the eigenvalue equation (2.32) for plane-stratified media. Now the reasoning in Secs. 2.2.3 to 2.5.2 can be applied without the integration over a (finite) stratification surface as in Sec. 2.6.2, but this holds only for media whose stratification surfaces are either planes or (circular) cylinders.

For media with other stratification surfaces we go back to Maxwell's equations (2.21) without sources

$$\mathsf{L}\,\mathbf{e} := [i\omega \mathsf{K} + \mathsf{D}]\mathbf{e} = 0 \tag{2.136}$$

The adjoint equation is

$$\bar{\mathsf{L}}\,\bar{\mathbf{e}} := [i\omega \mathsf{K}^T - \mathsf{D}]\bar{\mathbf{e}} = 0 \tag{2.137}$$

and the Lagrange identity, by analogy with (2.123) and (2.124), reads

$$\bar{\mathbf{e}}^T\,\mathsf{L}\,\mathbf{e} - \mathbf{e}^T\,\bar{\mathsf{L}}\,\bar{\mathbf{e}} = \mathsf{D} : \mathbf{e}\,\bar{\mathbf{e}}^T = \nabla \cdot \mathbf{P} \tag{2.138}$$

with the bilinear concomitant vector

$$\mathbf{P} := \mathbf{E} \times \overline{\mathbf{H}} + \overline{\mathbf{E}} \times \mathbf{H} \tag{2.139}$$

For $\mathsf{L}\,\mathbf{e} = 0$ and $\bar{\mathsf{L}}\,\bar{\mathbf{e}} = 0$, the left-hand side of (2.138) vanishes. To the right-hand side we apply Gauss' divergence theorem (in three dimensions)

$$\int \nabla \cdot \mathbf{P}\, d^3r = \oint \mathbf{P} \cdot d\mathbf{S} \tag{2.140}$$

with the surface elements $d\mathbf{S}$ directed outwards. The integration volume is bounded by two stratification surfaces, and between them by walls for which the boundary conditions (2.128) hold. The latter leave us with the contributions of the bounding stratification surfaces:

$$\int \bar{\mathbf{e}}^T \mathbf{U}_w \, \mathbf{e} \, dS \equiv \int P^w dS = \text{const} \tag{2.141}$$

where (2.125) has been used. This is the generalization of the result (2.39) for plane stratified media.

2.6.4 Scattering theorems

For the generalization of the scattering theorem (2.108) we introduce modal amplitudes, a_α and $\bar{a}_\alpha$, as in (2.74)

$$\mathbf{e}_\alpha = a_\alpha \hat{\mathbf{e}}_\alpha \qquad \bar{\mathbf{e}}_\alpha = \bar{a}_\alpha \hat{\bar{\mathbf{e}}}_\alpha \tag{2.142}$$

but with $\hat{\mathbf{e}}_\alpha$ and $\hat{\bar{\mathbf{e}}}_\alpha$ now normalized so that, cf. (2.132),

$$\int \hat{\bar{\mathbf{e}}}_\beta^T \, \mathbf{U}_w \, \hat{\mathbf{e}}_\alpha \, dS = \delta_{\alpha\beta} \, \text{sgn}(\alpha) \tag{2.143}$$

Putting this and the decompositions

$$\mathbf{e} = \sum_\alpha a_\alpha \hat{\mathbf{e}}_\alpha \qquad \bar{\mathbf{e}} = \sum_\beta \bar{a}_\beta \hat{\bar{\mathbf{e}}}_\beta \tag{2.144}$$

into (2.141), we obtain

$$\sum_\alpha \bar{a}_\alpha a_\alpha \, \text{sgn}(\alpha) = \text{const} \tag{2.145}$$

which is the same result as in (2.78) and (2.105) for plane-stratified media. This was the basis for the derivation of the scattering theorems (2.108) and (2.112) in Secs. 2.5.1 and 2.5.2, which need not be changed for the generalization to curved stratified media.

Chapter 3

Generalization of the scattering theorem

3.1 Scattering theorem with generalized base modes

3.1.1 Isotropic bounding media

In most problems of practical interest the plane-stratified anisotropic medium will be bounded by free space (or possibly by isotropic dielectric media), and one is then interested to express the elements of the scattering matrix in terms of incoming and outgoing plane waves having specific prescribed polarizations (linear or circular, for instance) in these bounding media. The difficulty is that the scattering theorem (2.112) has been defined only in terms of eigenmodes of the anisotropic medium which exhibit a biorthogonality relationship (2.49) with respect to the (well-defined) adjoint eigenmodes.

For simplicity we suppose that the bounding region is free space. (The ensuing discussion will be essentially the same for loss-free dielectric bounding media, except that the ratio of magnetic to electric wave fields will be increased by a factor n, the refractive index of the medium.) If we have a 'suitable' (as yet undefined) set of base modes, then equation (2.58) indicates that the adjoint modes in the bounding media can be formed by taking the complex conjugates of the given modes. In any case we can, *a posteriori*, test the validity of these base modes for use in the scattering theorem by demanding that they satisfy the orthogonality condition (2.58).

3.1.2 Determination of orthogonality conditions

We note that the biorthogonality condition (2.49), or (2.58) in loss-free media, involves only the x- and y-components of $\mathbf{e}_\alpha$ and $\bar{\mathbf{e}}_\beta$, since the elements of the third and sixth rows (and columns) of U_z (2.25) are all zero. Hence it will be convenient in the present application to express these conditions in a condensed form:

$$\bar{\mathbf{f}}_\beta^T \mathsf{U}_z^{(4)} \mathbf{f}_\alpha = \delta_{\alpha\beta}\, P_{z,\alpha} \tag{3.1}$$

$$\hat{\bar{\mathbf{f}}}_\beta^T \mathsf{U}_z^{(4)} \hat{\mathbf{f}}_\alpha = \delta_{\alpha\beta}\, \mathrm{sgn}(\alpha) \tag{3.2}$$

where $\mathbf{f}_\alpha$ and $\bar{\mathbf{f}}_\beta$, given by

$$\mathbf{f}_\alpha := [E_x,\, E_y,\, H_x,\, H_y]_\alpha^T, \qquad \bar{\mathbf{f}}_\beta := [\overline{E}_x\,, \overline{E}_y\,, \overline{H}_x, \overline{H}_y]_\beta^T$$

contain only four of the six components of $\mathbf{e}_\alpha$ and $\bar{\mathbf{e}}_\beta$ respectively, and

$$\mathsf{U}_z^{(4)} := \begin{bmatrix} 0 & 0 & 0 & 1 \\ 0 & 0 & -1 & 0 \\ 0 & -1 & 0 & 0 \\ 1 & 0 & 0 & 0 \end{bmatrix}$$

is formed from U_z (2.6) by removing the null rows and columns. In general applications we shall preserve the full 6-component wave fields in order to be able to distinguish clearly between reflection and rotation mappings.

The normalized eigenmodes, $\hat{\mathbf{f}}_\alpha$ or $\hat{\bar{\mathbf{f}}}_\alpha$, may be juxtaposed to form the 4×4 modal matrices, F or $\overline{\mathsf{F}}$, in analogy with the 4×6 matrices, E and $\overline{\mathsf{E}}$, in (2.100)

$$\mathsf{F} := \left[\hat{\mathbf{f}}_1\ \hat{\mathbf{f}}_2\ \hat{\mathbf{f}}_{-1}\ \hat{\mathbf{f}}_{-2}\right], \qquad \overline{\mathsf{F}} := \left[\hat{\bar{\mathbf{f}}}_1\ \hat{\bar{\mathbf{f}}}_2\ \hat{\bar{\mathbf{f}}}_{-1}\ \hat{\bar{\mathbf{f}}}_{-2}\right] \tag{3.3}$$

and will be used in this section to express the biorthogonality condition in a convenient form:

$$\overline{\mathsf{F}}^T \mathsf{U}_z^{(4)} \mathsf{F} = \begin{bmatrix} \mathsf{I}^{(2)} & 0 \\ 0 & -\mathsf{I}^{(2)} \end{bmatrix} \equiv \bar{\mathsf{I}}^{(4)} \tag{3.4}$$

To specify the base modes in free space we use an auxiliary coordinate system (x',y',z'), formed by rotating the (x,y,z) system through an azimuthal angle ϕ about the z $(= z')$ axis, so that the plane of incidence will coincide with the x'-z' plane, with the positive-going modal wave vectors having positive x' and z' components. The auxiliary conjugate system (x_c', y_c', z_c') is then formed by rotating the (x,y,z) frame through an angle $(\pi - \phi)$ about the z $(= z')$ axis. (Recall, cf. Sec. 2.1.4, that the x-axis was chosen so that the external magnetic field $\mathbf{b}$ lies in the x-z plane.)

The transverse wave polarizations, ρ_1 and ρ_2, are now defined in a coordinate system (ξ, η, ς), cf. (1.77), in which the ς-axis is tied to the wave normal (which is now at an angle of incidence θ), and the ξ-axis lies in the plane spanned by the wave normal and the external magnetic field **b**. In free space however (or in any isotropic bounding medium) the external magnetic field direction is irrelevant, and we are at liberty to choose the ξ-direction at our convenience. We choose it to lie in the plane of incidence (at an angle θ with respect to the x'-axis), which provides the only preferred direction. Then

$$\rho := \frac{E_\eta}{E_\xi} = -\frac{H_\xi}{H_\eta}$$

The modal wave fields, $\mathbf{e}_\alpha$, in the (ξ, η, ς) system, cf. (2.67), then have the form

$$\mathbf{e}_\alpha = [1,\ \rho_\alpha,\ 0;\ -Y_0\rho_\alpha,\ Y_0,\ 0]^T E_{\alpha\xi} \tag{3.5}$$

so that positive-going 4-element base vectors $\mathbf{f}_\alpha$ (α=1,2) in the (x',y',z') system are given by

$$\mathbf{f}_\alpha := \left[E_{x'}, E_{y'}, H_{x'}, H_{y'}\right]^T_\alpha = [\cos\theta,\ \rho_\alpha,\ -Y_0\rho_\alpha\cos\theta,\ Y_0]^T E_{\alpha\xi} \tag{3.6}$$

The negative-going base modes may be formed by reversing the signs of the H components (and thereby of the z-components of the Poynting flux), and the overall modal matrix becomes

$$\mathsf{F} = \begin{bmatrix} a\cos\theta & b\cos\theta & a\cos\theta & b\cos\theta \\ a\rho_1 & b\rho_2 & a\rho_1 & b\rho_2 \\ -a\rho_1 Y_0\cos\theta & -b\rho_2 Y_0\cos\theta & a\rho_1 Y_0\cos\theta & b\rho_2 Y_0\cos\theta \\ aY_0 & bY_0 & -aY_0 & -bY_0 \end{bmatrix} \tag{3.7}$$

where a and b are normalization constants. Setting $\overline{\mathsf{F}}=\mathsf{F}^\star$, we find

$$\overline{\mathsf{F}}^T \mathsf{U}_z^{(4)} \mathsf{F} = 2Y_0\cos\theta \begin{bmatrix} \mathsf{P} & 0 \\ 0 & -\mathsf{P} \end{bmatrix} \tag{3.8}$$

where

$$\mathsf{P} := \begin{bmatrix} a^\star a(1+\rho_1^\star\rho_1) & a^\star b(1+\rho_1^\star\rho_2) \\ ab^\star(1+\rho_1\rho_2^\star) & b^\star b(1+\rho_2^\star\rho_2) \end{bmatrix}$$

We note that whatever the values of ρ_1 and ρ_2, the positive- and negative-going modes are uncoupled, i.e. the Poynting-like products $\hat{\bar{\mathbf{f}}}_\alpha^T \mathsf{U}_z^{(4)} \hat{\mathbf{f}}_{-\beta}$ and $\hat{\bar{\mathbf{f}}}_{-\alpha}^T \mathsf{U}_z^{(4)} \hat{\mathbf{f}}_\beta$ (α,β=1,2) are zero, corresponding to the off-diagonal zeros in (3.8).

Furthermore, the off-diagonal terms in the sub-matrix P, corresponding to Poynting cross products of two positive-going, or two negative-going modes, are zero, as required by (3.4), provided that

$$\rho_1^\star \rho_2 = \rho_1 \rho_2^\star = -1 \tag{3.9}$$

If the polarizations are purely imaginary, this becomes

$$\rho_1 \rho_2 = 1 \tag{3.10}$$

corresponding to circularly polarized base modes, $\rho_1 = \pm i = -\rho_2$, or to elliptical polarizations, $\rho_1 = \pm i |\rho_1| = 1/\rho_2$, with the principal axes of the polarization ellipses lying on the ξ and η axes (i.e. in the plane of incidence and perpendicular to it), the two ellipses being mirror images with respect to a 45° line through the origin in the first and third quadrants.

As the eccentricity of the polarization ellipses increases indefinitely, (i.e. as they become increasingly narrow), so that in the limit

$$\rho_1 \to \pm i0, \qquad \rho_2 \to \mp i\infty, \qquad \rho_1 \rho_2 = 1 \tag{3.11}$$

these degenerate into straight lines (linear polarizations) with the electric wave vectors, $\mathbf{E}_1$ or $\mathbf{E}_2$, lying respectively parallel or perpendicular to the plane of incidence.

Normalization of the modes requires that

$$a = [2Y_0 \cos\theta(1 + \rho_1^\star \rho_1)]^{-\frac{1}{2}} \qquad b = [2Y_0 \cos\theta(1 + \rho_2^\star \rho_2)]^{-\frac{1}{2}} \tag{3.12}$$

which, together with (3.9), completes the specification of base modes in free space that satisfy the biorthogonality condition (3.4), and thereby qualify for use in the scattering theorem (2.112).

It should be noted that the prescription $\overline{\mathsf{F}} = \mathsf{F}^\star$, with the ensuing $\bar{\rho}_\alpha = \rho_\alpha^\star$, is not the only one which satisfies modal biorthogonality. We could heuristically have arrived at another prescription by imagining the isotropic region to have been formed from a gyrotropic one in the limit $\mathbf{b} \to 0$, or collision frequency ν becoming very large (as at the base of the ionosphere), and then have used the magnetic field reversal condition (2.66):

$$\bar{\rho}_\alpha = -\rho_\alpha$$

The adjoint eigenmodes so formed would also exhibit biorthogonality (note the structure of P in (3.8) with $-\rho_\alpha$ replacing $\rho_\alpha^\star$) provided that

$$\rho_1 \rho_2 = 1 \tag{3.13}$$

with ρ_1 and ρ_2 now possibly complex. This implies that *any pair* of elliptically polarized waves which are mirror images of each other with respect to a 45° line in the first and third quadrants passing through the origin, are also acceptable base modes. A more general formulation of the requirements on base modes to satisfy modal biorthogonality has been given by Altman and Suchy [13].

3.1.3 Linear and circular base modes

Linear base modes

Following the prescription of Sec. 3.1.1, we construct a set of linear modes with either electric or magnetic wave vectors along the positive y'-axis (perpendicular to the plane of incidence), project the other wave field (magnetic or electric) which is parallel to the plane of incidence, onto the x'-axis, and normalize so that each mode will eventually have unit Poynting flux (or to be more precise, will have $|\,\hat{\bar{\mathbf{f}}}^T_\alpha \mathbf{U}^{(4)}_z \bar{\mathbf{f}}_\alpha\,|=1$):

$$\mathsf{F} = \frac{1}{(2Y_0 \cos\theta)^{1/2}} \begin{bmatrix} \cos\theta & 0 & -\cos\theta & 0 \\ 0 & 1 & 0 & 1 \\ 0 & -Y_0\cos\theta & 0 & Y_0\cos\theta \\ Y_0 & 0 & Y_0 & 0 \end{bmatrix} \tag{3.14}$$

(The signs of the elements in the third column have been reversed, in order to conform with Budden's sign convention for linear modes [32, p.89] and thus to facilitate comparison with his reciprocity theorem [29,21] mentioned in Sec. 2.1.3.)

The matrix in (3.14) is purely real, and so we expect it to be self-adjoint:

$$\bar{\mathsf{F}} = \mathsf{F}^* = \mathsf{F} \tag{3.15}$$

Substitution of (3.14) and (3.15) into the matrix product $\bar{\mathsf{F}}^T \mathsf{U}^{(4)}_z \mathsf{F}$ produces $\bar{\mathsf{I}}^{(4)}$, as in (3.4), and confirms that the columns of the matrix F (3.14) constitute a valid orthonormal set of linear modes for use in the scattering theorem.

To construct the conjugate set of eigenmodes, F^c, we reflect the adjoint modes with respect to the magnetic meridian plane, y=0, and then time-reverse, cf. Secs. 2.4.1 and 2.4.2, which in the present context means reversing the direction of the (already reflected) magnetic wave fields, and thereby reversing the direction of propagation ($\mathbf{f}_\alpha \rightarrow \mathbf{f}_{-\alpha}$). The given and conjugate linear eigenmodes are depicted in Fig. 3.1, and we see that the conjugate eigenmodes in the (x'_c, y'_c, z'_c) coordinate system have the same components as the given eigenmodes in the (x', y', z') system:

$$\mathsf{F}^c(x'_c, y'_c) = \mathsf{F}(x', y') \tag{3.16}$$

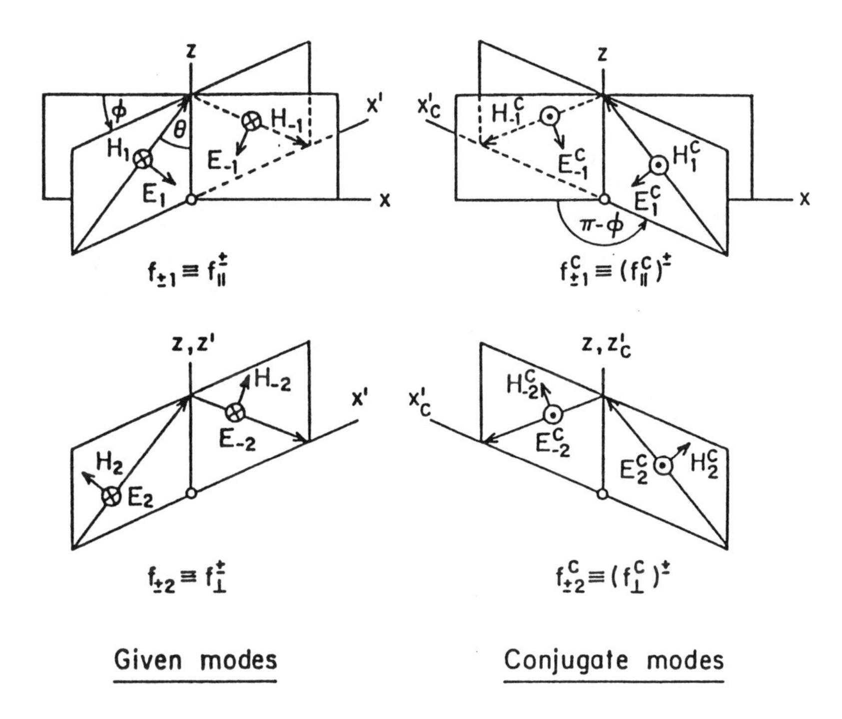

Figure 3.1: Given and conjugate linear modes.
$\mathbf{H}_{\pm 1}$ and $\mathbf{E}_{\pm 2}$ are along y', $\mathbf{H}^c_{\pm 1}$ and $\mathbf{E}^c_{\pm 2}$ are along y'_c. When the plane of incidence is normal to the magnetic meridian plane y=0 (i.e. when $\phi=\pi/2$), the given and conjugate modes coalesce.

To obtain this result formally, rather than with the hand-waving procedure just adopted, is a little tricky. The 4-component equivalent of (2.92) is

$$\mathsf{F}^c = \mathsf{Q}^c_{(4)}\mathsf{F}\mathsf{J}, \quad \mathsf{Q}^c_{(4)} := \begin{bmatrix} 1 & 0 & 0 & 0 \\ 0 & -1 & 0 & 0 \\ 0 & 0 & 1 & 0 \\ 0 & 0 & 0 & -1 \end{bmatrix}, \quad \mathsf{J} := \begin{bmatrix} 0 & 0 & 1 & 0 \\ 0 & 0 & 0 & 1 \\ 1 & 0 & 0 & 0 \\ 0 & 1 & 0 & 0 \end{bmatrix} \tag{3.17}$$

with $\mathsf{Q}^c_{(4)}$ effecting reflection with respect to the y=0 plane, and reversing the magnetic wave fields (in the sense that it treats them like polar vectors); J exchanges the upgoing eigencolumns with the downgoing. Putting $\overline{\mathsf{F}}$=F in (3.17), we obtain $\mathsf{F}^c(x,y) = -\mathsf{F}(x,y)$. But the eigenmodes are here defined in the *fixed* (x,y) system, tied to the magnetic meridian plane, and the results conform with the representation of Fig. 3.1.

Suppose that the wave fields $\mathbf{f}(\mathbf{k}_t)$ and $\mathbf{f}^c(\mathbf{k}^c_t)$ in the given and conjugate systems have been determined, by numerical analysis or otherwise, at the free-space boundaries, z and z', of a plane-stratified gyrotropic medium. These may be decomposed into linear base modes of amplitudes [cf. (2.77)]

$$a_{\parallel}^{\pm} = \pm \hat{\mathbf{f}}_{\pm 1}^{T} \mathbf{U}_{z}^{(4)} \mathbf{f}, \qquad a_{\perp}^{\pm} = \pm \hat{\mathbf{f}}_{\pm 2}^{T} \mathbf{U}_{z}^{(4)} \mathbf{f} \tag{3.18}$$

and similarly in the conjugate problem. The incoming and outgoing amplitudes will then be related by the respective scattering matrices, $\mathbf{S}(s_x, s_y; z, z')$ (2.103) and $\mathbf{S}^c(-s_x, s_y; z, z')$, (2.111). These in turn will be related by the scattering theorem (2.112) which, for linear modes, becomes

$$\begin{bmatrix} {}_{\parallel}R_{\parallel} & {}_{\parallel}R_{\perp} \\ {}_{\perp}R_{\parallel} & {}_{\perp}R_{\perp} \end{bmatrix}^{\pm} = \begin{bmatrix} {}_{\parallel}R_{\parallel}^{c} & {}_{\parallel}R_{\perp}^{c} \\ {}_{\perp}R_{\parallel}^{c} & {}_{\perp}R_{\perp}^{c} \end{bmatrix}^{\pm} \tag{3.19}$$

$$\begin{bmatrix} {}_{\parallel}T_{\parallel} & {}_{\parallel}T_{\perp} \\ {}_{\perp}T_{\parallel} & {}_{\perp}T_{\perp} \end{bmatrix}^{\pm} = \begin{bmatrix} {}_{\parallel}T_{\parallel}^{c} & {}_{\parallel}T_{\perp}^{c} \\ {}_{\perp}T_{\parallel}^{c} & {}_{\perp}T_{\perp}^{c} \end{bmatrix}^{\mp} \tag{3.20}$$

Eq. (3.19) is just the reciprocity theorem of Budden and Barron [29,21], and (3.20) is a generalization of it [13] for the transmission matrices. Special cases of (3.20) have been given by Heading [66].

Circular base modes

With the aid of (3.7), (3.9), (3.10) and (3.12) we find the modal matrix for left- and right-circular polarized modes in the (x',y') coordinate frame (i.e. the frame tied to the plane of incidence) to be

$$\mathsf{F}(x', y') = \frac{1}{(2Y_0 \cos\theta)^{\frac{1}{2}}} \begin{bmatrix} \cos\theta & \cos\theta & -\cos\theta & -\cos\theta \\ i & -i & -i & i \\ -iY_0 \cos\theta & iY_0\cos\theta & -iY_0\cos\theta & iY_0\cos\theta \\ Y_0 & Y_0 & Y_0 & Y_0 \end{bmatrix} \tag{3.21}$$

We have chosen the phases (signs) of all modes so that instantaneously, at t=0, the rotating field vectors, $\mathbf{E}_\alpha$ and $\mathbf{H}_\alpha$ ($\alpha = \pm 1, \pm 2$), point in the same direction as $\mathbf{E}_{\pm 1}$ and $\mathbf{H}_{\pm 1}$ in the linear base-mode representation of Fig. 3.1.

With $\bar{\mathsf{F}}$=F^*, straightforward calculation gives $\bar{\mathsf{F}}^T \mathsf{U}_z^{(4)} \mathsf{F} = \bar{\mathsf{I}}^{(4)}$, as in (3.4), confirming that F and $\bar{\mathsf{F}}$ constitute a valid orthonormal set of eigenmodes for use in the scattering theorem.

The conjugating transformation will instantaneously map the rotating wave fields into the configuration of $\mathbf{E}_{\pm 1}^c$ and $\mathbf{H}_{\pm 1}^c$ in Fig. 3.1. As to the mapped polarizations, we note that their signs are reversed twice: in the passage from $\mathsf{F} \equiv \mathsf{F}(\mathbf{b})$ to $\overline{\mathsf{F}} = \mathsf{F}^* = \mathsf{F}(-\mathbf{b})$, and in the reflection with respect to the y=0 magnetic meridian plane (corresponding to a reversal of the direction of $-\mathbf{b}$ to its original orientation). Hence

$$\mathsf{F}^c(x'_c, y'_c) = \mathsf{F}(x', y') \tag{3.22}$$

which, formally, is the same as (3.16) for linear base modes.

The formal derivation of (3.22) would require application of the transformation (3.17) to $\overline{\mathsf{F}}$, which yields

$$\mathsf{F}^c(x, y) = -\mathsf{F}(x, y) \tag{3.23}$$

and then to revert to the tagged coordinate systems which are tied to the respective planes of incidence. (This is most easily seen when the incidence planes are parallel to the magnetic meridian plane.)

In terms of these modes the scattering theorem, $\mathbf{S} = \tilde{\tilde{\mathbf{S}}}^c$, takes the form

$$\begin{bmatrix} {}_lR_l & {}_lR_r \\ {}_rR_l & {}_rR_r \end{bmatrix}^{\pm} = \begin{bmatrix} {}_lR_l^c & {}_rR_l^c \\ {}_lR_r^c & {}_rR_r^c \end{bmatrix}^{\pm} \tag{3.24}$$

$$\begin{bmatrix} {}_lT_l & {}_lT_r \\ {}_rT_l & {}_rT_r \end{bmatrix}^{\pm} = \begin{bmatrix} {}_lT_l^c & {}_rT_l^c \\ {}_lT_r^c & {}_rT_r^c \end{bmatrix}^{\mp} \tag{3.25}$$

which is a generalization of (3.19) and (3.20) to left (l) and right (r) circularly polarized modes. These could, of course, have been derived by forming the circularly polarized modes, $\mathbf{f}_l$ and $\mathbf{f}_r$, from the linear modes $\mathbf{f}_\parallel$ and $\mathbf{f}_\perp$:

$$\hat{\mathbf{f}}_l := \frac{1}{\sqrt{2}}(\hat{\mathbf{f}}_\parallel + i\hat{\mathbf{f}}_\perp), \qquad \hat{\mathbf{f}}_r := \frac{1}{\sqrt{2}}(\hat{\mathbf{f}}_\parallel - i\hat{\mathbf{f}}_\perp) \tag{3.26}$$

and applying the results for linear modes obtained earlier.

3.1.4 Magnetic field along the stratification: Heading's mirrored modes

When the external magnetic field $\mathbf{b}$ is in the plane of the stratification, the higher degree of spatial symmetry of the medium is expressed in the symmetry structure of the reflection and transmission matrices, $\mathbf{R}_\pm$ and $\mathbf{T}_\pm$. Whereas up to now the only non-trivial transformation that mapped the medium into itself was the 'conjugating transformation', the medium can now also be mapped into itself by means of reflection with respect to the $x=0$ plane. (The direction of the axial-vector field is unchanged in this case.)

Suppose that in the original problem we had a positive-going plane wave, $\mathbf{f}(\mathbf{k}_t)$, with linearly polarized modal components, $\mathbf{f}_1 := \mathbf{f}_\parallel^+$ and $\mathbf{f}_2 := \mathbf{f}_\perp^+$, incident on the stratified medium at $z=z^-$, which produced reflected modes, $\mathbf{f}_{-1}(z^-)$ and $\mathbf{f}_{-2}(z^-)$, at the lower boundary, and transmitted modes, $\mathbf{f}_1(z^+)$ and $\mathbf{f}_2(z^+)$, at the upper boundary. The amplitudes of these modes, $a_{\pm 1}$, $a_{\pm 2}$, are related by the reflection and transmission matrices, $\mathbf{R}_+$ and $\mathbf{T}_+$ [cf. (2.104)],

$$\boldsymbol{a}_-(z^-) = \mathbf{R}_+(z^-)\,\boldsymbol{a}_+(z^-), \qquad \boldsymbol{a}_+(z^+) = \mathbf{T}_+(z^+,z^-)\,\boldsymbol{a}_+(z^-) \tag{3.27}$$

or, written out in full,

$$\begin{bmatrix} a_{-1}(z^-) \\ a_{-2}(z^-) \end{bmatrix} = \begin{bmatrix} {}_\parallel R_\parallel^+ & {}_\parallel R_\perp^+ \\ {}_\perp R_\parallel^+ & {}_\perp R_\perp^+ \end{bmatrix} \begin{bmatrix} a_1(z^-) \\ a_2(z^-) \end{bmatrix} \tag{3.28}$$

$$\begin{bmatrix} a_1(z^+) \\ a_2(z^+) \end{bmatrix} = \begin{bmatrix} {}_\parallel T_\parallel^+ & {}_\parallel T_\perp^+ \\ {}_\perp T_\parallel^+ & {}_\perp T_\perp^+ \end{bmatrix} \begin{bmatrix} a_1(z^-) \\ a_2(z^-) \end{bmatrix} \tag{3.29}$$

Consider the linear modes $\mathbf{f}_{\pm 1}$, parallel to the plane of incidence, in Fig. 3.1. We see on inspection that reflection with respect to the $x=0$ plane produces another set, $\mathbf{f}_{\pm 1}^c$, of ingoing and outgoing mirrored modes in the conjugate plane of incidence, whereas reflection of the perpendicular modes, $\mathbf{f}_{\pm 2}$, produces $-\mathbf{f}_{\pm 2}^c$. Both sets so produced obey Maxwell's equations in the given medium, with

$$a_{\pm 1} = a_{\pm 1}^c, \qquad a_{\pm 2} = -a_{\pm 2}^c \tag{3.30}$$

(remember, as pointed out in Sec. 2.3.1, that the amplitude a_α carries the phase, and hence the sign, of the corresponding eigenmode) so that (3.28), for instance, becomes

$$\begin{bmatrix} a_{-1}^c(z^-) \\ -a_{-2}^c(z^-) \end{bmatrix} = \begin{bmatrix} {}_\parallel R_\parallel^+ & {}_\parallel R_\perp^+ \\ {}_\perp R_\parallel^+ & {}_\perp R_\perp^+ \end{bmatrix} \begin{bmatrix} a_1^c(z^-) \\ -a_2^c(z^-) \end{bmatrix}$$

or

$$\begin{bmatrix} a_{-1}^c \\ a_{-2}^c \end{bmatrix} = \begin{bmatrix} {}_\parallel R_\parallel^+ & -{}_\parallel R_\perp^+ \\ -{}_\perp R_\parallel^+ & {}_\perp R_\perp^+ \end{bmatrix} \begin{bmatrix} a_1^c \\ a_2^c \end{bmatrix} \tag{3.31}$$

Identifying the 2×2 matrix in (3.31) with $\mathbf{R}_+^c$, and noting that $\mathbf{R}_+^c = \widetilde{\mathbf{R}}_+$, by virtue of the scattering theorem (2.112), we find

$${}_\perp R_\parallel^+ = -{}_\parallel R_\perp^+, \qquad \text{and} \qquad {}_\perp R_\parallel^- = -{}_\parallel R_\perp^- \tag{3.32}$$

by analogy. A similar analysis yields

$$ {}_{\parallel}T_{\parallel}^{+} = {}_{\parallel}T_{\parallel}^{-}, \qquad {}_{\perp}T_{\perp}^{+} = {}_{\perp}T_{\perp}^{-}, \qquad {}_{\perp}T_{\parallel}^{\pm} = -{}_{\parallel}T_{\perp}^{\mp} \tag{3.33} $$

giving the specific symmetry of the reflection and transmission matrices which derive from the reflectional symmetry of this particular problem. These results, (3.32) and (3.33), were first given by Heading [66].

We may carry through a similar analysis with circular base modes, the important difference being the interchange of modes on mirroring in the x=0 plane, with α=±1 converting to α=±2 in the mirrored/conjugate system. The result, given by Altman and Suchy [13],

$$ \begin{aligned} {}_{l}R_{r}^{\pm} &= {}_{r}R_{l}^{\pm}, \qquad & {}_{r}R_{r}^{\pm} &= {}_{l}R_{l}^{\pm} \\ {}_{l}T_{r}^{\pm} &= {}_{r}T_{l}^{\mp}, \qquad & {}_{r}T_{r}^{\pm} &= {}_{l}T_{l}^{\mp} \end{aligned} \tag{3.34} $$

gives the specific symmetry of the scattering matrix in terms of circular base modes in the isotropic (free-space) bounding media.

The results found in this section, 3.1.4, do not strictly speaking belong to the category of 'reciprocity in k-space'. They were derived by a straightforward reflection transformation of the system, without time reversal (in contrast to the conjugating transformation discussed previously), and the results express what we shall later designate an 'equivalence' rather than a 'reciprocity' relation. The difference will be appreciated in Sec. 4.3, when fields and their sources will be considered together.

3.2 Scattering theorems: alternative derivations

Although the earlier reciprocity theorems are contained in the eigenmode scattering theorem (2.112), with its generalization to base modes in isotropic bounding regions (aside from Heading's mirror-mode relationships, discussed in Sec. 3.1.4, which derive from the reflection symmetry of the medium), it is nevertheless worthwhile to consider briefly other approaches which provide new insights and other points of view.

We consider three alternative approaches. The first (the 'classical' approach) is based on the symmetries of the Clemmow-Heading coupled differential equations (1.127). This approach usually by-passes some of the steps in our previous derivation, at the expense of some of the physical insight provided by it.

The second method considers the medium to vary step-wise in a large number of thin discrete layers, derives a scattering matrix and thence a scattering theorem for any interface separating two adjacent layers. The result is then extended by induction to the entire medium using a recursion process, which preserves the symmetry of the interface scattering relation.

The third approach is based on a matrizant formulation of the differential equations governing wave propagation in plane-stratified media, and following a procedure similar to that adopted in the previous chapter, derives a *transfer matrix* linking modal amplitudes at two different levels in the medium. This leads to a relationship between transfer matrices for the given and conjugate planes of incidence, which can easily be transformed into a scattering theorem relating incoming and outgoing modal amplitudes.

3.2.1 Bilinear concomitant and modal biorthogonality via the Clemmow-Heading equations

The Clemmow-Heading equations (1.127) give the z-variation of the transverse (x, y) components of the electric and magnetic fields, in the form

$$\frac{d\mathbf{g}}{dz} + ik_0 \mathbf{T}\mathbf{g} = 0 \tag{3.35}$$

[$\mathbf{T}$ should not be confused with the transmission matrix (2.104)] with

$$\mathbf{g} := (E_x, -E_y, \mathcal{H}_x, \mathcal{H}_y), \qquad \mathcal{H}_x := Z_0 H_x, \quad \mathcal{H}_y := Z_0 H_y \tag{3.36}$$

and $Z_0 \equiv 1/Y_0 := (\mu_0/\varepsilon_0)^{1/2}$ denoting the free-space impedance. This particular form of the field vector $\mathbf{g}$, viz. with one of the tangential components having a negative sign, has been chosen since, as may be seen from eq. (1.128), the elements of the propagation matrix $\mathbf{T}$ are then transposed with respect to its trailing diagonal when the elements of the electric permittivity tensor $\boldsymbol{\varepsilon}$ are transposed about its leading diagonal, $\boldsymbol{\varepsilon} \to \boldsymbol{\varepsilon}^T$. The use of magnetic field components, $\mathcal{H}_x$ and $\mathcal{H}_y$, having the same dimensions as the electric field, suppresses the impedance term Z_0 in the propagation matrix $\mathbf{T}$. The coordinate system is now tied to the x-z plane of incidence, so that $\mathbf{g} \sim \exp(-ik_0 sx)$ in both given and conjugate systems, whereas the direction cosines of the fixed magnetic field become

$$\hat{b}_x, \hat{b}_y, \hat{b}_z \to -\hat{b}_x, \hat{b}_y, \hat{b}_z \tag{3.37}$$

in passing to the conjugate system (see Fig. 3.1, with x' and x'_c replaced by x and x_c).

Because the permittivity matrix $\boldsymbol{\varepsilon}$, whose elements appear in $\mathbf{T}$, is gyrotropic (1.33), i.e.

$$\boldsymbol{\varepsilon}(-\mathbf{b}) = \boldsymbol{\varepsilon}^T(\mathbf{b}) \tag{3.38}$$

we may write the adjoint (magnetic-field reversed) propagation matrix $\overline{\mathbf{T}}$ as the transpose of $\mathbf{T}$ about its trailing diagonal,

$$\overline{\mathbf{T}} = \mathbf{U}\tilde{\mathbf{T}}\mathbf{U}, \qquad \mathbf{U} := \begin{bmatrix} 0 & 0 & 0 & 1 \\ 0 & 0 & 1 & 0 \\ 0 & 1 & 0 & 0 \\ 1 & 0 & 0 & 0 \end{bmatrix} = \mathbf{U}^T = \mathbf{U}^{-1} \tag{3.39}$$

In indicial notation [32, p.390] , [100], this reads

$$\overline{T}_{ij} = \sum_{j=1}^{4} T_{5-j,5-i}$$

If the medium is loss-free, $\boldsymbol{\varepsilon}$ is hermitian, i.e. $\boldsymbol{\varepsilon}^{\star}=\boldsymbol{\varepsilon}^T$, so that $\boldsymbol{\varepsilon}(-\mathbf{b})=\boldsymbol{\varepsilon}^{\star}(\mathbf{b})$ which gives $\mathbf{T}^{\star}=\overline{\mathbf{T}}$.

Let us write (3.35) in matrix-operator form

$$\left[\mathbf{U}\frac{d}{dz} + ik_0\mathbf{U}\mathbf{T}\right]\mathbf{g} \equiv \mathbf{H}\mathbf{g} = 0 \tag{3.40}$$

The adjoint equation is

$$\left[-\mathbf{U}\frac{d}{dz} + ik_0\mathbf{U}\overline{\mathbf{T}}\right]\bar{\mathbf{g}} \equiv \overline{\mathbf{H}}\bar{\mathbf{g}} = 0 \tag{3.41}$$

in which the sign of the differential operator has been reversed, (cf. Sec.2.2.4), and the matrix operator $\mathbf{U}\mathbf{T}$ has been replaced by its transpose $\tilde{\mathbf{T}}\mathbf{U}=\mathbf{U}\overline{\mathbf{T}}$ (3.39). The Lagrange identity for the operators $\mathbf{H}$ and $\overline{\mathbf{H}}$ leads to

$$\bar{\mathbf{g}}^T\mathbf{H}\mathbf{g} - \mathbf{g}^T\overline{\mathbf{H}}\bar{\mathbf{g}} = \frac{d}{dz}(\bar{\mathbf{g}}^T\mathbf{U}\mathbf{g}) = 0 \tag{3.42}$$

by analogy with (2.38), giving the z-component of the the bilinear concomitant vector

$$\bar{\mathbf{g}}^T\mathbf{U}\mathbf{g} = P_z = \text{const} \tag{3.43}$$

Now we assume local plane-wave solutions

$$\mathbf{g}_\alpha \sim \exp(-ik_0\, q_\alpha z), \qquad \bar{\mathbf{g}}_\beta \sim \exp(ik_0\, \bar{q}_\beta z) \tag{3.44}$$

yielding, as in (2.46), (2.49) and (2.50),

$$\bar{q}_\alpha = q_\alpha$$

and the modal biorthogonality relation

$$\bar{\mathbf{g}}_\beta^T\mathbf{U}\mathbf{g}_\alpha = \delta_{\alpha\beta}\, P_{z,\alpha} \tag{3.45}$$

With normalized eigenvectors, this gives

$$\hat{\bar{\mathbf{g}}}_\beta^T\mathbf{U}\hat{\mathbf{g}}_\alpha = \delta_{\alpha\beta}\,\mathrm{sgn}(\alpha) \tag{3.46}$$

which is equivalent to (3.2).

3.2.2 Biorthogonality of given and conjugate eigenmodes

The transformation of adjoint to conjugate wave fields is provided by the conjugating matrix $\mathbf{Q}^c_{(4)} = (\mathbf{Q}^c_{(4)})^{-1}$ (3.17), so that in general

$$\bar{\mathbf{g}} = -\mathbf{Q}^c_{(4)}\,\mathbf{g}^c \tag{3.47}$$

and, in the case of eigenmodes,

$$\bar{\mathbf{g}}_\beta = -\mathbf{Q}^c_{(4)}\,\mathbf{g}^c_{-\beta} \tag{3.48}$$

with a minus sign inserted on the right-hand side since the coordinate system is tied to the plane of incidence, cf. (3.22) and (3.23). Substitution of (3.47) into (3.43) gives the bilinear concomitant P_z in terms of *given and conjugate* wave fields

$$\tilde{\mathbf{g}}^c\mathbf{U}^c\mathbf{g} = P_z = \text{const} \tag{3.49}$$

and in the case of eigenmodes

$$\tilde{\mathbf{g}}^c_{-\beta}\mathbf{U}^c\mathbf{g}_\alpha = P_{z,\alpha}\delta_{\alpha\beta} \tag{3.50}$$

where

$$\begin{aligned}\mathbf{U}^c &= -\mathbf{Q}^c_{(4)}\mathbf{U} = \begin{bmatrix} 0 & 0 & 0 & -1 \\ 0 & 0 & 1 & 0 \\ 0 & -1 & 0 & 0 \\ 1 & 0 & 0 & 0 \end{bmatrix} \\ &= \mathbf{U}\mathbf{Q}^c_{(4)} = -\tilde{\mathbf{U}}^c = -(\mathbf{U}^c)^{-1}\end{aligned} \tag{3.51}$$

or, equivalently,

$$\sum_{j=1}^{4}(-1)^j(\mathbf{g}^c_{-\beta})_j\,(\mathbf{g}_\alpha)_{5-j} = P_{z,\alpha}\,\delta_{\alpha\beta} \tag{3.52}$$

Note that if only two forward, or two backward waves are present at any level z in the medium, one in the given and one in the conjugate system, then α and β have opposite signs and, from (3.50), we deduce that P_z=0 at that level, and hence throughout the medium.

With normalized eigenmodes, (3.50) reduces to

$$(\hat{\mathbf{g}}^c_{-\beta})^T\mathbf{U}^c\hat{\mathbf{g}}_\alpha = \delta_{\alpha\beta}\,\text{sgn}(\alpha) \tag{3.53}$$

For later use it will be convenient to gather the normalized eigenmodes into modal matrices, $\mathbf{G}$ and $\mathbf{G}^c$, by analogy with the modal matrix $\mathbf{F}$ (3.3):

$$\begin{aligned}[\hat{\mathbf{g}}_1\;\hat{\mathbf{g}}_2\;\hat{\mathbf{g}}_{-1}\;\hat{\mathbf{g}}_{-2}] &\equiv [\mathbf{G}_+\;\mathbf{G}_-] \equiv \mathbf{G} \\ [\hat{\mathbf{g}}^c_1\;\hat{\mathbf{g}}^c_2\;\hat{\mathbf{g}}^c_{-1}\;\hat{\mathbf{g}}^c_{-2}] &\equiv [\mathbf{G}^c_+\;\mathbf{G}^c_-] \equiv \mathbf{G}^c\end{aligned} \tag{3.54}$$

Eq. (3.53) can then be written, with the aid of $\bar{\mathbf{I}}$ (2.81), in the form

$$\mathbf{J}\tilde{\mathbf{G}}^c\mathbf{U}^c\mathbf{G} = \bar{\mathbf{I}} \tag{3.55}$$

with the matrix $\mathbf{J}$, as in (3.17), interchanging positive- and negative-going eigenmode rows in $\tilde{\mathbf{G}}^c$ (first and second with third and fourth). Then, multiplying from the left by $\mathbf{G}\bar{\mathbf{I}}=\mathbf{G}\bar{\mathbf{I}}^{-1}$ and from the right by $\mathbf{G}^{-1}(\mathbf{U}^c)^{-1}=-\mathbf{G}^{-1}\mathbf{U}^c$ we obtain, with

$$\bar{\mathbf{J}} := \bar{\mathbf{I}}\mathbf{J} = \begin{bmatrix} 0 & 0 & 1 & 0 \\ 0 & 0 & 0 & 1 \\ -1 & 0 & 0 & 0 \\ 0 & -1 & 0 & 0 \end{bmatrix} = -\mathbf{J}\bar{\mathbf{I}} = -\bar{\mathbf{J}}^T = -\bar{\mathbf{J}}^{-1} \tag{3.56}$$

the biorthogonality relation

$$\mathbf{G}\bar{\mathbf{J}}\tilde{\mathbf{G}}^c = [\mathbf{G}_+ \; \mathbf{G}_-] \begin{bmatrix} \tilde{\mathbf{G}}^c{}_- \\ -\tilde{\mathbf{G}}^c{}_+ \end{bmatrix} = -\mathbf{U}^c \tag{3.57}$$

We note finally that the propagation matrix $\mathbf{T}$ is transformed into $\mathbf{T}^c$ in the conjugate system by means of

$$\mathbf{T}^c = -\mathbf{U}^c\tilde{\mathbf{T}}\mathbf{U}^c, \qquad \text{or} \qquad \mathsf{T}^c_{ij} = \sum_j (-1)^j \, \mathsf{T}_{5-j,5-i} \tag{3.58}$$

This may be shown by constructing the Clemmow-Heading equations (3.35) for the conjugate system:

$$\left[\frac{d}{dz} + ik_0\mathbf{T}^c\right]\mathbf{g}^c = 0$$

with (3.47) substituted into (3.41) with the aid of (3.39), or directly [21,100], by substituting $\hat{b}_x \to -\hat{b}_x$ (3.37) in the matrix elements ε_{ij} (1.45) that appear in $\mathbf{T}$ (1.128).

3.2.3 Rederivation of Budden's reciprocity theorem

Although Budden's theorem was derived originally by use of modal biorthogonality (3.50) [29], or the relation between the propagation matrices, $\mathbf{T}$ and $\mathbf{T}^c$ (3.58) [21], it is instructive to follow the derivation due to Pitteway and Jespersen [100].

Consider two plane waves incident on the ionosphere from below, in the given and conjugate systems respectively, so that above the ionosphere only upgoing waves are present. Then P_z=0 at all heights in the ionosphere [see the remarks following (3.52)]. If we now let the upgoing incident wave have unit amplitude, with the electric wave vector parallel to the plane of incidence, we obtain reflected modes with parallel and perpendicular polarizations, and with amplitudes $_{\parallel}R_{\parallel}$ and $_{\perp}R_{\parallel}$ respectively. In the conjugate system we suppose, for instance, that the unit-amplitude incident wave has perpendicular polarization, so that the reflected modes will have amplitudes $_{\parallel}R^c_{\perp}$ and $_{\perp}R^c_{\perp}$. Application of (3.50) which gives non-zero products of upgoing modes in one system with downgoing modes in the conjugate system yields, with P_z=0, and with the aid of (2.105) and (2.109),

$$_{\parallel}R^+_{\perp} = ({}_{\perp}R^c_{\parallel})^+$$

Other combinations of linearly polarized incident modes yield the remaining equalities in (3.19).

Tsuruda [126] has noted that whether or not a conducting earth is present, P_z=0 below the ionosphere (and hence throughout the medium) for whistler modes incident on the ionosphere from above. Suppose we have two whistler modes, $\mathbf{g}_{-1}$ and $\mathbf{g}^c_{-2}$, incident from above on the high ionosphere (where $X \gg 1+Y$), in both the given and conjugate planes of incidence. Each will generate a reflected mode (the other mode cannot propagate in this region), as well as transmitted modes below the ionosphere which, for the moment, do not interest us. The method of modal biorthogonality just described, applied to the high ionosphere with P_z=0, yields

$$R^-_{11} = (R^c_{11})^- \tag{3.59}$$

a result which involves eigenmodes within the anisotropic medium, rather than base modes in free space or in isotropic bounding media.

3.2.4 Reciprocity with penetrating and non-penetrating modes

In the general framework of eigenmode scattering/reciprocity relations, the reciprocity theorem of Pitteway and Jespersen [100], which relates 'penetrating and non-penetrating modes', occupies a special place in that these 'modes' are not eigenmodes of the medium. They are, in fact, specific combinations of upgoing incident eigenmodes which maximize or minimize the transmission coefficients through the ionosphere. The theorem can be derived by equating the constant value of P_z high in the ionosphere with that in free space below [100]. We shall not follow this procedure here, but shall show instead that Pitteway's modes [98] can be decomposed into the eigenmodes of the medium in a specific combination, and then demonstrate that the theorem is a direct consequence of the eigenmode scattering theorem.

Let $\mathbf{g}_{\pm 1}$ represent up- and downgoing whistler modes in the ionosphere. The other modes, $\mathbf{g}_{\pm 2}$, are evanescent in the high ionosphere, but can propagate lower down, below the $X{=}1{+}Y$ cutoff (see Sec. 1.2). We express the non-penetrating and penetrating upward incident modes, $\mathbf{g}_n$ and $\mathbf{g}_p$, at a height z_0 at the base of the ionosphere, as a combination of the two eigenmodes

$$\mathbf{g}_n = \hat{\mathbf{g}}_1 + b\,\hat{\mathbf{g}}_2, \qquad \mathbf{g}_p = \hat{\mathbf{g}}_1 + c\,\hat{\mathbf{g}}_2 \tag{3.60}$$

where b and c are constants to be determined. As for the free-space eigenmodes, $\mathbf{g}_1$ and $\mathbf{g}_2$, we can consider them to be right- and left-circularly polarized modes (the electron-whistler mode has 'electronic polarization'—i.e. the electric wave vector circles the constant magnetic field in the same sense as free electrons do). For simplicity we have taken the amplitude of the whistler component $\hat{\mathbf{g}}_1$ to be unity at z_0 in the two modes, $\mathbf{g}_n$ and $\mathbf{g}_p$, since we shall be interested only in transmission coefficients which are independent of incident wave amplitudes. All field quantities in $\mathbf{g}_n$ and $\mathbf{g}_p$ vary as $\exp(-ik_0 sx)$, with x in the plane of incidence which is at an azimuthal angle ϕ with respect to the magnetic meridian plane (see Fig. 3.1).

The non-penetrating wave must, by definition, produce a zero-amplitude whistler mode at a height z', high in the ionosphere (the other mode $\mathbf{g}_2(z')$ becomes evanescent well below z', and so in any case has zero amplitude at z'). If $a_1(z)$ and $a_2(z)$ are the amplitudes of modes 1 and 2 at any level z, then the amplitude $a_1(z')$ of the whistler mode is given by

$$a_1(z') = T_{11}^+\, a_1(z_0) + T_{12}^+\, a_2(z_0) = 0 \tag{3.61}$$

in which $T_{ij}^+ \equiv T_{ij}^+(z', z_0; s, \phi)$, and the two eigenmodes that constitute the non-penetrating mode at z_0, cf. (3.60), have amplitudes $a_1(z_0){=}1$ and $a_2(z_0){=}b$. Substituting into (3.61), and using (3.60), we get

$$b = -T_{11}^+/T_{12}^+, \qquad \mathbf{g}_n(z_0) = \hat{\mathbf{g}}_1(z_0) - \frac{T_{11}^+}{T_{12}^+}\hat{\mathbf{g}}_2(z_0) \tag{3.62}$$

Next we determine the constant c in $\mathbf{g}_p$ (3.60), by using the property that the electric wave vectors in the penetrating and non-penetrating modes are hermitian orthogonal, as shown in Sec. 1.5.2. In the (ξ, η, ς) coordinate system, tied to the wave normal and the plane of incidence below the ionosphere (as in Sec. 3.1.2), this property, with $\rho_1{=}i$, $\rho_2{=}-i$, cf. (3.10), leads to

$$\left[(\hat{\xi} + i\hat{\eta}) + b(\hat{\xi} - i\hat{\eta})\right] \cdot \left[(\hat{\xi} + i\hat{\eta}) + c(\hat{\xi} - i\hat{\eta})\right]^* = 0$$

which gives, with (3.62),

$$c = -1/b^* = (T_{12}^+/T_{11}^+)^*$$

and

$$\mathbf{g}_p(z_0) = \hat{\mathbf{g}}_1(z_0) + \left(\frac{T_{12}^+}{T_{11}^+}\right)^* \hat{\mathbf{g}}_2(z_0) \tag{3.63}$$

The amplitude $a_1(z')$ of the penetrating mode at a level z', high in the ionosphere, then becomes

$$a_1(z') = T_{11}^+ + cT_{12}^+ = T_{11}^+ + \left(\frac{T_{12}^+}{T_{11}^+}\right)^* T_{12}^+$$

where $T_{ij}^+ \equiv T_{ij}^+(z', z_0)$, and the 4-element wave field is

$$\mathbf{g}_p(z') = a_1(z')\hat{\mathbf{g}}_1(z') = \left[T_{11}^+ + \left(\frac{T_{12}^+}{T_{11}^+}\right)^* T_{12}^+\right] \hat{\mathbf{g}}_1(z') \tag{3.64}$$

The ratio of transmitted to incident energy flux, $\tau_p^+(s,\phi)$, is then

$$\tau_p^+(s,\phi) = \frac{[\mathbf{g}_p(z')^*]^T \, \mathbf{U} \, \mathbf{g}_p(z')}{[\mathbf{g}_p(z_0)^*]^T \, \mathbf{U} \, \mathbf{g}_p(z_0)} \tag{3.65}$$

as in (2.63) and (3.43), with complex conjugate replacing adjoint modes in loss-free media. In view of the modal biorthogonality (3.46), this gives with the aid of (3.63) and (3.64)

$$\tau_p^+(s,\phi) = \frac{\left|T_{11}^+ + \left(\frac{T_{12}^+}{T_{11}^+}\right)^* T_{12}^+\right|^2}{1 + \left|\frac{T_{12}^+}{T_{11}^+}\right|^2} = |\, T_{11}^+ \,|^2 + |\, T_{12}^+ \,|^2$$

Next consider a downgoing whistler mode $\mathbf{g}_{-1}^c$ with unit amplitude, $a_{-1}^c(z') = 1$, in the conjugate plane. The transmitted wave field at the base of the ionosphere will be

$$\mathbf{g}(z_0) = a_{-1}^c(z_0)\,\hat{\mathbf{g}}_{-1}^c(z_0) + a_{-2}^c(z_0)\,\hat{\mathbf{g}}_{-2}^c(z_0)$$

which, with

$$a_{-1}^c(z_0) = (T_{11}^-)^c\, a_{-1}^c(z') = (T_{11}^-)^c, \qquad a_{-2}^c(z_0) = (T_{21}^-)^c$$

gives

$$\begin{aligned} \mathbf{g}(z_0) &= (T^-_{\bar{1}1})^c\, \hat{\mathbf{g}}^c_{-1}(z_0) + (T^-_{\bar{2}1})^c\, \hat{\mathbf{g}}^c_{-2}(z_0) \\ &= T^+_{\bar{1}1}\, \hat{\mathbf{g}}^c_{-1}(z_0) + T^+_{\bar{1}2}\, \hat{\mathbf{g}}^c_{-2}(z_0) \end{aligned} \tag{3.66}$$

by virtue of the eigenmode scattering theorem (2.112). The ratio of transmitted to incident energy flux is then

$$\tau^c_-(s, \pi - \phi) = | T^+_{\bar{1}1} |^2 + | T^+_{\bar{1}2} |^2 = \tau^+_p(s, \phi) \tag{3.67}$$

which is the first part of the of the reciprocity theorem of Pitteway and Jespersen. It states that the ratio of transmitted to incident energy flux for a downgoing whistler mode high in the ionsphere is equal to that of a 'penetrating mode' incident on the ionosphere from below in the conjugate direction.

The relation between the wave polarizations of the two waves at the base of the ionosphere (upgoing penetrating and transmitted downgoing whistler) can readily be found by performing a conjugating transformation on the two component eigenmodes in the penetrating mode (see, for instance, [14]). The result, however, is immediately evident when it is realized that the conjugate downgoing whistler in the high ionosphere is derived from the upgoing whistler in the orginal system by a reflection-time-reversal transformation. Hence the polarization ellipse produced by the two downgoing transmitted eigenmodes in the conjugate system will be a mirror image of the polarization ellipse produced by the two upgoing eigenmodes in the penetrating wave. The sense of rotation, however, will be the same in the two cases since it has been twice reversed — first by reflection and then by time reversal.

This then completes the proof of the theorem, which is seen to fit into the general framework of the eigenmode scattering theorem.

3.2.5 Derivation of the eigenmode scattering theorem without explicit use of adjoint wave fields

It will no doubt have been noted that all previous derivations of scattering (reciprocity) theorems, which were based in one way or another on the biorthogonality of the given and the *conjugate* (rather than given and adjoint) eigenmodes, gave results which were restricted to wave amplitudes in free space or in loss-free regions ("high in the ionosphere", for instance), where amplitudes could conveniently be defined through the z-component of a Poynting vector by means of complex-conjugate fields (as in Sec. 2.3.1). The reason, presumably, was that by-passing the adjoint wave fields, not only was physical insight lost as mentioned earlier, but one tended to overlook the formalism by which the amplitude of an eigenmode in a lossy medium could be expressed in a simple manner, and by which the energy flux in an arbitrary wave field could be decomposed into the sum of the contributions of each of the eigenmodes. We shall now show that the eigenmode scattering theorem can indeed be simply derived by making use of the given and conjugate eigenmodes only.

We decompose two wave fields, $\mathbf{g}$ and $\mathbf{g}^c$, propagating in the given and conjugate planes of incidence, into component eigenmodes, as in (2.61),

$$\mathbf{g} = \sum_\alpha a_\alpha \hat{\mathbf{g}}_\alpha , \qquad \mathbf{g}^c = \sum_\alpha a_\alpha^c \hat{\mathbf{g}}_\alpha^c \tag{3.68}$$

where

$$a_\alpha = (\mathbf{g}^c_{-\alpha})^T \mathbf{U}^c \, \mathbf{g} \, \mathrm{sgn}(\alpha), \qquad a_\alpha^c = \hat{\mathbf{g}}^T_{-\alpha} \mathbf{U}^c \, \mathbf{g}^c \, \mathrm{sgn}(\alpha) \tag{3.69}$$

with the aid of (3.53). With the wave fields $\mathbf{g}$ and $\mathbf{g}^c$ decomposed into eigenmodes at any level z in this way, the constant bilinear concomitant, $P_z = \tilde{\mathbf{g}}^c \mathbf{U}^c \, \mathbf{g}$ (3.49)), becomes with (3.68) and (3.69)

$$P_z = a^c_{-1} a_1 + a^c_{-2} a_2 - a^c_1 a_{-1} - a^c_2 a_{-2} \equiv \tilde{\boldsymbol{a}}^c_- \boldsymbol{a}_+ - \tilde{\boldsymbol{a}}^c_+ \boldsymbol{a}_- \tag{3.70}$$

in which modal amplitudes have been collected into 2-element column vectors, $\boldsymbol{a}_\pm$ and $\boldsymbol{a}^c_\pm$, as in (2.101).

Now equate $P_z(z^-) = P_z(z^+)$ at two levels, z^- and z^+ ($z^+ > z^-$), and collect incoming and outgoing modal amplitudes into 4-element columns, $\boldsymbol{a}_{in}$ and $\boldsymbol{a}_{out}$, as in (2.102):

$$\tilde{\boldsymbol{a}}^c_-(z^-)\boldsymbol{a}_+(z^-) + \tilde{\boldsymbol{a}}^c_+(z^+)\boldsymbol{a}_-(z^+) = \tilde{\boldsymbol{a}}^c_+(z^-)\boldsymbol{a}_-(z^-) + \tilde{\boldsymbol{a}}^c_-(z^+)\boldsymbol{a}_+(z^+)$$

which may be written as

$$\tilde{\boldsymbol{a}}^c_{out} \boldsymbol{a}_{in} = \tilde{\boldsymbol{a}}^c_{in} \boldsymbol{a}_{out} \tag{3.71}$$

Substituting $\boldsymbol{a}_{out} = \mathbf{S}\boldsymbol{a}_{in}$ and $\boldsymbol{a}^c_{out} = \mathbf{S}^c \boldsymbol{a}^c_{in}$ into (3.71), we get finally

$$\tilde{\boldsymbol{a}}^c_{in} \tilde{\mathbf{S}}^c \boldsymbol{a}_{in} = \tilde{\boldsymbol{a}}^c_{in} \mathbf{S} \, \boldsymbol{a}_{in} \tag{3.72}$$

from which we have $\mathbf{S} = \tilde{\mathbf{S}}^c$, which is just the eigenmode scattering theorem (2.112).

3.2.6 The scattering theorem in a multilayer medium

Quite a different approach is to derive the scattering theorem first for a plane interface separating two adjacent thin homogeneous layers, and then to extend the result by recursion to the multilayer structure that simulates the original continuously varying medium [12]. If we equate transverse (tangential) electric and magnetic wave-field components across the interface separating layers ν and $(\nu+1)$, we have

$$\mathbf{g}^{\nu} = \mathbf{g}^{\nu+1}$$

with g containing the four transverse wave-field components (3.36). Decomposing the wave fields into positive- and negative-going eigenvectors, we may express this relation by means of the eigenmode amplitudes, $\boldsymbol{a}^{\nu}$, as in (2.101), and the normalized eigenmode matrices $\mathbf{G}^{\nu}$ (3.54),

$$\mathbf{G}^{\nu}\boldsymbol{a}^{\nu} \equiv \left[\mathbf{G}_{+}^{\nu}\boldsymbol{a}_{+}^{\nu} + \mathbf{G}_{-}^{\nu}\boldsymbol{a}_{-}^{\nu}\right] = \left[\mathbf{G}_{+}^{\nu+1}\boldsymbol{a}_{+}^{\nu+1} + \mathbf{G}_{-}^{\nu+1}\boldsymbol{a}_{-}^{\nu+1}\right] \equiv \mathbf{G}^{\nu+1}\boldsymbol{a}^{\nu+1} \tag{3.73}$$

We now rearrange terms into outgoing and incoming eigenmodes,

$$\left[-\mathbf{G}_{-}^{\nu} \;\; \mathbf{G}_{+}^{\nu+1}\right]\begin{bmatrix} \boldsymbol{a}_{-}^{\nu} \\ \boldsymbol{a}_{+}^{\nu+1} \end{bmatrix} = \left[\mathbf{G}_{+}^{\nu} \;\; -\mathbf{G}_{-}^{\nu+1}\right]\begin{bmatrix} \boldsymbol{a}_{+}^{\nu} \\ \boldsymbol{a}_{-}^{\nu+1} \end{bmatrix} \tag{3.74}$$

or, in condensed notation,

$$\mathbf{G}_{out}\boldsymbol{a}_{out} = \mathbf{G}_{in}\boldsymbol{a}_{in} \tag{3.75}$$

With $\boldsymbol{a}_{out}=\mathbf{S}\,\boldsymbol{a}_{in}$ (2.103), this becomes

$$\mathbf{G}_{out}\mathbf{S}\,\boldsymbol{a}_{in} = \mathbf{G}_{in}\boldsymbol{a}_{in}$$

or

$$\mathbf{G}_{out}\mathbf{S} = \mathbf{G}_{in}, \qquad \mathbf{S} = \mathbf{G}_{out}^{-1}\,\mathbf{G}_{in} \tag{3.76}$$

This is an interesting result, showing that at an interface the scattering matrix can be determined from the incoming and outgoing modal polarizations only, and not necessarily from the incoming and outgoing modal amplitudes.

In order to relate given and conjugate scattering matrices and obtain an interface scattering theorem, we equate the constant matrix product $\mathbf{G}\bar{\mathbf{J}}\tilde{\mathbf{G}}^{c}=-\mathbf{U}^{c}$ (3.57) on both sides, ν and $(\nu+1)$, of the interface

$$\left[\mathbf{G}_{+}^{\nu} \;\; \mathbf{G}_{-}^{\nu}\right]\begin{bmatrix} \tilde{\mathbf{G}}_{-}^{c,\nu} \\ -\tilde{\mathbf{G}}_{+}^{c,\nu} \end{bmatrix} = \left[\mathbf{G}_{+}^{\nu+1} \;\; \mathbf{G}_{-}^{\nu+1}\right]\begin{bmatrix} \tilde{\mathbf{G}}_{-}^{c,\nu+1} \\ -\tilde{\mathbf{G}}_{+}^{c,\nu+1} \end{bmatrix}$$

Regrouping, we have

$$\left[-\mathbf{G}_{-}^{\nu} \;\; \mathbf{G}_{+}^{\nu+1}\right]\begin{bmatrix} \tilde{\mathbf{G}}_{+}^{c,\nu} \\ -\tilde{\mathbf{G}}_{-}^{c,\nu+1} \end{bmatrix} = \left[\mathbf{G}_{+}^{\nu} \;\; -\mathbf{G}_{-}^{\nu+1}\right]\begin{bmatrix} -\tilde{\mathbf{G}}_{-}^{c,\nu} \\ \tilde{\mathbf{G}}_{+}^{c,\nu+1} \end{bmatrix}$$

or, in condensed notation, cf. (3.74)–(3.75),

$$\mathbf{G}_{out}\tilde{\mathbf{G}}_{in}^{c} = \mathbf{G}_{in}\tilde{\mathbf{G}}_{out}^{c}$$

whence

$$\mathbf{G}_{out}^{-1}\mathbf{G}_{in} = \tilde{\mathbf{G}}_{in}^{c}\left[\tilde{\mathbf{G}}_{out}^{c}\right]^{-1} = \left[(\mathbf{G}_{out}^{c})^{-1}\mathbf{G}_{in}^{c}\right]^{T} \tag{3.77}$$

With (3.76), this yields the required ***interface eigenmode scattering theorem***

$$\mathbf{S} = \tilde{\mathbf{S}}^{c} \tag{3.78}$$

To extend this result to a multilayer system, we consider a multilayer slab, as in Sec. 1.5.3, bounded on either side by gyrotropic media which in general will also be part of (i.e. imbedded in) the multilayer structure. We denote the uppermost layer in the slab by ν, and the following two layers, outside the slab, by $(\nu+1)$ and $(\nu+2)$ (see Fig. 1.2). Let the 2×2 reflection and transmission matrices for the slab be denoted by $\mathbf{R}_{\pm}^{\nu}$ and $\mathbf{T}_{\pm}^{\nu}$, and for the composite slab formed by adding the additional layer $(\nu+1)$, $\mathbf{R}_{\pm}^{\nu+1}$ and $\mathbf{T}_{\pm}^{\nu+1}$. The corresponding matrices for the interface separating the layers $(\nu+1)$ and $(\nu+2)$ are denoted by $\mathbf{r}_{\pm}$ and $\mathbf{t}_{\pm}$. If $\mathbf{R}_{\pm}^{\nu}$, $\mathbf{T}_{\pm}^{\nu}$ and $\mathbf{r}_{\pm}$, $\mathbf{t}_{\pm}$ are known, then $\mathbf{R}_{\pm}^{\nu+1}$ and $\mathbf{T}_{\pm}^{\nu+1}$ are determined by the recursion relations (1.172) in Sec. 1.5.3. The phase matrices $\mathbf{\Delta}_{\pm}$ in these equations give the phase change of the eigenmodes in traversing the $(\nu+1)$th layer in terms of the eigenvalues $q_{\alpha}^{\nu+1}$, $(\alpha=\pm1,\pm2)$, and the layer thickness δz, cf. (1.166):

$$\mathbf{\Delta}_{\pm} = \begin{bmatrix} \exp(-ik_0 q_{\pm1}^{\nu+1}\delta z) & 0 \\ 0 & \exp(-ik_0 q_{\pm2}^{\nu+1}\delta z) \end{bmatrix}$$

A similar set of recursion relations may be written for the conjugate system, and it will be noted that since $q_{\alpha}=-q_{-\alpha}^{c}$ (2.91), the phase matrices in the conjugate system are

$$\mathbf{\Delta}_{\pm}^{c} = \mathbf{\Delta}_{\mp} \tag{3.79}$$

Given the interface scattering relations (3.78), it is easy to show from the recursion relations (1.172) that if the scattering theorem holds for the original slab, i.e. if

$$\mathbf{R}_{\pm}^{\nu} = \tilde{\mathbf{R}}_{\pm}^{c,\nu}, \qquad \mathbf{T}_{\pm}^{\nu} = \tilde{\mathbf{T}}_{\mp}^{c,\nu+1} \tag{3.80}$$

then it holds also for the composite slab, i.e.

$$\mathbf{R}_{\pm}^{\nu+1} = \tilde{\mathbf{R}}_{\pm}^{c,\nu+1}, \qquad \mathbf{T}_{\pm}^{\nu+1} = \tilde{\mathbf{T}}_{\mp}^{c,\nu+1} \tag{3.81}$$

Except for the additional relationship (3.79) which is now required, the proof, which is straightforward but tedious, is given in reference [7] for the case of normal incidence. The proof by induction for the system as a whole now commences with the slab which is just the first elementary layer, in which the matrices $\mathbf{R}_{\pm}^{\nu}$ and $\mathbf{T}_{\pm}^{\nu}(\nu = 1)$ are replaced by $\mathbf{r}_{\pm}$ and $\mathbf{t}_{\pm}$ for the first interface. These obey the scattering relations (3.78), and hence by induction the scattering relations apply to the system as a whole, or to any slab imbedded within the system.

It is interesting to note that in the scattering theorem of Altman and Postan [7], which applies to eigenmodes within (possibly absorbing) gyrotropic media, the modal amplitudes for normal incidence (motivated by the form of the WKB solutions [32, p.405]) were defined as

$$a_\alpha = {n_\alpha}^{1/2}(1 - {\rho_\alpha}^2)\, E_{\alpha,x} \qquad (\alpha = \pm 1, \pm 2) \tag{3.82}$$

which, aside from a numerical factor $\sqrt{2Y_0}$, is precisely the value obtained by the ratio

$$E_{\alpha\xi}/\widehat{E}_{\alpha\xi} = \mathbf{e}_\alpha/\hat{\mathbf{e}}_\alpha = a_\alpha$$

see (2.68), (2.74) and (2.75), when the (ξ, η, ς) and (x, y, z) systems coincide (normal incidence). Their proof of the interface scattering relation made use of the condition, $\rho_1\rho_2 = 1$, for two upgoing or two downgoing modes. Now this is precisely the (biorthogonality) conditon applying to eigenmodes propagating *in the same direction* in a magnetoplasma (1.79) (see also [104, p.17], [32, p.50]), as will occur *if they propagate in a direction normal to the stratification.* For oblique incidence the modal propagation vectors, $\mathbf{k}_\alpha$, are not in the same direction—only the transverse components, $\mathbf{k}_t$, remain equal—so that the condition $\rho_1\rho_2 = 1$ is no longer valid, and this explains why this method [7] could not be extended to oblique incidence until the general formulation of modal biorthogonality was applied.

3.2.7 Reciprocity via Maxwell's second order differential equations

Heading's [66] formulation of symmetries between electric wave fields, and between reflection and transmission coefficients, for modes propagating with different directions of incidence through plane-stratified magnetoplasmas, differs in approach from the other methods discussed until now, in that the starting point is one of Maxwell's second-order differential equations obtained by eliminating the magnetic field from the first-order equations. The procedure has a certain resemblance to that adopted by us, in which the application of a Lagrange identity (2.37) yields the divergence of a bilinear concomitant vector. We shall adapt Heading's notation to ours in order to appreciate similarities and some important differences.

Consider the Maxwell system, (2.21) and (2.22),

$$[i\omega\mathsf{K} + \mathsf{D}]\,\mathbf{e}(\mathbf{r}) = 0, \qquad \tilde{\mathbf{e}} := [\tilde{\mathbf{E}}, \tilde{\mathbf{H}}] \tag{3.83}$$

in a source-free ($\mathbf{j}(\mathbf{r})$=0) magnetoplasma where, as in (2.35),

$$\mathsf{K} = \begin{bmatrix} \boldsymbol{\varepsilon} & 0 \\ 0 & \mu_0\,\mathsf{I}^{(3)} \end{bmatrix} = \begin{bmatrix} \varepsilon_0(\mathsf{I}^{(3)} + \chi) & 0 \\ 0 & \mu_0\,\mathsf{I}^{(3)} \end{bmatrix}$$

in which the susceptibility tensor χ, as used by Heading, is related to the conductivity tensor $\boldsymbol{\sigma}$ (1.33) through

$$\boldsymbol{\varepsilon} = \varepsilon_0\mathsf{I}^{(3)} - i\boldsymbol{\sigma}/\omega\,, \qquad \chi = -i\boldsymbol{\sigma}/\varepsilon_0\omega$$

If we operate on (3.83) with the differential operator D (2.22)

$$\mathsf{D}\,[i\omega\mathsf{K} + \mathsf{D}]\,\mathbf{e}(\mathbf{r}) = 0 \tag{3.84}$$

the upper part of this matrix equation yields

$$-\,i\omega\mu_0\boldsymbol{\nabla}\times\mathbf{H} - \boldsymbol{\nabla}\times(\boldsymbol{\nabla}\times\mathbf{E}) = 0 \tag{3.85}$$

since

$$\mathsf{D}^2 \equiv \mathsf{DD} = -\begin{bmatrix} \boldsymbol{\nabla}\times(\boldsymbol{\nabla}\times\mathsf{I}^{(3)}) & 0 \\ 0 & \boldsymbol{\nabla}\times(\boldsymbol{\nabla}\times\mathsf{I}^{(3)}) \end{bmatrix} \tag{3.86}$$

Now substitute the upper part of (3.83),

$$i\omega\varepsilon_0(\mathsf{I}^{(3)} + \chi)\mathbf{E} - \boldsymbol{\nabla}\times\mathbf{H} = 0 \tag{3.87}$$

into (3.85) to give, with $k_0 = \omega/c$,

$$\mathsf{L}_E\mathbf{E} := \left[k_0{}^2(\mathsf{I}^{(3)} + \chi) - \boldsymbol{\nabla}\times(\boldsymbol{\nabla}\times\mathsf{I}^{(3)})\right]\mathbf{E} = 0 \tag{3.88}$$

If we were to construct an equation adjoint to (3.88), the differential operator, which is second order in $\boldsymbol{\nabla}$, would remain unchanged in sign. The susceptibility tensor χ, on the other hand, would be replaced by its transpose χ^T, and would be eliminated on application of the Lagrange identity. Heading however does not work with the Lagrange identity, nor with the adjoint media and fields. Instead, he considers two different local plane-wave solutions, $\mathbf{E}_1$ and $\mathbf{E}_2$, of the Maxwell system (3.88), for two different directions of incidence:

$$\begin{aligned} &\mathbf{E}_2^T\mathsf{L}_E\mathbf{E}_1 - \mathbf{E}_1^T\mathsf{L}_E\mathbf{E}_2 \\ = \;& k_0{}^2(\mathbf{E}_2^T\chi\mathbf{E}_1 - \mathbf{E}_1^T\chi\mathbf{E}_2) + \mathbf{E}_1\cdot\{\boldsymbol{\nabla}\times(\boldsymbol{\nabla}\times\mathbf{E}_2)\} - \mathbf{E}_2\cdot\{\boldsymbol{\nabla}\times(\boldsymbol{\nabla}\times\mathbf{E}_1)\} \\ = \;& k_0{}^2\mathbf{E}_2^T\left[\chi - \chi^T\right]\mathbf{E}_1 - \boldsymbol{\nabla}\cdot\{\mathbf{E}_1\times(\boldsymbol{\nabla}\times\mathbf{E}_2) - \mathbf{E}_2\times(\boldsymbol{\nabla}\times\mathbf{E}_1)\} \\ = \;& 0 \end{aligned} \tag{3.89}$$

Had this been a Lagrange identity, with $\mathbf{E}_2 \rightarrow \overline{\mathbf{E}}_1$, the first term in the third line of (3.89) would have been eliminated, as mentioned, and the second term would have given the divergence of a bilinear concomitant vector. In Heading's version the divergence term is equal to

$$k_0{}^2 \mathbf{E}_2^T \left[\chi - \chi^T\right] \mathbf{E}_1 = k_0{}^2 \mathbf{E}_2^T \left[\varepsilon - \varepsilon^T\right] \mathbf{E}_1 / \varepsilon_0 = -2ik_0{}^2 D \hat{\mathbf{b}} \cdot \mathbf{E}_1 \times \mathbf{E}_2$$

rather than zero, as may be seen by use of the form of $\varepsilon/\varepsilon_0$ for a gyrotropic medium in (1.38). With local plane-wave x-dependence

$$\mathbf{E}_j \sim \exp(-ik_0\, s_j x), \qquad j = 1,2$$

in a coordinate system tied to the plane of incidence, the divergence operator has two non-zero components, $\nabla = (-ik_0 s_j,\, 0,\, \partial/\partial z)$, and (3.89) can be integrated over z from below the ionosphere to above it [66]. For specialized directions of incidence, the expressions simplify to yield relations between scattering coefficients, some of which, (3.20) and (3.33), have been cited earlier.

This method requires more mathematical manipulation than others, and yields relations governing wave fields in the isotropic bounding regions only. One of its merits is that insofar as the directions of incidence are arbitrary, one is at liberty, at the end of the z-integration, to choose all specialized directions of incidence that will yield useful identities. The mirror-mode identities, discussed in Sec. 3.1.4, are examples of such identities that were missed by most of the other methods described. This method lends itself to establishing generalized reciprocity theorems relating to wave propagation governed by second order differential equations containing n independent variables [67,68], and equations with self-adjoint differential operators of order $2n$ [69].

3.3 Matrizants and transfer matrices

The mapping of a plane-stratified medium into itself by a conjugating transformation has been shown to lead to a reciprocity relation (2.112) between the scattering matrices in the given and conjugate problems. This is a natural way to express the symmetry between the two systems when the solution of the propagation problem is aimed at obtaining reflection and transmission matrices as its end products. Another approach, discussed in Sec. 1.5.4 [78,128,71,23], is to produce as output matrizants that relate the *wave fields*, $\mathbf{g}(z^-)$ and $\mathbf{g}(z^+)$, at two levels, z^- and z^+, or matrizants (termed transfer matrices) that relate the eigenmode *amplitudes*, $\mathbf{a}(z^-)$ and $\mathbf{a}(z^+)$, at these levels. As would be expected, there is again a simple symmetry relationship between such matrices in the given and conjugate problems [119].

The four-component wave fields, $\mathbf{g}(z^-)$ and $\mathbf{g}(z^+)$ at z^- and z^+ (3.36), will be related in general by a 4×4 matrix, or matrizant, $\mathsf{M}(z^+,z^-)$:

$$\mathbf{g}(z^+) = \mathsf{M}(z^+, z^-)\,\mathbf{g}(z^-) \tag{3.90}$$

and similarly for the adjoint problem

$$\bar{\mathbf{g}}(z^+) = \overline{\mathsf{M}}(z^+, z^-)\,\bar{\mathbf{g}}(z^-) \tag{3.91}$$

with

$$\mathsf{M}(z^+, z^-) = \mathsf{M}^{-1}(z^-, z^+) \tag{3.92}$$

The bilinear concomitant (3.43)

$$\bar{\mathbf{g}}^T(z^+)\mathsf{U}\,\mathbf{g}(z^+) = \bar{\mathbf{g}}^T(z^-)\mathsf{U}\,\mathbf{g}(z^-) = \text{const} \tag{3.93}$$

yields, with (3.90) and (3.91),

$$\bar{\mathbf{g}}^T(z^-)\overline{\mathsf{M}}^T(z^+, z^-)\mathsf{U}\,\mathsf{M}(z^+, z^-)\mathbf{g}(z^-) = \bar{\mathbf{g}}^T(z^-)\mathsf{U}\,\mathbf{g}(z^-) \tag{3.94}$$

leading to

$$\overline{\mathsf{M}}^T(z^+, z^-)\mathsf{U}\,\mathsf{M}(z^+, z^-) = \mathsf{U} \tag{3.95}$$

or, with (3.92), and recalling that $\mathsf{U}=\mathsf{U}^T=\mathsf{U}^{-1}$ (3.39),

$$\overline{\mathsf{M}}(z^+, z^-) = \mathsf{U}\,\mathsf{M}^T(z^-, z^+)\mathsf{U} \tag{3.96}$$

This means that $\overline{\mathsf{M}}(z^+,z^-)$ is just the transpose of $\mathsf{M}(z^-,z^+)$ with respect to its trailing diagonal [cf. (3.39)].

If in (3.91) we transform from adjoint to conjugate wave fields, as in (3.47),

$$\bar{\mathbf{g}} = -\mathsf{Q}^c_{(4)}\mathbf{g}^c$$

with $\mathsf{Q}^c_{(4)}=\left(\mathsf{Q}^c_{(4)}\right)^{-1}$ defined in (3.17), we have

$$\mathbf{g}^c(z^+) = \mathsf{Q}^c_{(4)}\,\overline{\mathsf{M}}(z^+, z^-)\,\mathsf{Q}^c_{(4)}\mathbf{g}^c(z^-) \tag{3.97}$$

so that

$$\mathsf{M}^c(z^+, z^-) = \mathsf{Q}^c_{(4)}\,\overline{\mathsf{M}}(z^+, z^-)\,\mathsf{Q}^c_{(4)} \tag{3.98}$$

When this is substituted into (3.96), with

$$\mathsf{U}^c := -\mathsf{Q}^c_{(4)}\,\mathsf{U} = \mathsf{U}\,\mathsf{Q}^c_{(4)}$$

cf. (3.51), we have finally

$$\mathbf{M}^c(z^+,z^-) = -\mathbf{U}^c\,\mathbf{M}^T(z^-,z^+)\,\mathbf{U}^c, \quad M^c_{ij}(z^+,z^-) = \sum_j (-1)^j M_{5-j,5-i}(z^-,z^+) \tag{3.99}$$

in analogy with the relation (3.58) between $\mathbf{T}^c$ and $\mathbf{T}$. This is the *matrizant theorem*, relating matrizants in the given and conjugate problems.

Next we decompose the wave fields in (3.90) into eigenmodes, as in (3.68), with the aid of $\mathbf{G}$ (3.54),

$$\mathbf{g}(z^+) = \mathbf{G}(z^+)\,\boldsymbol{a}(z^+) = \mathbf{M}(z^+,z^-)\,\mathbf{g}(z^-) = \mathbf{M}(z^+,z^-)\,\mathbf{G}(z^-)\boldsymbol{a}(z^-) \tag{3.100}$$

Hence

$$\boldsymbol{a}(z^+) = \left[\mathbf{G}^{-1}(z^+)\,\mathbf{M}(z^+,z^-)\,\mathbf{G}(z^-)\right]\boldsymbol{a}(z^-) = \mathbf{P}(z^+,z^-)\,\boldsymbol{a}(z^-) \tag{3.101}$$

defining thereby the *transfer matrix* [128] or *propagator* [85], $\mathbf{P}(z^+,z^-)$. Substituting (2.101) we obtain

$$\begin{aligned} \boldsymbol{a}(z^+) := \begin{bmatrix} a_+(z^+) \\ a_-(z^+) \end{bmatrix} &= \begin{bmatrix} \mathbf{P}_{++}(z^+,z^-) & \mathbf{P}_{+-}(z^+,z^-) \\ \mathbf{P}_{-+}(z^+,z^-) & \mathbf{P}_{--}(z^+,z^-) \end{bmatrix} \begin{bmatrix} a_+(z^-) \\ a_-(z^-) \end{bmatrix} \\ &\equiv \mathbf{P}(z^+,z^-)\,\boldsymbol{a}(z^-) \end{aligned} \tag{3.102}$$

with

$$\mathbf{P}(z^-,z^+) = \mathbf{P}^{-1}(z^+,z^-) \tag{3.103}$$

The transfer matrix, defined by (3.101), is seen to be related to the matrizant through

$$\mathbf{P}(z^+,z^-) = \mathbf{G}^{-1}(z^+)\,\mathbf{M}(z^+,z^-)\,\mathbf{G}(z^-) \tag{3.104}$$

or

$$\mathbf{M}(z^+,z^-) = \mathbf{G}(z^+)\,\mathbf{P}(z+,z^-)\,\mathbf{G}^{-1}(z^-) \tag{3.105}$$

There is a similar relation in the conjugate system

$$\mathbf{M}^c(z^+,z^-) = \mathbf{G}^c(z^+)\,\mathbf{P}^c(z^+,z^-)\,\left[\mathbf{G}^c(z^-)\right]^{-1} \tag{3.106}$$

We now substitute (3.105), with z^+ and z^- interchanged, and (3.106) into (3.99), and rearrange terms using $(\mathbf{U}^c)^{-1}{=}-\mathbf{U}^c$ (3.51), to get

$$\left[\mathbf{G}^T(z^+)\,\mathbf{U}^c\,\mathbf{G}^c(z^+)\right]\,\mathbf{P}^c(z^+,z^-) = \mathbf{P}^T(z^-,z^+)\left[\mathbf{G}^T(z^-)\,\mathbf{U}^c\,\mathbf{G}^c(z^-)\right] \tag{3.107}$$

yielding, with the aid of (3.55), and with $\widetilde{\mathbf{U}}^c{=}-\mathbf{U}^c$, $\bar{\mathbf{J}}^T{=}\bar{\mathbf{J}}^{-1}{=}-\bar{\mathbf{J}}$ (3.56),

$$\bar{\mathbf{J}}\,\mathbf{P}^c(z^+,z^-) = \mathbf{P}^T(z^-,z^+)\,\bar{\mathbf{J}} \tag{3.108}$$

It will be convenient at this point to condense the notation by using

$$\mathbf{P}(z^+, z^-) \to \overrightarrow{\mathbf{P}}, \qquad \mathbf{P}(z^-, z^+) \to \overleftarrow{\mathbf{P}} \tag{3.109}$$

and (3.108) can be put in the form

$$\overrightarrow{\mathbf{P}}^c \equiv \begin{bmatrix} \overrightarrow{\mathbf{P}}^c_{++} & \overrightarrow{\mathbf{P}}^c_{+-} \\ \overrightarrow{\mathbf{P}}^c_{-+} & \overrightarrow{\mathbf{P}}^c_{--} \end{bmatrix} = \begin{bmatrix} \overleftarrow{\mathbf{P}}^T_{--} & -\overleftarrow{\mathbf{P}}^T_{+-} \\ -\overleftarrow{\mathbf{P}}^T_{-+} & \overleftarrow{\mathbf{P}}^T_{++} \end{bmatrix} \equiv -\bar{\mathbf{J}}\,\overleftarrow{\mathbf{P}}^T\,\bar{\mathbf{J}} \tag{3.110}$$

which is the *transfer theorem*, analogous to the scattering theorem (2.112) derived in Sec. 2.5.2. If we equate submatrices such as

$$\overrightarrow{\mathbf{P}}^c_{++} = \overleftarrow{\mathbf{P}}^T_{--}$$

we get typically (in obvious notation)

$$\overrightarrow{P}^c_{1,1} = \overleftarrow{P}_{-1,-1}, \qquad \overrightarrow{P}^c_{1,2} = \overleftarrow{P}_{-2,-1} \tag{3.111}$$

Eq. (3.102), which relates $\mathbf{a}(z^-)$ to $\mathbf{a}(z^+)$, or its inverse, using (3.103), gives

$$\mathbf{a}(z^-) = \overrightarrow{\mathbf{P}}^{-1}\,\mathbf{a}(z^+) = \overleftarrow{\mathbf{P}}\,\mathbf{a}(z^+)$$

This can be reorganized so as to relate incoming and outgoing amplitudes, $\mathbf{a}_{in}$ and $\mathbf{a}_{out}$ (2.102) [128, 105, 119], thus enabling us to express reflection and transmission matrices in terms of the sub-matrices of $\mathbf{P}$. One such symmetric form

$$\mathbf{S} \equiv \begin{bmatrix} \mathbf{R}_+ & \mathbf{T}_- \\ \mathbf{T}_+ & \mathbf{R}_- \end{bmatrix} = \begin{bmatrix} \overleftarrow{\mathbf{P}}_{-+}\,\overleftarrow{\mathbf{P}}^{-1}_{++} & \overrightarrow{\mathbf{P}}^{-1}_{--} \\ \overleftarrow{\mathbf{P}}^{-1}_{++} & \overrightarrow{\mathbf{P}}_{+-}\,\overrightarrow{\mathbf{P}}^{-1}_{--} \end{bmatrix} \tag{3.112}$$

can be understood by means of a simple example. Consider a positive-going wave, of amplitude $\mathbf{a}_+(z^-)$, incident on the lower boundary z^- of a plane-stratified slab (see Fig. 3.2). A reflected and a transmitted wave of amplitudes $\mathbf{a}_-(z^-)$ and $\mathbf{a}_+(z^+)$ are generated at the lower and upper boundaries respectively:

$$\mathbf{a}_-(z^-) = \mathbf{R}_+\,\mathbf{a}_+(z^-), \qquad \mathbf{a}_+(z^+) = \mathbf{T}_+\mathbf{a}_+(z^-) \tag{3.113}$$

Clearly,

$$\mathbf{a}_+(z^-) = \overleftarrow{\mathbf{P}}_{++}\,\mathbf{a}_+(z^+), \qquad \mathbf{a}_-(z^-) = \overleftarrow{\mathbf{P}}_{-+}\,\mathbf{a}_+(z^+)$$

since no negative-going wave is incident on the upper boundary z^+, and hence

$$\mathbf{a}_+(z^+) = \overleftarrow{\mathbf{P}}^{-1}_{++}\,\mathbf{a}_+(z^-) \tag{3.114}$$

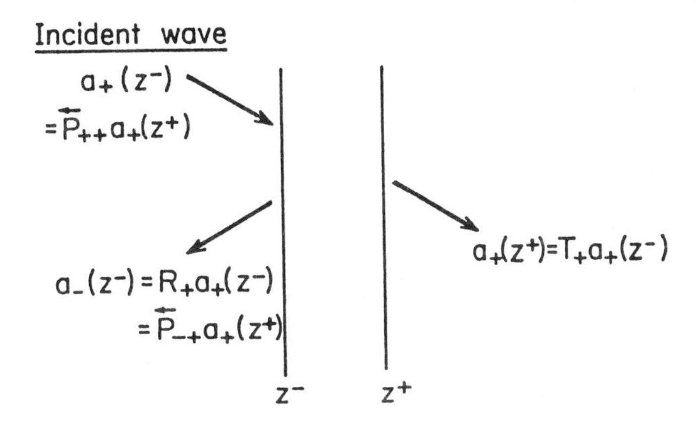

Figure 3.2: Relations between scattering and transfer sub-matrices

$$a_-(z^-) = \overleftarrow{\mathsf{P}}_{-+}\, a_+(z^+) = \overleftarrow{\mathsf{P}}_{-+}\, \overleftarrow{\mathsf{P}}^{-1}_{++} a_+(z^-) \tag{3.115}$$

in which we have substituted (3.114) into (3.115).

Comparison of the last two equations with (3.113) yields immediately

$$\mathsf{R}_+ = \overleftarrow{\mathsf{P}}_{-+}\, \overleftarrow{\mathsf{P}}^{-1}_{++}, \qquad \mathsf{T}_+ = \overleftarrow{\mathsf{P}}^{-1}_{++} \tag{3.116}$$

The other two equalities in (3.112) can be checked by a similar simple construction. Application of the transfer theorem (3.110) to (3.112), in the given and conjugate problems, leads after some manipulation [119], to the scattering theorem (2.112), as could be expected.

3.4 The Lorentz-adjoint system

3.4.1 The adjoint Maxwell system: alternative formulations

It was mentioned in Sec. 2.2.4 that the prescription we adopted for forming a set of equations adjoint to the Maxwell set ($\mathsf{K} \to \mathsf{K}^T$, $\mathsf{D} \to -\mathsf{D}^T$) was not unique, and that any other prescription that satisfies a Lagrange identity such as (2.37) is also valid. Consider the Maxwell system (2.40)

$$\mathsf{L}\,\mathsf{e} := [i\omega \mathsf{K} + \mathsf{D}]\,\mathsf{e}(\mathbf{r}) = \left[i\omega\mathsf{K} + \mathsf{U}_x \frac{\partial}{\partial x} + \mathsf{U}_y \frac{\partial}{\partial y} + \mathsf{U}_z \frac{\partial}{\partial z}\right] \mathsf{e}(\mathbf{r}) = 0 \tag{3.117}$$

The adjoint equation, (2.41):

$$\bar{\mathsf{L}}\,\bar{\mathsf{e}} := \left[i\omega\mathsf{K}^T - \mathsf{D}^T\right] \bar{\mathsf{e}}(\mathbf{r}) = 0 \tag{3.118}$$

defined an 'adjoint medium', characterized by the transposed constitutive tensor K^Twhich, in the case of a magnetoplasma, was no more than the magnetic-field reversed medium. Application of the Lagrange identity (2.37) yielded the concomitant vector $\mathbf{P}$ (2.42):

$$\mathbf{P} = \overline{\mathbf{E}} \times \mathbf{H} + \mathbf{E} \times \overline{\mathbf{H}}, \qquad \nabla \cdot \mathbf{P} = 0 \tag{3.119}$$

Suppose however that we rewrite the Maxwell system in the form

$$\mathsf{L}'\mathbf{e} := \bar{\mathsf{I}}\,\mathsf{L}\,\mathbf{e} = \bar{\mathsf{I}}\left[i\omega\mathsf{K} + \mathsf{D}\right]\mathbf{e}(\mathbf{r}) = 0 \tag{3.120}$$

with $\bar{\mathsf{I}} = \bar{\mathsf{I}}^T$ (2.81). The equation formally adjoint to it (constructed, it will be recalled, by transposing the matrix operators and changing the signs of the linear differential operators)

$$\bar{\mathsf{L}}'\mathbf{e}^{(L)} := \left[i\omega\mathsf{K}^T - \mathsf{D}^T\right]\bar{\mathsf{I}}\,\mathbf{e}^{(L)} = 0 \tag{3.121}$$

will satisfy a Lagrange identity

$$\tilde{\mathbf{e}}^{(L)}\mathsf{L}'\,\mathbf{e} - \tilde{\mathbf{e}}\bar{\mathsf{L}}'\,\mathbf{e}^{(L)} = \nabla \cdot \left(\mathbf{E}^{(L)} \times \mathbf{H} - \mathbf{E} \times \mathbf{H}^{(L)}\right) \tag{3.122}$$

with $\mathbf{e}^{(L)} := (\mathbf{E}^{(L)}, \mathbf{H}^{(L)})$, which with (3.120) and (3.121) gives

$$\nabla \cdot \mathbf{P}^{(L)} = 0, \qquad \mathbf{P}^{(L)} := \mathbf{E}^{(L)} \times \mathbf{H} - \mathbf{E} \times \mathbf{H}^{(L)} \tag{3.123}$$

Note the diference in form—the minus sign before the second term—between $\mathbf{P}^{(L)}$ (3.123) and $\mathbf{P}$ (3.119). Premultiplying (3.121) by $\bar{\mathsf{I}}$ gives

$$\left[i\omega\mathsf{K}^{(L)} + \mathsf{D}\right]\mathbf{e}^{(L)} = 0 \tag{3.124}$$

with

$$\bar{\mathsf{I}}\,\mathsf{D}^T\,\bar{\mathsf{I}} = -\mathsf{D}, \qquad \mathsf{K}^{(L)}(\mathbf{r}) := \bar{\mathsf{I}}\,\mathsf{K}^T(\mathbf{r})\,\bar{\mathsf{I}} \tag{3.125}$$

For reasons which will become clear in the next section, we shall employ the term 'Lorentz-adjoint' for the medium characterized by the constitutive tensor $\mathsf{K}^{(L)}$ (3.125). Kerns [81] has termed such a medium the 'adjoint medium', Kong and Cheng [84] have called it the 'complementary medium'. We shall show in Chap. 6 that in fact any orthogonal spatial mapping (rotation, reflection, inversion) of the Lorentz-adjoint medium will serve as a reciprocal medium, in which transformed currents and fields will exhibit a Lorentz-type reciprocity relation with respect to the original currents and fields in the given medium. In the case of a magnetoplasma the tensors K^T and $\mathsf{K}^{(L)}$ will be identical, but will differ for bianisotropic media, see (2.20), with

$$\mathsf{K}^T = \begin{bmatrix} \varepsilon^T & \eta^T \\ \xi^T & \mu^T \end{bmatrix}, \qquad \mathsf{K}^{(L)} = \begin{bmatrix} \varepsilon^T & -\eta^T \\ -\xi^T & \mu^T \end{bmatrix} \tag{3.126}$$

In order to avoid confusion we shall henceforth reserve the name 'adjoint', or 'formally adjoint', to describe the medium characterized by the transposed constitutive tensor K^T. The Lorentz-adjoint medium $\mathsf{K}^{(L)}$ will be identified in Chap. 7 with the 'time-reversed medium'.

We shall call $\mathbf{e}^{(L)}$, (3.122) and (3.124), the 'Lorentz-adjoint' or 'Lorentz-reversed' field, and such fields are used in the next section to derive a version of the eigenmode scattering theorem different to that found in our earlier treatment. Comparison of (3.118) and (3.121) gives

$$\mathbf{e}^{(L)}(\mathbf{r}) = \bar{\mathsf{I}}\,\bar{\mathbf{e}}(\mathbf{r}) \tag{3.127}$$

We note that $\mathbf{e}^{(L)}$ obeys Maxwell's equations (3.124) in a medium $\mathsf{K}^{(L)}$ which could in principle be a physically realizable medium. The 'non-physicality' of $\bar{\mathbf{e}}$, as opposed to $\mathbf{e}^{(L)}$, is expressed by the fact that its direction of propagation, as defined by its plane-wave ansatz $\exp(ik_0 q_\alpha z)$, is 'wrong', i.e. in a direction opposite to that of its Poynting vector. The Poynting vector of the field $\mathbf{e}^{(L)} = \bar{\mathsf{I}}\bar{\mathbf{e}}$ (3.127), on the other hand, has been reversed by the matrix $\bar{\mathsf{I}}$, which changes the sign of the magnetic field components, and thereby restores its 'physicality' (see the discussion of the restricted time-reversal procedure in Sec. 2.4.2). We recall that the *conjugate* modal wave fields $\mathbf{e}^c_\alpha$, discussed earlier in this chapter, are also 'physical', for the same reason, being related to the adjoint modal fields by the transformation (2.92)

$$\mathbf{e}^c_{-\alpha}(-s_x, s_y) = \mathbf{Q}_y\,\bar{\mathsf{I}}\,\bar{\mathbf{e}}_\alpha(s_x, s_y) \tag{3.128}$$

in which $\bar{\mathsf{I}}$ 'restores physicality', and $\mathbf{Q}_y$ provides a reflection mapping of the resultant Lorentz-reversed modes.

In anticipation of the spatial mappings of vector and tensor fields which will be discussed systematically in Chap. 6, we remark that the spatial inversion mappings of the transposed tensor $\mathsf{K}^T(\mathbf{r})$ and the adjoint field $\bar{\mathbf{e}}(\mathbf{r})$ resemble those generated by the Poynting-vector reversing operator $\bar{\mathsf{I}}$ in (3.125) and (3.127). The inversion transformations will be shown to be

$$\mathsf{K}'(\mathbf{r}') = (-\bar{\mathsf{I}})\,\mathsf{K}^T(\mathbf{r})(-\bar{\mathsf{I}}) = \bar{\mathsf{I}}\,\mathsf{K}^T(\mathbf{r})\,\bar{\mathsf{I}}, \qquad \mathbf{e}'(\mathbf{r}') = -\bar{\mathsf{I}}\,\bar{\mathbf{e}}(\mathbf{r}), \qquad \mathbf{r}' = -\mathbf{r} \tag{3.129}$$

giving, inter alia, $\mathbf{E}'(\mathbf{r}') = -\overline{\mathbf{E}}(\mathbf{r})$, $\mathbf{H}'(\mathbf{r}') = \overline{\mathbf{H}}(\mathbf{r})$. These indeed resemble the transformations just derived (besides the unimportant sign difference in the transformed fields, which can be regarded as a 180° phase shift), which is not surprising, since the Poynting vector is reversed in direction in the inversion transformation too. The important difference, however, is that in the inversion mapping (3.129), the fields are mapped into an inverse space, $\mathbf{r}' = -\mathbf{r}$, whereas the Lorentz-reversed fields, (3.127), are mapped into the same space.

It should be remarked in conclusion that Tai [122] has obtained a reciprocity theorem through a *rotational* transformation of the magnetic-field reversed medium, which restores the magnetic field **b** to its original direction, while the fields and currents are rotated by 180° about an axis perpendicular to the field **b**. If the medium is plane stratified, the rotated field-reversed medium will in general no longer coincide with the original medium, as does the conjugate (field-reversed, reflected) plane-statified medium discussed in this chapter, and hence will be of limited interest. In other geometries however, see for instance Figs. 4.2 and 4.4, this may be a useful 'conjugating transformation', yielding Lorentz-type reciprocity relations.

3.4.2 Lorentz-adjoint scattering theorem: the eigenmode generalization

We now consider Kerns' scattering theorem, discussed in Sec. 2.1.1, in the restricted form it takes (2.14) when the scattering object becomes a plane-stratified slab, and the scattered angular spectrum reduces to a single reflected, and a single transmitted pair of base modes (each with parallel and perpendicular wave polarizations). We shall consider it, however, in the more general context of eigenmodes within the medium [10], and not restrict the discussion to linearly polarized base modes in the free space bounding the stratified medium.

We start with the eigenmode equation (2.29) in a source-free medium, $\mathbf{j}(\mathbf{k}_t,z)=0$,

$$\mathsf{L}\,\mathbf{e}_\alpha := ik_0[c\mathsf{K} - s_x\mathsf{U}_x - s_y\mathsf{U}_y - q_\alpha\mathsf{U}_z]\,\mathbf{e}_\alpha = 0$$

with $\mathsf{K}\equiv\mathsf{K}(\mathbf{b})$, $\mathbf{e}_\alpha\equiv\mathbf{e}_\alpha(\mathbf{b};s_x,s_y)$ and U_x, U_y and U_z defined in (2.25). In the field-reversed medium, with $\mathsf{K}(-\mathbf{b})=\mathsf{K}^T$ (2.36), and with the aid of (2.44) and (2.46), we have

$$\bar{\mathsf{L}}\,\bar{\mathbf{e}}_\alpha := ik_0\left[c\mathsf{K}^T - s_x\mathsf{U}_x - s_y\mathsf{U}_y - q_\alpha\mathsf{U}_z\right]\bar{\mathbf{e}}_\alpha = 0 \tag{3.130}$$

Now multiply from the left with $\bar{\mathsf{I}}$, using $\bar{\mathsf{I}}=\bar{\mathsf{I}}^{-1}$ (2.81),

$$\left[\bar{\mathsf{I}}\,\bar{\mathsf{L}}\,\bar{\mathsf{I}}\right]\left[\bar{\mathsf{I}}\,\bar{\mathbf{e}}\right] = 0 \tag{3.131}$$

and since $\bar{\mathsf{I}}\,\mathsf{U}_i\bar{\mathsf{I}}=-\mathsf{U}_i$ $(i=x,y,z)$, and K for a magnetoplasma is unaffected by the transformation, i.e. $\bar{\mathsf{I}}\,\mathsf{K}^T\bar{\mathsf{I}}=\mathsf{K}^T$, the last two equations give

$$[c\mathsf{K}(-\mathbf{b}) + s_x\mathsf{U}_x + s_y\mathsf{U}_y + q_\alpha\mathsf{U}_z]\,\bar{\mathsf{I}}\,\bar{\mathbf{e}}_\alpha = 0$$

by analogy with (3.124). We may thus identify the 'Lorentz-adjoint' or 'Lorentz reversed' modes (see Sec. 3.4.1)

$$\mathbf{e}_{-\alpha}^{(L)}(-\mathbf{b};-s_x,-s_y)=\bar{\mathbf{I}}\,\bar{\mathbf{e}}_\alpha,\qquad \bar{\mathbf{e}}_{-\alpha}^{(L)}(-\mathbf{b};-s_x,-s_y)=\bar{\mathbf{I}}\,\mathbf{e}_\alpha \tag{3.132}$$

to obtain [cf. (2.92), (2.94) and (2.97)]

$$a_\alpha^{(L)}=\bar{a}_{-\alpha},\qquad \bar{a}_\alpha^{(L)}=a_{-\alpha} \tag{3.133}$$

This relation between the 'Lorentz-adjoint' and the given modal amplitudes is thus identical with that between the conjugate and the given modal amplitudes. This is not surprising since the conjugate and Lorentz-adjoint eigenvectors are related by a simple reflection transformation, as may be seen by comparing (2.92) and (3.132),

$$\mathbf{e}_\alpha^c=\mathbf{Q}_y\,\bar{\mathbf{I}}\,\bar{\mathbf{e}}_{-\alpha}=\mathbf{Q}_y\,\mathbf{e}_\alpha^{(L)}$$

with

$$\mathbf{e}_\alpha^c=\mathbf{e}_\alpha(\mathbf{b};\,-s_x,s_y),\qquad \mathbf{e}_\alpha^{(L)}=\mathbf{e}_\alpha(-\mathbf{b};-s_x,-s_y)$$

where the diagonal matrix $\mathbf{Q}_y$ defined in (2.85), performs a reflection transformation with respect to the y=0 magnetic-meridian plane. The corresponding scattering matrices are therefore equal, and hence with (2.112)

$$\mathbf{S}^{(L)}=\mathbf{S}^c=\mathbf{S}^T \tag{3.134}$$

giving the Lorentz-adjoint reflection and transmission matrices

$$\mathbf{R}_\pm^{(L)}(-\mathbf{b};\,-s_x,-s_y)=\tilde{\mathbf{R}}_\pm(\mathbf{b};\,s_x,s_y)$$

$$\mathbf{T}_\pm^{(L)}(-\mathbf{b};\,-s_x,-s_y)=\tilde{\mathbf{T}}_\mp(\mathbf{b};\,s_x,s_y)$$

This result is not strictly in the same category as the previously derived scattering theorem (2.112), although closely related to it, in that here the modal amplitudes are in two different media—the given and the adjoint (magnetic-field reversed) media. These reciprocity (scattering) relations have proved useful, and have been employed for instance by Bahar and Agrawal [19,20] to check the consistency of numerically computed scattering coefficients for plane-stratified magnetoplasmas.

Chapter 4

Reciprocity in media with sources

4.1 Plane-stratified uniaxial media

Before generalizing the discussion to media with sources (currents), we extend the scattering theorem to more general plane-stratified uniaxial anisotropic or bianisotropic media. This has a dual purpose. First, these media exhibit reflection symmetry, which will enable us to demonstrate the different treatment of reflection transformations in the case of eigenmodes with a prescribed local plane-wave variation in source-free media, and in the general case of wave fields associated with arbitrary current distributions. Second, the stratified structure permits the decomposition of wave fields into well-defined eigenmodes, which is a necessary condition for the derivation and formulation of a scattering theorem. In the subsequent discussion, involving currents and fields in arbitrary media having reflection symmetry, there will no longer be any need to restrict the discussion to plane-stratified configurations.

4.1.1 The constitutive tensors in uniaxial media

In uniaxial anisotropic (crystalline), gyrotropic or bianisotropic media, the tensorial character of all constitutive tensors depends on a single unit vector, representing the symmetry axis of the medium. In gyrotropic media, considered in previous chapters, this was the vector $\hat{\mathbf{b}}$, the direction of the external magnetic field. This symmetry axis, together with the normal to the stratification, defines a plane of symmetry—the 'meridian' plane—to which we may tie the cartesian coordinate system (the y=0 plane), and with respect to which we may perform symmetry transformations such as reflection. A comprehensive discussion of such, and other, media is to be found in the book of Kong [83].

Uniaxial anisotropic media

Under this heading we consider non-magnetic crystalline media which are described in general by symmetric permittivity tensors, $\boldsymbol{\varepsilon}=\boldsymbol{\varepsilon}^T$, $\boldsymbol{\mu}=\mu_0\mathbf{I}^{(3)}$. When $\boldsymbol{\varepsilon}$ is real (lossless), a real principal-axis coordinate system can always be found in which $\boldsymbol{\varepsilon}$ is diagonal. In cubic crystals the three eigenvalues ε_i (i=1,2,3) are equal, and the medium is isotropic (scalar permittivity). In tetragonal, hexagonal and rhombohedral crystals, two of the three eigenvalues ε_i are equal. Such crystals are *uniaxial*, and the symmetry axis is called the *optic axis*. The constitutive tensor has the form

$$\boldsymbol{\varepsilon}=\begin{bmatrix}\varepsilon & 0 & 0\\ 0 & \varepsilon & 0\\ 0 & 0 & \varepsilon'\end{bmatrix}, \qquad \mathsf{K}=\begin{bmatrix}\boldsymbol{\varepsilon} & 0\\ 0 & \mu_0\mathbf{I}^{(3)}\end{bmatrix}=\mathsf{K}^T \tag{4.1}$$

in which $\varepsilon_1=\varepsilon_2=\varepsilon$, $\varepsilon_3=\varepsilon'$. In orthorhombic, monoclinic and triclinic crystals, all three eigenvalues are different and the medium is *biaxial*.

Gyrotropic media

These unsymmetric, uniaxial media which have been discussed in the previous chapters, may be *gyroelectric* (magnetoplasmas) or *gyromagnetic* (ferrites), and the respective tensors, $\boldsymbol{\varepsilon}$ or $\boldsymbol{\mu}$, depend on the external field, $\mathbf{b}$, cf. (2.35). In the absence of absorption these tensors are hermitian.

Magnetoelectric bianisotropic media

When placed in an electric or a magnetic field the medium becomes both polarized and magnetized. Such a medium, in which the permanent electric dipoles also have magnetic moments, was conceived by Tellegen [124] as the basis of a new network element, the 'gyrator'. The existence of bianisotropic magnetoelectric materials was predicted, on theoretical grounds, by Dzyaloshinski [50], and many antiferromagnetic crystals, such as chromium oxide, as well as ferromagnetic crystals like gallium iron oxide, have been found experimentally. Dzyaloshinski indicated that in substances like antiferromagnetic chromium oxide, the 3×3 tensors $\boldsymbol{\varepsilon}$, $\boldsymbol{\mu}$, $\boldsymbol{\xi}$ and $\boldsymbol{\eta}$, (2.20), would all have the same symmetry axis, which we shall take as the z-axis, with $\boldsymbol{\xi}=\boldsymbol{\eta}$, yielding symmetric uniaxial bianisotropic constitutive tensors of the form

$$\mathsf{K}=\begin{bmatrix}\boldsymbol{\varepsilon} & \boldsymbol{\xi}\\ \boldsymbol{\xi} & \boldsymbol{\mu}\end{bmatrix}=\mathsf{K}^T \tag{4.2}$$

with

$$\boldsymbol{\varepsilon} = \begin{bmatrix} \varepsilon & 0 & 0 \\ 0 & \varepsilon & 0 \\ 0 & 0 & \varepsilon' \end{bmatrix}, \qquad \boldsymbol{\mu} = \begin{bmatrix} \mu & 0 & 0 \\ 0 & \mu & 0 \\ 0 & 0 & \mu' \end{bmatrix}, \qquad \boldsymbol{\xi} = \begin{bmatrix} \xi & 0 & 0 \\ 0 & \xi & 0 \\ 0 & 0 & \xi' \end{bmatrix}$$

In the discussion which follows we consider plane-stratified systems consisting of uniaxial media, as in (4.1) and (4.2). If the direction cosines of the axis of symmetry are $(\hat{b}_x, 0, \hat{b}_z)$, then the matrices $\boldsymbol{\varepsilon}$, $\boldsymbol{\mu}$ and $\boldsymbol{\xi}$, (4.1) and (4.2), have the typical form

$$\boldsymbol{\varepsilon} = \begin{bmatrix} \hat{b}_x^2\varepsilon' + \hat{b}_z^2\varepsilon & 0 & \hat{b}_x\hat{b}_z(\varepsilon' - \varepsilon) \\ 0 & \varepsilon & 0 \\ \hat{b}_x\hat{b}_z(\varepsilon' - \varepsilon) & 0 & \hat{b}_x^2\varepsilon + \hat{b}_z^2\varepsilon' \end{bmatrix} = \boldsymbol{\varepsilon}^T \tag{4.3}$$

Moving media—the Lorentz bianisotropy

For a medium which is isotropic in the rest frame, i.e. $\boldsymbol{\varepsilon}(0)=\varepsilon\,\mathsf{I}^{(3)}$, $\boldsymbol{\mu}(0)=\mu\,\mathsf{I}^{(3)}$, $\boldsymbol{\xi}(0)=0=\boldsymbol{\eta}(0)$, then in the laboratory frame in which the medium is moving with a velocity $\boldsymbol{v}$ in the x-direction, the electric and magnetic fields are coupled by the relativistic Lorentz transformations. The constitutive tensor then becomes bianisotropic [83, Sec. 2.3c]:

$$\mathsf{K} = \left[\begin{array}{ccc:ccc} \varepsilon & \cdot & \cdot & \cdot & \cdot & \cdot \\ \cdot & \varepsilon' & \cdot & \cdot & \cdot & -\xi \\ \cdot & \cdot & \varepsilon' & \cdot & \xi & \cdot \\ \hdashline \cdot & \cdot & \cdot & \mu & \cdot & \cdot \\ \cdot & \cdot & \xi & \cdot & \mu' & \cdot \\ \cdot & -\xi & \cdot & \cdot & \cdot & \mu' \end{array}\right] = \mathsf{K}^T \tag{4.4}$$

with

$$\frac{\varepsilon'}{\varepsilon} = \frac{\mu'}{\mu} := \frac{1 - v^2\varepsilon_0\mu_0}{1 - v^2\varepsilon\mu}, \qquad \xi := \frac{v}{c^2}\,\frac{c^2\varepsilon\mu - 1}{1 - v^2\varepsilon\mu}$$

It will be found convenient later to consider the constitutive tensor K also in a coordinate-free representation. It may be shown [83, Sec. 2.3], [36, eqs. (8.13)–(8.16)] that K has the form

$$\mathsf{K} = \begin{bmatrix} \varepsilon'\,(\mathsf{I}^{(3)} - \hat{\boldsymbol{v}}\hat{\boldsymbol{v}}^T) + \varepsilon\,\hat{\boldsymbol{v}}\hat{\boldsymbol{v}}^T & \xi(\hat{\boldsymbol{v}} \times \mathsf{I}^{(3)}) \\ -\xi(\hat{\boldsymbol{v}} \times \mathsf{I}^{(3)}) & \mu'\,(\mathsf{I}^{(3)} - \hat{\boldsymbol{v}}\hat{\boldsymbol{v}}^T) + \mu\,\hat{\boldsymbol{v}}\hat{\boldsymbol{v}}^T \end{bmatrix} \tag{4.4a}$$

with ε', μ' and ξ defined above. The operator $\hat{\boldsymbol{v}}\hat{\boldsymbol{v}}^T$ projects $\mathbf{E}$ (or $\mathbf{H}$) onto the unit vector $\hat{\boldsymbol{v}}$; $(\mathsf{I}^{(3)} - \hat{\boldsymbol{v}}\hat{\boldsymbol{v}}^T)$ gives the transverse (to $\hat{\boldsymbol{v}}$) projection of $\mathbf{E}$ (or $\mathbf{H}$):

$$\mathbf{E}_{\parallel} = \hat{\boldsymbol{v}}\hat{\boldsymbol{v}}^T\,\mathbf{E}\,, \qquad \mathbf{E}_{\perp} = (\mathsf{I}^{(3)} - \hat{\boldsymbol{v}}\hat{\boldsymbol{v}}^T)\,\mathbf{E}$$

Insofar as we are dealing with stratified media, the simplest physical model is to assume a gradient in the z-direction of the fluid-velocity vector, $\mathbf{v}=v\,\hat{\mathbf{x}}$, which imposes a stratification on the system even when the constitutive parameters are isotropic and constant in the local rest frame. In Chap. 7 we shall encounter other types of media, viz. the compressible magnetoplasma and the isotropic chiral medium.

4.1.2 Transformation of Maxwell's equations

Suppose, for simplicity, that the medium is crystalline (anisotropic) or magnetoelectric (bianisotropic), as in (4.1) or (4.2). The medium is assumed to be source-free and to vary in the z-direction only. The axis of symmetry of the medium is assumed to be parallel to the y=0 plane, and the form of the 3×3 constitutive tensor is as in (4.3). Maxwell's equations, when Fourier analysed in the transverse (stratification) plane, (2.27), with fields having harmonic, $\exp(i\omega t)$, time dependence, yield (2.29)

$$\mathsf{L}\,\mathbf{e}(\mathbf{k}_t,z) := ik_0\left[c\mathsf{K} - s_x\mathsf{U}_x - s_y\mathsf{U}_y - \frac{i}{k_0}\mathsf{U}_z\frac{d}{dz}\right]\mathbf{e}(\mathbf{k}_t,z) = 0 \qquad (4.5)$$

with $\mathbf{k}_t$=$k_0(s_x,s_y)$ (2.26), where s_x and s_y, by Snell's law, are constants of the propagation.

The equation formally adjoint to (4.5) will be, as in (2.34) and (2.43),

$$\bar{\mathsf{L}}\,\bar{\mathbf{e}}(\mathbf{k}_t,z) := ik_0\left[c\mathsf{K}^T - s_x\mathsf{U}_x^T - s_y\mathsf{U}_y^T + \frac{i}{k_0}\mathsf{U}_z^T\frac{d}{dz}\right]\bar{\mathbf{e}}(\mathbf{k}_t,z) = 0 \qquad (4.6)$$

with K=K^T, (4.1–4.3), and U_i=U_i^T (i=x,y,z), (2.25). These two equations, (4.5) and (4.6), are just those found for gyrotropic media, (2.30) and (2.34). Assumption of a local plane-wave ansatz for $\mathbf{e}$ and $\bar{\mathbf{e}}$,

$$\mathbf{e}_\alpha(\mathbf{k}_t,z) = \mathbf{e}_\alpha(\mathbf{k}_t)\exp(-ik_0\,q_\alpha z) \qquad \bar{\mathbf{e}}_\beta(\mathbf{k}_t,z) = \bar{\mathbf{e}}_\beta(\mathbf{k}_t)\exp(ik_0\,\bar{q}_\beta z) \quad (4.7)$$

yields eigenmode equations, as in (2.32) and (2.44),

$$\mathsf{L}\,\mathbf{e}_\alpha := ik_0\left[c\mathsf{K} - s_x\mathsf{U}_x - s_y\mathsf{U}_y - q_\alpha\mathsf{U}_z\right]\mathbf{e}_\alpha = 0 \qquad (4.8)$$

$$\bar{\mathsf{L}}\,\bar{\mathbf{e}}_\beta := ik_0\left[c\mathsf{K}^T - s_x\mathsf{U}_x^T - s_y\mathsf{U}_y^T - \bar{q}_\beta\mathsf{U}_z^T\right]\bar{\mathbf{e}}_\beta = 0 \qquad (4.9)$$

Since K=K^T and U_i=U_i^T, ($i = x,y,z$), we conclude that

$$\bar{q}_\beta = q_\beta \qquad \text{and} \qquad \bar{\mathbf{e}}_\beta(\mathbf{k}_t) = \mathbf{e}_\beta(\mathbf{k}_t) \qquad (4.10)$$

i.e. that the given and adjoint eigenmodes have the same polarization (unlike the corresponding relations—(2.67) and (2.68)—in gyrotropic media), but different z-dependence (4.7). Application of the Lagrange identity (2.37) gives the modal biorthogonality relation, cf. (2.50),

$$\hat{\mathbf{e}}_\beta^T \, \mathbf{U}_z \, \hat{\mathbf{e}}_\alpha = \delta_{\alpha\beta} \, \mathrm{sgn}(\alpha), \qquad \alpha, \beta = \pm 1, \pm 2 \tag{4.11}$$

and definition of modal amplitudes in the given and adjoint systems, as in (2.77), leads finally to a relationship between the respective scattering matrices, $\mathbf{S}$ and $\overline{\mathbf{S}}$,

$$\overline{\mathbf{S}}^T \mathbf{S} = \mathbf{I}^{(4)} = \mathbf{S} \, \overline{\mathbf{S}}^T \tag{4.12}$$

as in (2.108).

We now introduce the reflection matrices $\mathbf{Q}_i$ ($i = x$, y or z), which generate reflection with respect to the x, y or $z = 0$ plane), (2.85) and (2.84),

$$\mathbf{Q}_i := \begin{bmatrix} \mathbf{q}_i & 0 \\ 0 & (\det \mathbf{q}_i) \, \mathbf{q}_i \end{bmatrix} = \begin{bmatrix} \mathbf{q}_i & 0 \\ 0 & -\mathbf{q}_i \end{bmatrix} = \mathbf{Q}_i{}^T = \mathbf{Q}_i{}^{-1} \tag{4.13}$$

where

$$\mathbf{q}_x := \begin{bmatrix} -1 & 0 & 0 \\ 0 & 1 & 0 \\ 0 & 0 & 1 \end{bmatrix} \qquad \mathbf{q}_y := \begin{bmatrix} 1 & 0 & 0 \\ 0 & -1 & 0 \\ 0 & 0 & 1 \end{bmatrix} \qquad \mathbf{q}_z := \begin{bmatrix} 1 & 0 & 0 \\ 0 & 1 & 0 \\ 0 & 0 & -1 \end{bmatrix}$$

with $\mathbf{q}_i = \mathbf{q}_i{}^T = \mathbf{q}_i{}^{-1}$. The adjoint-reflection (conjugating) matrix, $\overline{\mathbf{Q}}_i$, is similarly defined

$$\overline{\mathbf{Q}}_i := \mathbf{Q}_i \, \bar{\mathbf{I}} = \begin{bmatrix} \mathbf{q}_i & 0 \\ 0 & -(\det \mathbf{q}_i) \, \mathbf{q}_i \end{bmatrix} = \begin{bmatrix} \mathbf{q}_i & 0 \\ 0 & \mathbf{q}_i \end{bmatrix} = \overline{\mathbf{Q}}_i{}^T = \overline{\mathbf{Q}}_i{}^{-1} \tag{4.14}$$

and $\overline{\mathbf{Q}}_i = \overline{\mathbf{Q}}_y = \mathbf{Q}_y \bar{\mathbf{I}} \equiv \mathbf{Q}_y^c$, when $\mathbf{q}_i = \mathbf{q}_y$, cf. (2.93). The term $\det \mathbf{q}_y = -1$ has been inserted in the matrix $\mathbf{Q}_y$ (4.13) (see the discussion in Sec. 2.4) since $\mathbf{Q}_y$ operates on a mixed polar-axial electomagnetic field $\mathbf{e}$,

$$\mathbf{e}' := \begin{bmatrix} \mathbf{E}' \\ \mathbf{H}' \end{bmatrix} = \mathbf{Q}_y \begin{bmatrix} \mathbf{E} \\ \mathbf{H} \end{bmatrix} \equiv \mathbf{Q}_y \mathbf{e} \tag{4.15}$$

and the sign (direction) of the mapped field

$$\mathbf{H}' = (\det \mathbf{q}_y) \, \mathbf{q}_y \, \mathbf{H} \tag{4.16}$$

must be reversed under reflection (or inversion) when $\det \mathbf{q} = -1$. The adjoint matrix, $\overline{\mathbf{Q}}_y$, has been derived from $\mathbf{Q}_y$ via the matrix $\bar{\mathbf{I}}$ (2.81) which reverses the sign of $\mathbf{H}'$, and hence of the Poynting vector, $\mathbf{E}'\times\mathbf{H}'$. In Sec. 2.4.2 we identified $\overline{\mathbf{Q}}_y=\mathbf{Q}^c$ as a (restricted-)time-reversal-cum-reflection operator, although it still remains to be seen to what extent the adjoint fields are, in general, no more than spatially-mapped time-reversed fields.

Now apply $\overline{\mathbf{Q}}_y = \overline{\mathbf{Q}}_y^{-1}$ to (4.9)

$$\overline{\mathbf{Q}}_y\overline{\mathbf{L}}\,\bar{\mathbf{e}}_\alpha = \left[\overline{\mathbf{Q}}_y\overline{\mathbf{L}}\,\overline{\mathbf{Q}}_y\right]\overline{\mathbf{Q}}_y\bar{\mathbf{e}}_\alpha = ik_0\left[c\mathbf{K} + s_x\mathbf{U}_x - s_y\mathbf{U}_y + q_\alpha\mathbf{U}_z\right]\overline{\mathbf{Q}}_y\bar{\mathbf{e}}_\alpha$$
$$= ik_0\left[c\mathbf{K} + s_x\mathbf{U}_x - s_y\mathbf{U}_y - q^c_{-\alpha}\mathbf{U}_z\right]\mathbf{e}^c_{-\alpha} = 0 \quad (4.17)$$

since $\mathbf{U}_i$ $(i = x, y, z)$, (2.25), is transformed as

$$\overline{\mathbf{Q}}_y\mathbf{U}_x\overline{\mathbf{Q}}_y = -\mathbf{U}_x, \quad \overline{\mathbf{Q}}_y\mathbf{U}_y\overline{\mathbf{Q}}_y = \mathbf{U}_y, \quad \overline{\mathbf{Q}}_y\mathbf{U}_z\overline{\mathbf{Q}}_y = -\mathbf{U}_z, \quad \mathbf{U}_i = \mathbf{U}_i^T \quad (i = x, y, z) \quad (4.18)$$

and $\mathbf{K}$, whose assumed structure is given by (4.2) and (4.3), is transformed as

$$\overline{\mathbf{Q}}_y\mathbf{K}\,\overline{\mathbf{Q}}_y = \mathbf{K} = \mathbf{K}^T \quad (4.19)$$

We have identified the first result in (4.17), by analogy with (2.90), as the eigenmode equation for the conjugate system, and comparison with (4.8) gives

$$-q^c_{-\alpha} = q_\alpha = \bar{q}_\alpha \;; \qquad \mathbf{e}^c_{-\alpha}(-s_x, s_y) = \overline{\mathbf{Q}}_y\bar{\mathbf{e}}_\alpha(s_x, s_y) \quad (4.20)$$

as in (2.91). The relation between given and conjugate eigenvectors are depicted in Fig. 2.2. Following the same analysis as in Sec. 2.5, we find the scattering relation ('reciprocity in k-space'), $\mathbf{S}^c=\tilde{\mathbf{S}}$ (2.112), relating outgoing to incoming eigenmode amplitudes in the given and conjugate systems.

4.1.3 Reciprocity and equivalence in k-space

In the case of crystalline, anisotropic media, (4.1), we could have arrived at another relation between eigenmodes in the two systems by operating with $\mathbf{Q}_y$ on (4.8) rather than with $\overline{\mathbf{Q}}_y$ on (4.9), i.e. by a straightforward reflection mapping without time reversal:

$$\mathbf{Q}_y\mathbf{L}\,\mathbf{e}_\alpha = \left[\mathbf{Q}_y\mathbf{L}\,\mathbf{Q}_y\right]\mathbf{Q}_y\mathbf{e}_\alpha = ik_0\left[c\mathbf{K} - s_x\mathbf{U}_x + s_y\mathbf{U}_y - q_\alpha\mathbf{U}_z\right]\mathbf{Q}_y\mathbf{e}_\alpha = 0 \quad (4.21)$$

Maxwell's equations in this reflected system are simply

$$\left[c\mathbf{K} - s'_x\mathbf{U}_x - s'_y\mathbf{U}_y - q'_\alpha\mathbf{U}_z\right]\mathbf{e}'_\alpha = 0 \quad (4.22)$$

with

$$\mathbf{e}'_\alpha(s'_x, s'_y) = \mathbf{Q}_y\mathbf{e}_\alpha(s_x, s_y), \qquad q'_\alpha = q_\alpha, \qquad s'_x = s_x, \;\; s'_y = -s_y$$

The modes $\mathbf{e}_\alpha$ and $\mathbf{e}'_\beta$ in the given and reflected systems are no longer biorthogonal, their amplitudes are not linked by a reciprocity relation, $\mathbf{S}=\tilde{\mathbf{S}}^c$ (2.112), but rather by a simple *equivalence* relation, $\mathbf{S}=\mathbf{S}'$. [The term 'equivalence' is sometimes used in another sense in electromagnetics. Two sources producing the same fields within a region of space are said to be *equivalent* within that region [62, Sec. 3-5]. In the case of simple geometries, for instance, this is achieved by replacing current or charge distributions induced on bounding surfaces of a region, by electric or magnetic images outside that region.]

This equivalence via a reflection mapping could not have been achieved with a magnetoelectric bianisotropic medium, with the symmetry axis in the y=0 plane, (4.2) and (4.3), since the reflected medium is different from the original medium: the off-diagonal matrix ξ changes sign under reflection, and so

$$\mathbf{K} = \mathbf{K}^T \qquad \text{but} \qquad \mathbf{Q}_y \mathbf{K} \mathbf{Q}_y \neq \mathbf{K} \tag{4.23}$$

However, under reflection *and* time reversal, the original medium, as we have seen (4.19), is unchanged, i.e. $\overline{\mathbf{Q}}_y \mathbf{K} \overline{\mathbf{Q}}_y = \mathbf{K}$, which can be understood physically since ξ relates $\mathbf{E}$ fields, which are *even* under reflection and under time reversal, to $\mathbf{H}$ fields which are *odd* both under reflection and under time reversal. (We are using the term 'odd under reflection' to describe the property of a physically reflected axial vector whose direction is opposite to that of its geometrical reflection. A polar vector, by analogy, will be 'even under reflection'.)

Another way of looking at these properties of magnetoelectric media is to recall that these consist of elements or domains possessing both magnetic-dipole and electric-dipole moments. Under reflection the magnetic dipoles are reversed, but not the electric dipoles. When the reflected medium, however, is also time reversed, the magnetic dipoles are restored to their original orientations.

We consider, finally, transformations of moving media, with the motion in the x-direction, parallel to the stratification, and $\mathbf{K}$ given by (4.4). Because of the higher degree of of spatial symmetry of the medium (cf. the discussion of Heading's mirrored modes, Sec. 3.1.4, in which the external magnetic field $\mathbf{b}$ is parallel to the stratification), the medium can be mapped into itself by means of the reflection matrix $\mathbf{Q}_y$, with $\mathbf{q}_i$=$\mathbf{q}_y$, (4.13), or the conjugating matrix $\overline{\mathbf{Q}}_x$, (4.14), with $\mathbf{q}_i$=$\mathbf{q}_x$. This can be seen by inspection if $\mathbf{K}$ is given by (4.4):

$$\mathbf{Q}_y \mathbf{K}\, \mathbf{Q}_y = \mathbf{K}, \qquad \overline{\mathbf{Q}}_x \mathbf{K}\, \overline{\mathbf{Q}}_x = \mathbf{K} = \mathbf{K}^T \tag{4.24}$$

The first (reflection) transformation yields an equivalence relation, the second (conjugating) transformation yields a scattering theorem. If the eigenmodes

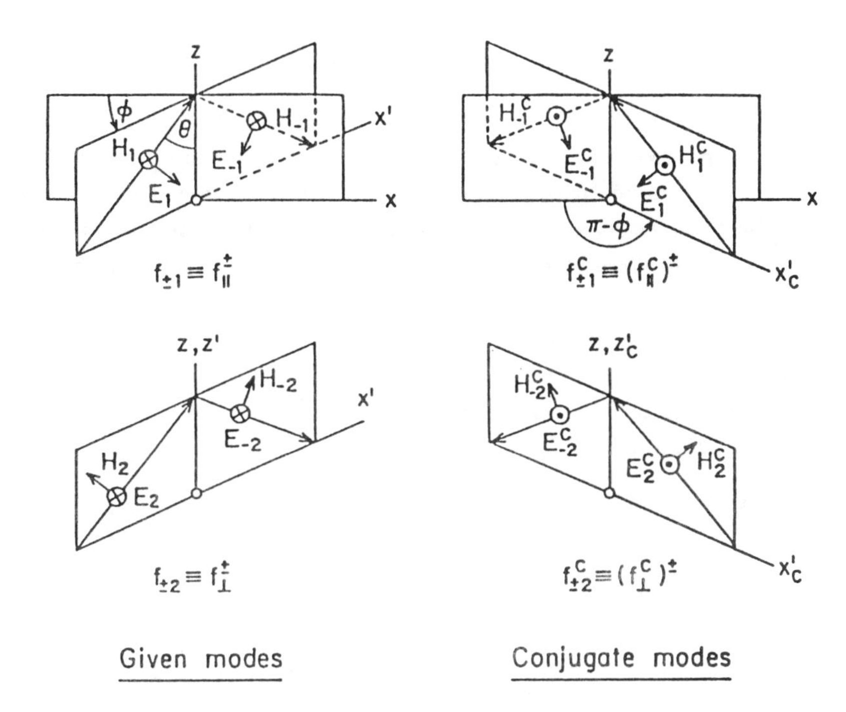

Figure 4.1: Given and conjugate linear modes for moving media. The velocity vector is in the x-direction. 'Plane stratification' is provided by a velocity gradient in the z-direction.

are linearly polarized, parallel ($\parallel$) or perpendicular ($\perp$) to the plane of incidence in the free-space bounding medium, as in Fig. 3.1, the scattering theorem (2.112) becomes

$$\mathbf{S}^c(s_x, -s_y) = \tilde{\mathbf{S}}(s_x, s_y), \qquad \begin{bmatrix} \mathbf{R}_+^c & \mathbf{T}_-^c \\ \mathbf{T}_+^c & \mathbf{R}_-^c \end{bmatrix} = \begin{bmatrix} \tilde{\mathbf{R}}_+ & \tilde{\mathbf{T}}_+ \\ \tilde{\mathbf{T}}_- & \tilde{\mathbf{R}}_- \end{bmatrix} \tag{4.25}$$

where typically, cf. (3.19) and (3.20),

$$\left({}_{\parallel}R_{\parallel}^{c}\right)^{\pm} = {}_{\parallel}R_{\parallel}{}^{\pm}, \qquad ({}_{\parallel}R_{\perp}^{c})^{\pm} = {}_{\perp}R_{\parallel}{}^{\pm}$$

$$({}_{\perp}T_{\perp}^{c})^{\pm} = {}_{\perp}T_{\perp}{}^{\mp}, \qquad \left({}_{\perp}T_{\parallel}^{c}\right)^{\pm} = {}_{\parallel}T_{\perp}{}^{\mp}$$

The given and conjugate eigenmodes are depicted in Fig. 4.1. (Compare these conjugate eigenmodes with those formed by a conjugating transformation with respect to the y=0 plane, Fig. 3.1.)

The equivalence relation

$$\mathbf{S}'(s_x, -s_y) = \mathbf{S}(s_x, s_y) \tag{4.26}$$

is analogous to that obtained with Heading's mirrored modes, Sec. 3.1.4, but the mirroring of modes is now with respect to the y=0 plane (rather than the x=0 plane in the case of Heading's modes) in order to map the polar velocity vector--the symmetry axis—into itself. When the electric wave vector is parallel to the plane of incidence, the mirrored mode coincides with the conjugate mode, depicted in Fig. 4.1. When the electric vector is perpendicular to the plane of incidence, its direction is opposite to that of the corresponding conjugate eigenmode. Comparison of (4.25) and (4.26) thus yields the specific symmetry of the scattering matrix $\mathbf{S}(s_x,s_y)$, in analogy with that derived in Sec. 3.1.4, cf. (3.32) and (3.33),

$${}_\perp R_\parallel^\pm = -{}_\parallel R_\perp^\pm, \qquad {}_\parallel T_\parallel^+ = {}_\parallel T_\parallel^-, \quad {}_\perp T_\perp^+ = {}_\perp T_\perp^-, \quad {}_\perp T_\parallel^\pm = -{}_\parallel T_\perp^\mp \tag{4.27}$$

4.2 Unbounded media with sources

4.2.1 The bilinear concomitant vector at infinity

Consider the Maxwell system (2.21) for an electromagnetic field $\mathbf{e}(\mathbf{r})$ generated by, or associated with, a current distribution $\mathbf{j}(\mathbf{r})$, that is confined to a finite region of space

$$\mathbf{L}\,\mathbf{e} := [i\omega\mathbf{K} + \mathbf{D}]\,\mathbf{e}(\mathbf{r}) = -\mathbf{j}(\mathbf{r}) \tag{4.28}$$

$\mathbf{K}$, $\mathbf{D}$, $\mathbf{e}$ and $\mathbf{j}$ are defined in (2.20), (2.22) and (2.23). The formally adjoint equation, (2.41), for a second, confined current system, $\bar{\mathbf{j}}(\mathbf{r})$, and its associated field, $\bar{\mathbf{e}}(\mathbf{r})$, may be written as

$$\bar{\mathbf{L}}\bar{\mathbf{e}} := [i\omega\mathbf{K}^T - \mathbf{D}]\,\bar{\mathbf{e}}(\mathbf{r}) = -\bar{\mathbf{j}}(\mathbf{r}) \tag{4.29}$$

Application of the Lagrange identity

$$\bar{\mathbf{e}}^T\mathbf{L}\,\mathbf{e} - \mathbf{e}^T\bar{\mathbf{L}}\bar{\mathbf{e}} = \nabla\cdot\mathbf{P}$$

or

$$-\,\bar{\mathbf{e}}\cdot\mathbf{j} + \mathbf{e}\cdot\bar{\mathbf{j}} = \nabla\cdot\mathbf{P} = \nabla\cdot(\overline{\mathbf{E}}\times\mathbf{H} + \mathbf{E}\times\overline{\mathbf{H}}) \tag{4.30}$$

gives a reciprocity relation when integrated over all space, the divergence term giving a surface integral (Gauss) at infinity:

$$\int(\mathbf{e}\cdot\bar{\mathbf{j}} - \bar{\mathbf{e}}\cdot\mathbf{j})\,d^3r = \int(\mathbf{E}\times\overline{\mathbf{H}} + \overline{\mathbf{E}}\times\mathbf{H})\cdot d\mathbf{S} \tag{4.31}$$

For this result to be physically useful, we would require the right-hand side to equal zero, and then be left with a relation between currents and fields. Indeed, if $\mathbf{e}(\mathbf{r})$ were to represent a single outgoing 'eigenmode' of the homogeneous equation, $\mathsf{L}\,\mathbf{e} = 0$, at a point $\mathbf{r}$ on S, described by a locally-plane wave of the form $\mathbf{e} \sim \exp(-i\mathbf{k}\cdot\mathbf{r})$, then $\bar{\mathbf{e}} \sim \exp(i\mathbf{k}\cdot\mathbf{r})$ would give the spatial variation of the adjoint field. In loss-free media this adjoint field would be simply the complex conjugate of the original field, (cf. the discussion in Sec. 2.3.1), and the surface integral in (4.31) would be no more than the net outward energy flow — a conserved, non-zero quantity. Indeed, in Chaps. 2 and 3, we used the constancy of (the z-component of) the integrand, the bilinear concomitant vector $\mathbf{P}$, cf. (2.55) and (2.56), as the basis for the ensuing discussion. Here, however, we would require the bilinear concomitant vector, $\mathbf{P}$, to be zero at a large distance from the source, or at least to give a zero surface integral.

4.2.2 The Lorentz-adjoint wave fields at infinity

To obtain a bilinear concomitant vector, $\mathbf{P}$, with vanishing surface flux at infinity, we could use the Lorentz-adjoint system, (3.120), described in Sec. 3.4,

$$\mathsf{L}'\,\mathbf{e} \equiv \bar{\mathsf{I}}\,\mathsf{L}\,\mathbf{e} := \bar{\mathsf{I}}\,[i\omega\mathsf{K} + \mathsf{D}]\,\mathbf{e}(\mathbf{r}) = -\bar{\mathsf{I}}\,\mathbf{j}(\mathbf{r}) \tag{4.32}$$

The formally adjoint system, (3.121), becomes

$$\mathsf{L}^{(L)}\,\mathbf{e}^{(L)} \equiv \bar{\mathsf{L}}'\,\mathbf{e}^{(L)} := [i\omega\mathsf{K}^T - \mathsf{D}^T]\,\bar{\mathsf{I}}^T\mathbf{e}^{(L)} = -\bar{\mathsf{I}}\,\mathbf{j}^{(L)}(\mathbf{r}) \tag{4.33}$$

with $\mathsf{D}^T{=}\mathsf{D}$ and $\bar{\mathsf{I}}^T{=}\bar{\mathsf{I}}$, (2.22) and (2.81). Comparison of (4.33) with (4.29) indicates the simple relationship [cf. (3.132)] between the Lorentz-adjoint fields and currents, $\mathbf{e}^{(L)}$ and $\mathbf{j}^{(L)}$, and the corresponding adjoint quantities, $\bar{\mathbf{e}}$ and $\bar{\mathbf{j}}$,

$$\mathbf{e}^{(L)}(\mathbf{r}) = \bar{\mathsf{I}}\,\bar{\mathbf{e}}(\mathbf{r}), \qquad \mathbf{j}^{(L)}(\mathbf{r}) = \bar{\mathsf{I}}\,\bar{\mathbf{j}}(\mathbf{r}) \tag{4.34}$$

with $\bar{\mathsf{I}}{=}\bar{\mathsf{I}}^T{=}\bar{\mathsf{I}}^{-1}$. Application of the Lagrange identity, (3.122), to (4.32) and (4.33) yields

$$\tilde{\mathbf{e}}^{(L)}\mathsf{L}'\,\mathbf{e} - \tilde{\mathbf{e}}\,\bar{\mathsf{L}}'\,\mathbf{e}^{(L)} = \boldsymbol{\nabla}\cdot\mathbf{P}^{(L)}, \qquad \mathbf{P}^{(L)} := \mathbf{E}^{(L)}\times\mathbf{H} - \mathbf{E}\times\mathbf{H}^{(L)} \tag{4.35}$$

and, on integration over all space,

$$\int(-\tilde{\mathbf{e}}^{(L)}\bar{\mathsf{I}}\,\mathbf{j} + \tilde{\mathbf{e}}\,\bar{\mathsf{I}}\,\mathbf{j}^{(L)})\,d^3r = \int\mathbf{P}^{(L)}\cdot d\mathbf{S} = \int(\mathbf{E}^{(L)}\times\mathbf{H} - \mathbf{E}\times\mathbf{H}^{(L)})\cdot d\mathbf{S} \tag{4.36}$$

To consider the behaviour of $\mathbf{P}^{(L)}$ at large distances from the source, we need to know the behaviour of the fields $\mathbf{E}^{(L)}$ and $\mathbf{H}^{(L)}$. Premultiplication of (4.33) by $\bar{\mathsf{I}}$ yields, as in (3.124),

$$[i\omega\mathsf{K}^{(L)} + \mathsf{D}]\,\mathsf{e}^{(L)} = -\mathsf{j}^{(L)}(\mathbf{r}), \qquad \mathsf{K}^{(L)} := \bar{\mathsf{I}}\,\mathsf{K}^T\,\bar{\mathsf{I}} \tag{4.37}$$

This is just the Maxwell system, cf. (4.28), for fields and currents in a medium described by the tensor $\mathsf{K}^{(L)}$, which could in fact represent a physical medium. Suppose, for simplicity, that we restrict ourselves for the present to media which may be anisotropic, but *not bianisotropic*. In anisotropic crystalline media, for instance, the tensor K is symmetric, so that $\mathsf{K}^{(L)}$=K^T represents the original medium. In magnetoplasmas or ferrites, $\mathsf{K}^{(L)} = \mathsf{K}^T$ represents a physical medium in which the direction of the external magnetic field is opposite to that in the given medium. Thus $\mathsf{e}^{(L)}$ (4.37) represents in principle a physical field, generated in a physical (transposed) medium by a localized current distribution, $\mathsf{j}^{(L)}(\mathbf{r})$. The constituent fields, $\mathbf{E}^{(L)}$ and $\mathbf{H}^{(L)}$, which appear in (4.35) and (4.36) therefore describe 'outgoing' fields at large distances on S, as do of course the given fields, $\mathbf{E}$ and $\mathbf{H}$, in the given medium.

The behaviour of the concomitant vector $\mathbf{P}^{(L)}$ at infinity may be inferred by application of the so-called *radiation condition* [111, p. 189], [72, p.429]. For isotropic media this stipulates that the wave fields at a large distance $\mathbf{r}$, measured from any point within the region containing the localized current distribution, will be of the form of an outgoing spherical wave

$$|\mathsf{e}| \sim \frac{\exp(-ikr)}{r} + \mathrm{O}\left(\frac{1}{r^2}\right) \tag{4.38}$$

Mathematically this may be expressed in the form

$$\lim_{r\to\infty} r\left(\frac{\partial A}{\partial r} - ikA\right) = 0 \tag{4.39}$$

where A represents any field component transverse to $\mathbf{r}$, and $k = \omega\sqrt{\varepsilon\mu}$ [53, p.87], [83, p.245]. In view of the local plane-wave nature of the fields as $r \to \infty$, the relation between $\mathbf{E}$ and $\mathbf{H}$ will be simply

$$\hat{\mathbf{r}} \times \mathbf{E} = \sqrt{\frac{\mu}{\varepsilon}}\,\mathbf{H} \tag{4.40}$$

The normal component of the concomitant vector, $\mathbf{P}^{(L)} \cdot \hat{\mathbf{r}}$ (4.35), becomes

$$\sqrt{\frac{\varepsilon}{\mu}}\left\{\mathbf{E} \times (\hat{\mathbf{r}} \times \mathbf{E}^{(L)}) - \mathbf{E}^{(L)} \times (\hat{\mathbf{r}} \times \mathbf{E})\right\} \cdot \hat{\mathbf{r}} = 0 \tag{4.41}$$

and the surface integral in (4.36), taken over a spherical surface of radius $r \to \infty$, vanishes identically for isotropic media.

For anisotropic or gyrotropic media the radiation condition stipulates outward power flow (outward ray directions) at infinity, rather than 'outgoing wave propagation' [53, p.748]. At any point on the surface, S, more than one ray may intersect, implying that a number of different eigenmodes, each with a local plane-wave spatial variation, $\exp(-i\,\mathbf{k}_\alpha \cdot \mathbf{r})$, may be superimposed. The directions of the characteristic propagation vectors, $\mathbf{k}_\alpha$, will in general differ from those of the rays, and certainly from one another. Outgoing rays could also be associated with backward, incoming waves ($\mathbf{k}_\alpha \cdot \hat{\mathbf{r}} < 0$). Application of the 'radiation condition' is thus by no means a trivial problem. We shall show however in the next section that here too the surface integral $\int \mathbf{P}^{(L)} \cdot d\mathbf{S}$ will vanish as $\mathbf{r} \to \infty$.

4.2.3 Refractive-index surfaces for given and Lorentz-adjoint eigenmodes

Consider local plane-wave solutions of (4.32) and (4.33) in the source-free ($\mathbf{j}$=0) far field ($\mathbf{r} \to \infty$). With the plane-wave ansatz

$$\mathbf{e}(\mathbf{r}) \sim \exp(-i\,\mathbf{k} \cdot \mathbf{r}) \qquad \text{and} \qquad \mathbf{e}^{(L)}(\mathbf{r}) \sim \exp(-i\,\mathbf{k}' \cdot \mathbf{r})$$

we get

$$\mathsf{D}\,\mathbf{e} \to -i \begin{bmatrix} 0 & -\mathbf{k} \times \mathsf{I} \\ \mathbf{k} \times \mathsf{I} & 0 \end{bmatrix} \mathbf{e} \equiv -i\mathcal{K}\mathbf{e} = -i\mathcal{K}^T\mathbf{e}$$

$$\mathsf{D}\,\mathbf{e}^{(L)} \to -i\mathcal{K}'\mathbf{e}^{(L)} = -i\mathcal{K}'^T\mathbf{e}^{(L)} \tag{4.42}$$

defining thereby the symmetric matrices $\mathcal{K}$ and $\mathcal{K}'$, by means of which (4.32) and (4.33) become algebraic equations:

$$\mathsf{L}(\mathbf{k})\mathbf{e}(\mathbf{k}) := i[\omega\mathsf{K} - \mathcal{K}]\,\mathbf{e}(\mathbf{k}) = 0 \tag{4.43}$$

$$\overleftarrow{\mathsf{L}}(\mathbf{k}')\,\mathbf{e}^{(L)}(\mathbf{k}') := i[\omega\mathsf{K}^T + \mathcal{K}']\,\bar{\mathsf{I}}\,\mathbf{e}^{(L)}(\mathbf{k}') = 0 \tag{4.44}$$

Premultiplication of (4.44) by $\bar{\mathsf{I}}$ gives

$$i[\omega\mathsf{K}^{(L)} - \mathcal{K}']\,\mathbf{e}^{(L)}(\mathbf{k}') = 0 \tag{4.45}$$

Suppose that the propagation vectors, $\mathbf{k}$ and $\mathbf{k}'$, are in a specified direction, $\hat{\mathbf{k}}$

$$\mathbf{k} = k\,\hat{\mathbf{k}}, \quad \mathcal{K} = k\hat{\mathcal{K}}; \qquad \mathbf{k}' = k'\,\hat{\mathbf{k}}, \quad \mathcal{K}' = k'\hat{\mathcal{K}}$$

The eigenvalue equations for k and k', with $\widehat{\mathcal{K}}=\widehat{\mathcal{K}}^T$, are seen from (4.43) and (4.45) to be identical for anisotropic media in which $\mathbf{K}^{(L)}=\mathbf{K}^T$:

$$\det[\omega\mathbf{K} - k\widehat{\mathcal{K}}] = 0, \qquad \det[\omega\mathbf{K}^T - k'\widehat{\mathcal{K}}^T] = 0 \tag{4.46}$$

yielding quartic equations in k or k' (as becomes evident if one of the coordinate axes is taken in the direction of $\hat{\mathbf{k}}$). If, as in the present discussion, the media are not bianisotropic, the quartic is in fact quadratic in k^2, or in n^2, where n is the refractive index: $k=nk_0$. This is easily seen if $[\omega\mathbf{K} - k\widehat{\mathcal{K}}]$ in (4.43) is premultiplied and postmultiplied by $\bar{\mathbf{I}}$, the only resultant change being a reversal of the sign of k. (For cold magnetoplasmas this quadratic is just the Appleton-Hartree-Lassen formula (1.58), derived in Chap. 1).

The two roots, ${k_1}^2$ and ${k_2}^2$, of the eigenvalue equation, with $k_\alpha = -k_{-\alpha}$, $\alpha=1$ or 2, will correspond to left- or right-handed modes, L or R, if the wave polarization is elliptic or circular, and to ordinary or extraordinary, O or X, if the transverse polarization is linear. As the linear polarization passes over continuously into elliptic, the designation, O or X, is often retained in this range too. For a given wave frequency, ω, there will thus be two sets of $\mathbf{k}$-surfaces (refractive-index surfaces), one for each modal type or polarization, which will be symmetric with respect to the origin, $\mathbf{k}=0$. The equality of the eigenvalues, k and k' (4.46), means that that the refractive-index surfaces will be identical for the given and Lorentz-adjoint wave fields (i.e. for local plane-wave solutions in the given and transposed media). It should be noted that had our starting point been the adjoint equation (4.29), rather than the Lorentz-adjoint equation (4.33), we would have obtained the same eigenvalue equation (4.46) with the plane-wave ansatz $\bar{\mathbf{e}} \sim \exp(i\,\mathbf{k}\cdot\mathbf{r})$, thereby justifying a remark made in Sec. 4.2.1 that the adjoint (as distinct from the Lorentz-adjoint) far field would vary as an *incoming* plane wave. For a given value of $\mathbf{k}$, i.e. for a given point on one of the refractive-surfaces, the ray direction will be normal to the surface at that point [33, Sec. 5.3], and conversely, for a given ray direction we could generally have up to three wave-normal directions, $\hat{\mathbf{k}}$ [33, Sec. 5.4].

It is necessary, at this stage, to specify physically acceptable behaviour of the media at large distances. In the case of a magnetoplasma, characterized by an external magnetic field which is generated by a localized current system outside of the source region (in which we are interrelating currents and fields), the external magnetic field tends to zero at infinity, and the medium becomes isotropic, with consequences that were discussed in the last section.

In the case of anisotropic media, or gyrotropic media such as ferrites, we may assume that beyond a certain large distance r_0, $(r > r_0)$, the medium becomes homogeneous. In the case of ferrites this would require a constant magnetization. In such a region the rays would travel in straight lines, and in the limit, $r \to \infty$, the outgoing ray directions would be normal to the spherical bounding surface S at infinity ($r >> r_0$). In other words, there would be a single outgoing ray direction at each point on the bounding surface, but several local plane waves or eigenmodes, with different values of $\mathbf{k}_\alpha$, could be superimposed at that point. This then is the form taken by the 'radiation condition' in anisotropic media. In view of the previous discussion, we may infer that the same characteristic wave vectors will also be present in the Lorentz-adjoint wave fields, i.e. $\mathbf{k'}_\alpha = \mathbf{k}_\alpha$, although with different relative intensities as determined by the localized current distributions, $\mathbf{j}^{(L)}(\mathbf{r}) \neq \mathbf{j}(\mathbf{r})$, which generate the fields.

4.2.4 The bilinear concomitant vector in the far field

Let $\mathbf{k}_\alpha$ represent the propagation vector of one of the local plane-wave eigenmodes (4.43) at a point $\mathbf{r}$ on S in the given problem, and $\mathbf{k'}_\beta$, one of the corresponding eigenmodes in the Lorentz-adjoint problem (4.44). Both have the same ray direction in common. Then, with $\bar{\mathsf{I}}^T = \bar{\mathsf{I}}$ (2.81) and $\mathcal{K'}^T = \mathcal{K'}$ (4.42), we obtain from (4.43) and (4.44)

$$\begin{aligned}
[\bar{\mathsf{I}}\, \mathbf{e}_\beta^{(L)}]^T \mathsf{L}(\mathbf{k})\, \mathbf{e}_\alpha - \mathbf{e}_\alpha^T\, \bar{\mathsf{L}}'(\mathbf{k}')\, \mathbf{e}_\beta^{(L)} &= -i \left[\tilde{\mathbf{e}}_\beta^{(L)}\, \bar{\mathsf{I}}^T (\mathcal{K}_\alpha + \mathcal{K'}_\beta)\mathbf{e}_\alpha\right] \\
&= i(\mathbf{k}_\alpha + \mathbf{k'}_\beta) \cdot (\mathbf{E}_\alpha \times \mathbf{H}_\beta^{(L)} - \mathbf{E}_\beta^{(L)} \times \mathbf{H}_\alpha] \\
&= i(\mathbf{k}_\alpha + \mathbf{k'}_\beta) \cdot \mathbf{P}_{\alpha\beta}^{(L)} = 0 \qquad (4.47)
\end{aligned}$$

We wish to show that this implies that the bilinear concomitant vector, (4.35),

$$\mathbf{P}_{\alpha\beta}^{(L)} := \mathbf{E}_\beta^{(L)} \times \mathbf{H}_\alpha - \mathbf{E}_\alpha \times \mathbf{H}_\beta^{(L)} \qquad (4.48)$$

vanishes for all α and β.

Note first that the possibility, $\mathbf{k}_\alpha + \mathbf{k'}_\beta = 0$, is ruled out. If $\mathbf{k'}_\beta = -\mathbf{k}_\alpha$, then there is also a solution, $\mathbf{k'}_\beta = \mathbf{k}_\alpha$ [see the discussion following (4.46)]. If both the given and the Lorentz adjoint wave vectors generate outgoing ray directions, then clearly $\mathbf{k'}_\beta = \mathbf{k}_\alpha$ is the relevant choice.

We now consider two other possibilities. In some media, such as magnetoplasmas, the wave field vectors, $\mathbf{E}$ and $\mathbf{H}$, will describe ellipses in the transverse (to $\mathbf{k}$) plane, while the electric vector will also have a longitudinal component. The result is that $\mathbf{E}$ and $\mathbf{H}$ will move in ellipses in different planes, while the instantaneous Poynting vector, $\mathbf{E}\times\mathbf{H}$, will rotate along the surface of a cone, so that only its mean direction (along the axis of the cone) will correspond to the direction of the ray [104, p.191]. The Lorentz-adjoint wave fields too, will move in different planes, the $\mathbf{H}_\beta^{(L)}$ vector in a plane transverse to $\mathbf{k}'$, the $\mathbf{E}_\beta^{(L)}$ vector in a tilted plane. Consequently, each vector product on the right-hand side of (4.48) will change its direction continuously, as would their difference, if not equal to zero. But in that case (4.47) could not be satisfied at all times, unless $\mathbf{P}_{\alpha\beta}^{(L)} = 0$.

The other possibility is that the Poynting vector for each mode is in a fixed direction, the ray direction, $\hat{\mathbf{r}}$, which would then also be the direction of the concomitant vector, $\mathbf{P}^{(L)}$, with the wave fields, $\mathbf{E}_\alpha$ and $\mathbf{H}_\alpha$, thus lying in a plane normal to the ray direction, i.e. tangential to the bounding surface S. Application of (4.47) would then imply that $(\mathbf{k}_\alpha + \mathbf{k}'_\beta) \cdot \hat{\mathbf{r}} = 0$, requiring either that both wave normals be perpendicular to the ray direction, or else that one of the wave normals lie in the backward direction, $\mathbf{k} \cdot \hat{\mathbf{r}} < 0$.

Now backward waves are indeed quite common in hot plasmas having spatial dispersion, as for instance the cyclotron-harmonic Bernstein modes, but such media are not treated here. If we then exclude the two 'pathological cases' just mentioned, we may conclude that the concomitant vector for any pair of local plane waves, or eigenmodes, vanishes identically at large distances:

$$\mathbf{P}_{\alpha\beta}^{(L)} = 0, \qquad \mathbf{r} \to \infty \tag{4.49}$$

In all other cases we could solve our problem with the simple, if inelegant, device of assuming that the medium is 'slightly absorbing' at infinity, i.e. that the $\mathbf{k}$-vectors have a small negative imaginary part. Then all outgoing field vectors, both given and Lorentz-adjoint, will vanish at infinity, including the 'pathological cases' just discussed. (Note, however, that the adjoint wave fields $\bar{\mathbf{e}}$ (4.29),—as opposed to the 'Lorentz-adjoint' fields $\mathbf{e}^{(L)}$ (4.37) that we are now discussing—would blow up as $\mathbf{r} \to \infty$ in absorbing media, and the concomitant vector $\mathbf{P}$ would remain constant, as mentioned in Sec. 4.2.1.)

If, finally, we consider the overall concomitant vector, $\mathbf{P}^{(L)}$ (4.35), on the bounding surface S, in which the wave fields consist possibly of a number of superimposed eigenmodes,

$$\mathbf{E}(\mathbf{r}) = \sum_\alpha \mathbf{E}_\alpha(\mathbf{r}_0) \exp(-i\,\mathbf{k}_\alpha \cdot \mathbf{r}), \qquad \mathbf{E}^{(L)}(\mathbf{r}) = \sum_\beta \mathbf{E}_\beta^{(L)}(\mathbf{r}_0) \exp(-i\,\mathbf{k}'_\beta \cdot \mathbf{r})$$

then the concomitant vector for each pair of eigenmodes, α, β, in the cross products (4.48), will vanish, and so

$$\mathbf{P}^{(L)} = \sum_\alpha \sum_\beta \mathbf{P}_{\alpha,\beta}^{(L)} = 0 \tag{4.50}$$

The important consequence, from our point of view, will then be that (4.36) yields

$$\int (\tilde{\mathbf{e}}\,\bar{\mathbf{I}}\,\mathbf{j}^{(L)} - \tilde{\mathbf{e}}^{(L)}\bar{\mathbf{I}}\,\mathbf{j})d^3r \equiv \int \{(\mathbf{E}\cdot\mathbf{J}_\mathrm{e}^{(L)} - \mathbf{H}\cdot\mathbf{J}_\mathrm{m}^{(L)}) - (\mathbf{E}^{(L)}\cdot\mathbf{J}_\mathrm{e} - \mathbf{H}^{(L)}\cdot\mathbf{J}_\mathrm{m})\}\,d^3r$$
$$= 0 \qquad (4.51)$$

which is the form of the Lorentz reciprocity theorem that will be generalized in this chapter.

4.3 Boundary conditions at impedance walls

4.3.1 Surface impedance boundaries

One often encounters problems in which the currents and fields are confined to bounded regions of space, enclosed or separated from other regions of space by 'impedance walls'. On these walls it is sometimes possible to define a dyadic *surface impedance*, $\mathsf{Z}_s(\mathbf{r}_s)$, that characterizes *in an approximate way* the relation between the tangential electric and magnetic wave fields at that point, $\mathbf{r}_s$, on the surface [53, p.16]. In mixed vector-matrix notation this may be written as

$$\hat{\mathbf{n}}\times\mathbf{E} = \mathsf{Z}_s\mathbf{H} = \mathsf{Z}_s\mathbf{H}_t \qquad \text{with} \qquad \mathsf{Z}_s\hat{\mathbf{n}} = 0 \qquad (4.52)$$

where $\hat{\mathbf{n}}$ is an outward unit normal vector at $\mathbf{r}_s$ on the surface S. When the surface-impedance concept is applicable, it simplifies the solution of boundary-value problems by circumventing the need to evaluate the fields beyond the surface boundaries.

The formal structure of Z_s, which operates only on the transverse component of $\mathbf{H}$, is

$$\mathsf{Z}_s = \mathsf{Z}_t\left[\mathsf{I}^{(3)} - \hat{\mathbf{n}}\,\hat{\mathbf{n}}^T\right] \qquad (4.53)$$

where the matrix $[\mathsf{I}^{(3)} - \hat{\mathbf{n}}\,\hat{\mathbf{n}}^T]$, operating on any vector $\mathbf{A}$, projects it onto the plane transverse to $\hat{\mathbf{n}}$:

$$\left[\mathsf{I}^{(3)} - \hat{\mathbf{n}}\,\hat{\mathbf{n}}^T\right]\mathbf{A} = \mathbf{A}_t, \qquad \hat{\mathbf{n}}\cdot\mathbf{A}^T = 0 \qquad (4.54)$$

The matrix Z_t, operating on a vector in the plane transverse to $\hat{\mathbf{n}}$, produces another vector in that plane. The surface impedance is a 'scalar' if $\mathsf{Z}_t = Z_t\mathsf{I}^{(3)}$, so that

$$\mathsf{Z}_s = Z_t\left[\mathsf{I}^{(3)} - \hat{\mathbf{n}}\,\hat{\mathbf{n}}^T\right], \qquad \mathsf{Z}_s = \mathsf{Z}_s{}^T \qquad (4.55)$$

where Z_t is a complex scalar impedance, and $\mathbf{H}_t$ is then in the direction of $\hat{\mathbf{n}} \times \mathbf{E}$.

In some cases such as, for instance, the problem of diffraction by a partially conducting wedge [53, Sec. 6.6], whose surfaces satisfy 'homogeneous boundary conditions', the separability of the solutions at the boundary requires that the surface impedance increase linearly, or decrease inversely, as the distance from the linear edge, depending on whether the electric or magnetic wave fields, respectively, are parallel to the edge of the wedge [53, eqs. (3a) and (4a)]. The wall impedance can take on a tensor character, even with (scalar) resistance walls. If the walls are banded or corrugated, for instance, then a tangential electric field, $\mathbf{E}_t$, at an angle to the bands or corrugation, will generate currents in a direction different to that of the field.

We now examine the requirements on the dyadic (tensor) surface impedance for Lorentz reciprocity to hold in the bounded region.

4.3.2 Surface impedance and its Lorentz adjoint

Consider the surface integral in (4.36)

$$\begin{aligned}\int \mathbf{P}^{(L)} \cdot d\mathbf{S} &\equiv \int (\mathbf{E}^{(L)} \times \mathbf{H} - \mathbf{E} \times \mathbf{H}^{(L)}) \cdot \hat{\mathbf{n}}\, dS \\ &= \int \{(\hat{\mathbf{n}} \times \mathbf{E}^{(L)}) \cdot \mathbf{H} - (\hat{\mathbf{n}} \times \mathbf{E}) \cdot \mathbf{H}^{(L)}\}\, dS \qquad (4.56)\end{aligned}$$

at any point $\mathbf{r}_s$ on the impedance surface. With the aid of (4.52), and a similar relation for the Lorentz-adjoint wave fields,

$$\hat{\mathbf{n}} \times \mathbf{E}^{(L)} = \mathsf{Z}_s^{(L)} \mathbf{H}^{(L)} \qquad \text{with} \qquad \mathsf{Z}_s^{(L)} \hat{\mathbf{n}} = 0 \qquad (4.57)$$

the right-hand side of (4.56) becomes, in matrix notation,

$$\int \left\{ \mathbf{H}^T \mathsf{Z}_s^{(L)} \mathbf{H}^{(L)} - \tilde{\mathbf{H}}^{(L)} \mathsf{Z}_s \mathbf{H} \right\} dS = \int \mathbf{H}^T [\mathsf{Z}_s^{(L)} - \mathsf{Z}_s^T] \mathbf{H}^{(L)}\, dS$$

with the second term in the integrand on the left transposed. The surface integral vanishes identically if

$$\mathsf{Z}_s^{(L)} = \mathsf{Z}_s^T \qquad (4.58)$$

a relation which is satisfied if, for instance, the surface impedances are scalars, as in (4.55).

In summary, we compare solutions of Maxwell's equations within a volume V, bounded by a surface S. In a given medium with a constitutive tensor

$\mathsf{K}(\mathbf{r})$ and a surface impedance $\mathsf{Z}_s(\mathbf{r}_s)$, ($\mathbf{r} \in V$, $\mathbf{r}_s \in S$), the fields and current distributions are $\mathrm{e}(\mathbf{r})$ and $\mathrm{j}(\mathbf{r})$. In an associated Lorentz-adjoint problem in the same region, in which 'physical' fields propagate in the Lorentz-adjoint medium $\mathsf{K}^{(L)}(\mathbf{r})$ bounded by a surface impedance $\mathsf{Z}_s^{(L)}(\mathbf{r}_s)$, the fields and currents are $\mathrm{e}^{(L)}(\mathbf{r})$ and $\mathrm{j}^{(L)}(\mathbf{r})$. Then the currents and fields will obey the Lorentz reciprocity relation, cf. (4.36) and (4.58),

$$\int(\tilde{\mathrm{e}}\,\bar{\mathrm{I}}\,\mathrm{j}^{(L)} - \tilde{\mathrm{e}}^{(L)}\bar{\mathrm{I}}\,\mathrm{j})d^3r \equiv \int\{(\mathbf{E}\cdot\mathbf{J}_{\mathrm{e}}^{(L)} - \mathbf{H}\cdot\mathbf{J}_{\mathrm{m}}^{(L)}) - (\mathbf{E}^{(L)}\cdot\mathbf{J}_{\mathrm{e}} - \mathbf{H}^{(L)}\cdot\mathbf{J}_{\mathrm{m}})\}\,d^3r = 0 \tag{4.59}$$

provided that (4.58)

$$\mathsf{Z}_s^{(L)} = \mathsf{Z}_s^T.$$

4.4 Uniaxial media with sources

In this section we consider uniaxial media containing sources (currents) and a plane of symmetry with respect to a reflection transformation, $\mathcal{R}$, or a conjugating transformation, $\mathcal{RT}$. We consider initially a gyrotropic magnetoplasma, where we can clearly distinguish between the given and transposed media, K and K^T. We no longer restrict ourselves to plane-stratified media, and so the external field b is not necessarily constant, nor parallel to the symmetry plane.

4.4.1 Transformation of gyrotropic media with sources

Suppose that the plasma is spherically symmetric, i.e. the plasma parameters, such as electron or ion densities and collision frequencies, are functions of the radial distance r only. Let the plasma be immersed in a magnetic field generated by a magnetic dipole at the origin parallel to the z-axis (see Fig. 4.2). This may be considered as an idealized model of the earth's ionosphere. Recalling that the magnetic field is an axial vector, we note that $z=0$ defines a plane of reflection symmetry: the field $\mathrm{b}' \equiv \mathrm{b}(\mathbf{r}') = \mathrm{b}(x, y, -z)$ in the lower half space, $z < 0$, is a reflection of the field $\mathrm{b}(\mathbf{r}) \equiv \mathrm{b}(x, y, z)$ in the upper half space, $z > 0$,

$$\mathrm{b}' \equiv \mathrm{b}(\mathbf{r}') \equiv \mathcal{R}\mathrm{b}(\mathbf{r}) = (\det \mathsf{q}_z)\mathsf{q}_z\mathrm{b}(\mathbf{r}) = -\mathsf{q}_z\mathrm{b}(\mathbf{r}), \qquad \mathbf{r}' = \mathsf{q}_z\mathbf{r} \tag{4.60}$$

with, cf. (4.13),

$$\mathsf{q}_z := \begin{bmatrix} 1 & 0 & 0 \\ 0 & 1 & 0 \\ 0 & 0 & -1 \end{bmatrix}, \qquad \mathrm{b}' = (-b_x, -b_y, b_z), \quad \mathbf{r}' = (x, y, -z)$$

The y=0 plane, on the other hand, is a plane of symmetry under $\mathcal{RT}$, the conjugating transformation. Time reversal changes the sign of $\mathbf{b}$, so that

$$\mathbf{b}^c \equiv \mathbf{b}(\mathbf{r}') \equiv \mathcal{RT}\mathbf{b}(\mathbf{r}) = -(\det \mathbf{q}_y)\mathbf{q}_y\mathbf{b}(\mathbf{r}) = \mathbf{q}_y\mathbf{b}(\mathbf{r}), \qquad \mathbf{r}' = \mathbf{q}_y\mathbf{r} \quad (4.61)$$

with $\mathbf{q}_y$, as in (4.13), given by

$$\mathbf{q}_y := \begin{bmatrix} 1 & 0 & 0 \\ 0 & -1 & 0 \\ 0 & 0 & 1 \end{bmatrix} \qquad \mathbf{b}^c = (b_x, -b_y, b_z), \quad \mathbf{r}' = (x, -y, z)$$

Transformation of the gyrotropic dielectric tensor $\boldsymbol{\varepsilon}\{\mathbf{b}(\mathbf{r}),\mathbf{r}\}$ in $\mathbf{K}(\mathbf{r})$ (2.35), with $\mathbf{q}_z$ or $\mathbf{q}_y$, gives

$$\mathbf{q}_z\boldsymbol{\varepsilon}\{\mathbf{b}(\mathbf{r}),\mathbf{r}\}\mathbf{q}_z = \boldsymbol{\varepsilon}(\mathbf{b}',\mathbf{r}'), \qquad \mathbf{r}' = (x, y, -z) \quad (4.62)$$

or

$$\mathbf{q}_y\boldsymbol{\varepsilon}\{\mathbf{b}(\mathbf{r}),\mathbf{r}\}\mathbf{q}_y = \boldsymbol{\varepsilon}(-\mathbf{b}^c,\mathbf{r}') = \boldsymbol{\varepsilon}^T(\mathbf{b}^c,\mathbf{r}'), \qquad \mathbf{r}' = (x, -y, z) \quad (4.63)$$

with $\mathbf{b}'$, $\mathbf{b}^c$, $\mathbf{q}_y$ and $\mathbf{q}_z$ defined above in (4.60) and (4.61).

We consider next the transformation of the Maxwell system (4.28) by means of the reflection operator $\mathbf{Q}_z$ and the adjoint reflection operator $\overline{\mathbf{Q}}_y$ where, cf. (4.13) and (4.14),

$$\begin{aligned} \mathbf{Q}_z &= \begin{bmatrix} \mathbf{q}_z & 0 \\ 0 & (\det \mathbf{q}_z)\mathbf{q}_z \end{bmatrix} = \begin{bmatrix} \mathbf{q}_z & 0 \\ 0 & -\mathbf{q}_z \end{bmatrix}, \\ \overline{\mathbf{Q}}_y &= \begin{bmatrix} \mathbf{q}_y & 0 \\ 0 & -(\det \mathbf{q}_y)\mathbf{q}_y \end{bmatrix} = \begin{bmatrix} \mathbf{q}_y & 0 \\ 0 & \mathbf{q}_y \end{bmatrix} \end{aligned} \quad (4.64)$$

In view of (4.62) and (4.63), we note that the constitutive tensor $\mathbf{K}(\mathbf{b},\mathbf{r})$, (2.35), transforms as

$$\mathbf{Q}_z\mathbf{K}(\mathbf{b},\mathbf{r})\mathbf{Q}_z = \mathbf{K}(\mathbf{b}',\mathbf{r}'), \qquad \mathbf{r}' = \mathbf{q}_z\mathbf{r}, \quad \mathbf{b}' = \mathbf{b}(\mathbf{r}') = \mathcal{R}\mathbf{b}(\mathbf{r}) \quad (4.65)$$

$$\begin{gathered} \overline{\mathbf{Q}}_y\mathbf{K}(\mathbf{b},\mathbf{r})\overline{\mathbf{Q}}_y = \bar{\mathbf{I}}\,\mathbf{K}(-\mathbf{b}^c,\mathbf{r}')\,\bar{\mathbf{I}} = \mathbf{K}^T(\mathbf{b}^c,\mathbf{r}') = \mathbf{K}^T\{\mathbf{b}(\mathbf{r}'),\mathbf{r}'\} \\ \mathbf{r}' = \mathbf{q}_y\mathbf{r}, \qquad \mathbf{b}^c = \mathbf{b}(\mathbf{r}') = \mathcal{RT}\mathbf{b}(\mathbf{r}) \end{gathered} \quad (4.66)$$

To represent these results schematically we could specify that $\mathbf{K}(\mathbf{r})$, in the first quadrant in Fig. 4.2, represents the 'given medium'; then the fourth quadrant would contain the 'reflected medium', and the second quadrant, the conjugate

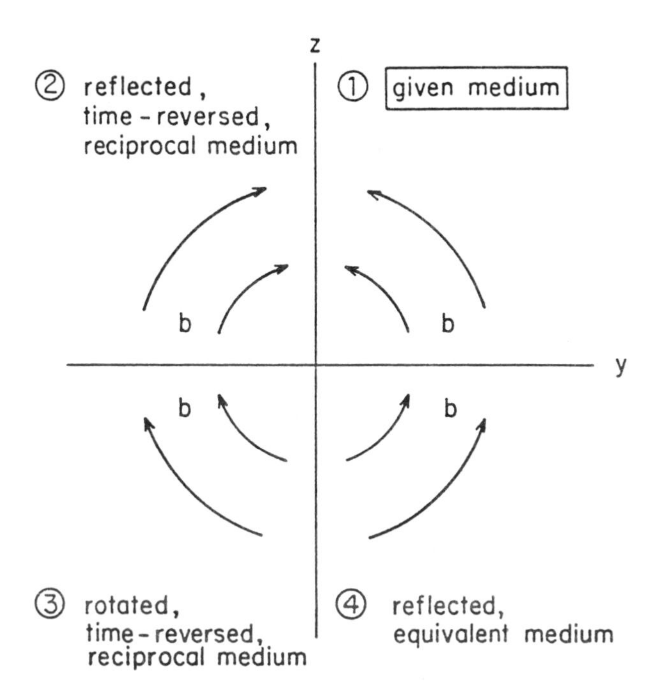

Figure 4.2: Mapping a gyrotropic medium (a magnetoplasma) into itself. **b** is the external magnetic field. Currents and fields in the half-spaces $y<0$ and $y>0$ are reciprocal. Those in $z<0$ and $z>0$ are 'equivalent'.

(reflected and time-reversed) medium. The medium in the third quadrant could similarly be described as rotated (i.e. twice reflected) and time reversed. But we could, just as well, have taken any part of the overall medium, or all of it—i.e. all of space—as the 'given medium'. In the latter case the reflected or conjugate media would also extend over all space, and would coincide of course with the original medium in all of space.

The transformation of the differential operator $\mathsf{D}\equiv\mathsf{D}(\mathbf{r})$, (2.22) and (2.24), requires special care. The operation $\mathbf{Q}_i\mathsf{D}\mathbf{Q}_i$ or $\overline{\mathbf{Q}}_i\mathsf{D}\overline{\mathbf{Q}}_i$ will change the sign of some of the elements of the 6×6 matrix $\mathsf{D}(\mathbf{r})$. However, the partial derivatives, (2.24), in D will no longer operate on the given field $\mathbf{e}(\mathbf{r})$ (2.23), but on the transformed field,

$$\mathbf{e}'(\mathbf{r}') = \mathbf{Q}_i\,\mathbf{e}(\mathbf{r}), \qquad \mathbf{r}' = \mathbf{q}_i\mathbf{r} \qquad (i = x, y \text{ or } z) \tag{4.67}$$

The transformed field $\mathbf{e}'(\mathbf{r}')$ has not only a different direction to that of the original field $\mathbf{e}(\mathbf{r})$, but a different spatial structure. Hence

$$\begin{aligned}\mathbf{Q}_z\mathbf{D}(\mathbf{r})\mathbf{Q}_z &= \mathbf{Q}_z\left[\mathbf{U}_x\frac{\partial}{\partial x}+\mathbf{U}_y\frac{\partial}{\partial y}+\mathbf{U}_z\frac{\partial}{\partial z}\right]\mathbf{Q}_z \\ &= \left[\mathbf{U}_x\frac{\partial}{\partial x}+\mathbf{U}_y\frac{\partial}{\partial y}-\mathbf{U}_z\frac{\partial}{\partial z}\right] \\ &= \left[\mathbf{U}_x\frac{\partial}{\partial x'}+\mathbf{U}_y\frac{\partial}{\partial y'}+\mathbf{U}_z\frac{\partial}{\partial z'}\right] \\ &= \mathbf{D}(\mathbf{r}') \end{aligned} \tag{4.68}$$

since $\mathbf{r}'=\mathbf{q}_z\mathbf{r}=(x,\,y,\,-z)$. Similarly, with $\overline{\mathbf{Q}}_y=\mathbf{Q}_y\bar{\mathbf{I}}=\bar{\mathbf{I}}\,\mathbf{Q}_y$, (4.14), we have

$$\overline{\mathbf{Q}}_y\mathbf{D}(\mathbf{r})\overline{\mathbf{Q}}_y=\bar{\mathbf{I}}\,\mathbf{D}(\mathbf{r}')\bar{\mathbf{I}}=-\mathbf{D}(\mathbf{r}') \tag{4.69}$$

as in (3.125), with $\mathbf{r}'=\mathbf{q}_y\mathbf{r}=(x,-y,\,z)$.

Reflection mapping of the Maxwell system

The spatial symmetries of the medium just described, are expressed in the corresponding transformations of Maxwell's equations. The reflection transformation with respect to the z=0 plane gives, with $\mathbf{Q}_z=\mathbf{Q}_z^{-1}$,

$$\begin{aligned}\mathbf{Q}_z\mathbf{L}\,\mathbf{e}(\mathbf{r}) &= [\mathbf{Q}_z\mathbf{L}\mathbf{Q}_z]\,[\mathbf{Q}_z\mathbf{e}(\mathbf{r})] \\ &= \mathbf{Q}_z\left[i\omega\mathbf{K}(\mathbf{b},\mathbf{r})+\mathbf{D}(\mathbf{r})\right]\mathbf{Q}_z\cdot\mathbf{Q}_z\mathbf{e}(\mathbf{r})=-\mathbf{Q}_z\mathbf{j}(\mathbf{r})\end{aligned}$$

or, with the aid of (4.60), (4.65), (4.67) and (4.68),

$$\left[i\omega\mathbf{K}(\mathbf{b}',\mathbf{r}')+\mathbf{D}(\mathbf{r}')\right]\mathbf{e}'(\mathbf{r}')=-\mathbf{j}'(\mathbf{r}') \tag{4.70}$$

with

$$\mathbf{e}'(\mathbf{r}')=\mathbf{Q}_z\mathbf{e}(\mathbf{r}),\quad \mathbf{j}'(\mathbf{r}')=\mathbf{Q}_z\mathbf{j}(\mathbf{r}),\quad \mathbf{b}'\equiv\mathbf{b}(\mathbf{r}')=-\mathbf{q}_z\mathbf{b}(\mathbf{r}),\quad \mathbf{r}'=\mathbf{q}_z\mathbf{r}$$

The reflected wave field, $\mathbf{e}'(\mathbf{r}')$, and the reflected current distribution, $\mathbf{j}'(\mathbf{r}')$, are thus seen to satisfy Maxwell's equations in the reflected medium, as expected.

Adjoint mapping of the adjoint system

Consider next the mapping of the adjoint quantities by means of the 'adjoint operator' $\overline{\mathbf{Q}}_y$. We premultiply the adjoint system (4.29)

$$\bar{\mathbf{L}}\bar{\mathbf{e}}(\mathbf{r}):=\left[i\omega\mathbf{K}^T\{\mathbf{b}(\mathbf{r}),\mathbf{r}\}-\mathbf{D}(\mathbf{r})\right]\,\bar{\mathbf{e}}(\mathbf{r})=-\bar{\mathbf{j}}(\mathbf{r}) \tag{4.71}$$

by $\overline{\mathbf{Q}}_y = \mathbf{Q}_y\,\bar{\mathbf{I}} = \overline{\mathbf{Q}}_y^{\,-1}$, and make use of $\bar{\mathbf{I}} = \bar{\mathbf{I}}^{-1}$, to obtain

$$\mathbf{Q}_y\left[\bar{\mathbf{I}}\,\bar{\mathbf{L}}\,\bar{\mathbf{I}}\right]\bar{\mathbf{I}}\,\bar{\mathbf{e}}(\mathbf{r}) = \mathbf{Q}_y\left[i\omega\,\bar{\mathbf{I}}\,\mathbf{K}^T\,\bar{\mathbf{I}} + \mathbf{D}(\mathbf{r})\right]\bar{\mathbf{I}}\,\bar{\mathbf{e}}(\mathbf{r}) = -\mathbf{Q}_y\,\bar{\mathbf{I}}\,\bar{\mathbf{j}}(\mathbf{r}) \qquad (4.72)$$

Making use of (3.125) and (4.37), and recalling that $\mathbf{Q}_y = \mathbf{Q}_y^{\,-1}$, we see that this is no more than a reflection mapping of the Lorentz-adjoint system,

$$\mathbf{Q}_y\left[i\omega\mathbf{K}^{(L)}(\mathbf{r}) + \mathbf{D}(\mathbf{r})\right]\mathbf{Q}_y \cdot \mathbf{Q}_y\,\mathbf{e}^{(L)}(\mathbf{r}) = -\mathbf{Q}_y\,\mathbf{j}^{(L)}(\mathbf{r}) \qquad (4.73)$$

To perform the reflection mapping it is convenient to revert to (4.72), and with the aid of (4.66) and (4.69) this becomes

$$\left[i\omega\bar{\mathbf{I}}\,\mathbf{K}(\mathbf{b}^c,\mathbf{r}')\,\bar{\mathbf{I}} + \mathbf{D}(\mathbf{r}')\right]\mathbf{e}^c(\mathbf{r}') = -\mathbf{j}^c(\mathbf{r}'), \quad \mathbf{r}' = \mathbf{q}_y\mathbf{r}, \quad \mathbf{b}^c := \mathcal{R}\mathcal{T}\mathbf{b}(\mathbf{r}) = \mathbf{b}(\mathbf{r}') \qquad (4.74)$$

with $\mathbf{e}^c(\mathbf{r}')$ and $\mathbf{j}^c(\mathbf{r}')$, the conjugate fields and currents, satisfying Maxwell's equations in the conjugate medium

$$\mathbf{K}^c(\mathbf{r}') \equiv \bar{\mathbf{I}}\,\mathbf{K}(\mathbf{b}^c,\mathbf{r}')\,\bar{\mathbf{I}} = \mathbf{Q}_y\mathbf{K}^{(L)}(\mathbf{r})\mathbf{Q}_y \qquad (4.75)$$

Comparison with (4.73), together with (4.34), gives

$$\mathbf{e}^c(\mathbf{r}') = \mathbf{Q}_y\mathbf{e}^{(L)}(\mathbf{r}) = \overline{\mathbf{Q}}_y\bar{\mathbf{e}}(\mathbf{r}), \qquad \mathbf{j}^c(\mathbf{r}') = \mathbf{Q}_y\mathbf{j}^{(L)}(\mathbf{r}) = \overline{\mathbf{Q}}_y\bar{\mathbf{j}}(\mathbf{r}) \qquad (4.76)$$

We draw attention to our use of $\bar{\mathbf{I}}\,\mathbf{K}(\mathbf{b}^c,\mathbf{r}')\,\bar{\mathbf{I}}$ to define the conjugate medium, rather than $\mathbf{K}(\mathbf{b}^c,\mathbf{r}')$. The two tensors are identical as far as gyrotropic or other anisotropic media are concerned. They will differ however for bianisotropic media, in the same way that the adjoint and the Lorentz-adjoint media, discussed in Sec. 3.4.1, differed from one another. This particular formulation will later be seen to be consistent with our definition of the conjugate medium as derived from a reflection-cum-time-reversal transformation.

Adjoint mapping of the Maxwell system

We finally apply the adjoint operator $\overline{\mathbf{Q}}_y$ to the given (Maxwell) system, rather than to the adjoint system,

$$\left[\overline{\mathbf{Q}}_y\,\mathbf{L}\,\overline{\mathbf{Q}}_y\right]\overline{\mathbf{Q}}_y\,\mathbf{e}(\mathbf{r}) \equiv \overline{\mathbf{Q}}_y\left[i\omega\mathbf{K}(\mathbf{b},\mathbf{r}) + \mathbf{D}(\mathbf{r})\right]\overline{\mathbf{Q}}_y \cdot \overline{\mathbf{Q}}_y\,\mathbf{e}(\mathbf{r}) = -\overline{\mathbf{Q}}_y\mathbf{j}(\mathbf{r})$$

and obtain with the help of (4.66) and (4.69)

$$\left[i\omega\,\bar{\mathbf{I}}\,\mathbf{K}^T(\mathbf{b}^c,\mathbf{r}')\,\bar{\mathbf{I}} - \mathbf{D}(\mathbf{r}')\right]\bar{\mathbf{e}}'(\mathbf{r}') = -\bar{\mathbf{j}}'(\mathbf{r}') \qquad (4.77)$$

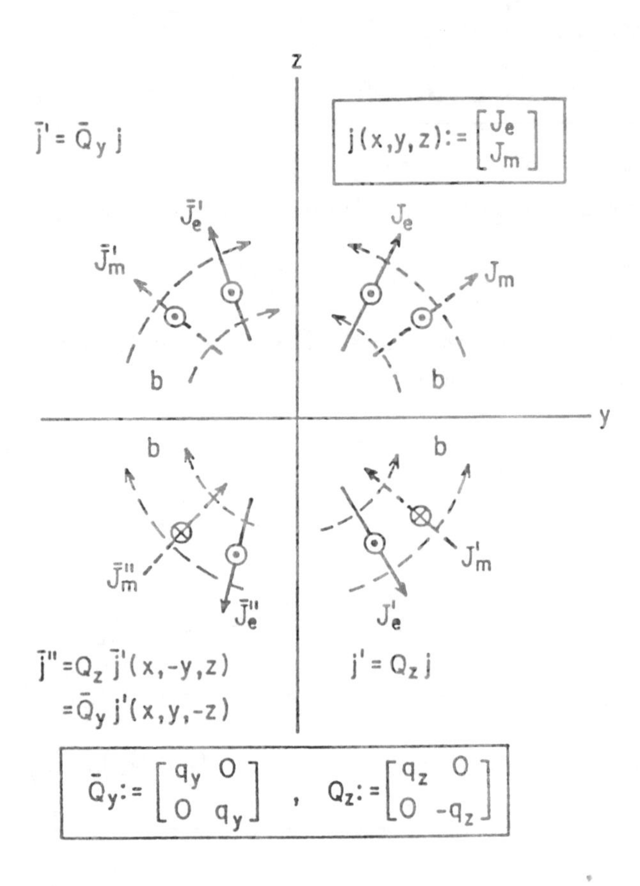

Figure 4.3: Mapping of electric and magnetic currents.
Each current is shown with parallel and perpendicular components with respect to the $y-z$ plane. Currents are mapped from one quadrant to another by means of $\mathbf{Q}_i$ or $\overline{\mathbf{Q}}_i$. If a 'quadrant' includes all of space, then 'another quadrant' will include all of space too.

which is just the adjoint equation for the conjugate medium. We may thus identify $\bar{\mathrm{e}}'$ and $\bar{\mathrm{j}}'$ with the adjoint conjugate quantities, $\bar{\mathrm{e}}^c$ and $\bar{\mathrm{j}}^c$, and remembering that $\overline{\mathbf{Q}}_i = \bar{\mathsf{I}}\,\mathbf{Q}_i$ we find, with (4.67),

$$\bar{\mathrm{e}}'(\mathbf{r}') \equiv \bar{\mathrm{e}}^c(\mathbf{r}') = \overline{\mathbf{Q}}_y\, \mathrm{e}(\mathbf{r}) = \bar{\mathsf{I}}\, \mathrm{e}'(\mathbf{r}'), \quad \bar{\mathrm{j}}'(\mathbf{r}') \equiv \bar{\mathrm{j}}^c(\mathbf{r}') = \overline{\mathbf{Q}}_y\, \mathrm{j}(\mathbf{r}) = \bar{\mathsf{I}}\, \mathrm{j}'(\mathbf{r}') \tag{4.78}$$

representing adjoint fields and currents in the transposed conjugate medium, $\bar{\mathsf{I}}\,\mathsf{K}^T(\mathbf{r}')\,\bar{\mathsf{I}} = \mathsf{K}\{-\mathrm{b}(\mathbf{r}'), \mathbf{r}'\}$.

Some of the mappings described in this section are illustrated in Fig. 4.3, in which a current distribution, $\mathrm{j}(\mathbf{r}) = [\mathbf{J}_e(\mathbf{r}), \mathbf{J}_m(\mathbf{r})]$, in the first quadrant of the model magnetoplasma, is mapped into the other three quadrants.

4.4.2 Notation and some concepts summarized and systematized

It will be convenient to pause at this stage, in order to summarize and systematize some of the concepts and the notation we are using. For the sake of completeness we shall anticipate some results that will be derived in later chapters.

- $\mathbf{e}(\mathbf{r})$ and $\mathbf{j}(\mathbf{r})$ are wave fields and currents that obey Maxwell's equations in a given medium characterized by a constitutive tensor $\mathsf{K}(\mathbf{r})$ (4.28).

- $\mathbf{e}'(\mathbf{r}')$, $\mathbf{j}'(\mathbf{r}')$ and $\mathsf{K}'(\mathbf{r}')$ are derived by mapping (reflecting, rotating or inverting) the above wave fields, currents and constitutive tensor from a region V $(\mathbf{r} \in V)$ into a region V' $(\mathbf{r}' \in V')$ in which they too satisfy Maxwell's equations. (The regions V and V' may overlap or be distinct.) In the case of reflection transformations, cf. (4.65), (4.67), (4.68) and (4.70)
$\mathsf{K}'(\mathbf{r}')=\mathsf{Q}_i\mathsf{K}(\mathbf{r})\mathsf{Q}_i$, $\mathsf{D}(\mathbf{r}')=\mathsf{Q}_i\mathsf{D}(\mathbf{r})\mathsf{Q}_i$, $\mathbf{e}'(\mathbf{r}')=\mathsf{Q}_i\mathbf{e}(\mathbf{r})$, $\mathbf{r}'=\mathbf{q}_i\mathbf{r}$ $(i=x, y$ or $z)$

- $\bar{\mathbf{e}}(\mathbf{r})$, $\bar{\mathbf{j}}(\mathbf{r})$ and $\overline{\mathsf{K}}(\mathbf{r}) \equiv \mathsf{K}^T(\mathbf{r})$, $(\mathbf{r} \in V)$, are the adjoint fields and currents, and the adjoint medium, which obey the adjoint Maxwell equations, derived from the given Maxwell equations by a change of sign of the differential operators and transposition of the constitutive tensor (4.29). If the currents are 'physical', then the fields will be 'unphysical', insofar as they obey a non-physical, adjoint set of equations. The transposed tensor $\mathsf{K}^T(\mathbf{r})$ may be physically realizable, but will differ in general from that in the original medium.

- $\mathbf{e}^{(L)}(\mathbf{r})$ and $\mathbf{j}^{(L)}(\mathbf{r})$ are the Lorentz-adjoint (Lorentz-reversed) fields and currents which obey the physical Maxwell equations in the Lorentz-adjoint medium, $\mathsf{K}^{(L)}(\mathbf{r}) \equiv \bar{\mathsf{I}}\,\mathsf{K}^T\,\bar{\mathsf{I}}$ (4.37). They are related to the adjoint fields and currents by the operator $\bar{\mathsf{I}}$, which reverses the direction of the Poynting vector (3.127) and (4.34): $\mathbf{e}^{(L)}(\mathbf{r}) = \bar{\mathsf{I}}\;\bar{\mathbf{e}}(\mathbf{r})$ and $\mathbf{j}^{(L)}(\mathbf{r}) = \bar{\mathsf{I}}\;\bar{\mathbf{j}}(\mathbf{r})$. All quantities—fields, currents and the constitutive tensor—are related to the corresponding quantities in the given medium by straightforward time reversal: $\mathbf{e}^{(L)}(\mathbf{r}) = \mathcal{T}\mathbf{e}(\mathbf{r})$, $\mathsf{K}^{(L)}(\mathbf{r}) = \mathcal{T}\,\mathsf{K}(\mathbf{r})\ldots$

- $\mathbf{e}^c(\mathbf{r}')$ and $\mathbf{j}^c(\mathbf{r}')$, the conjugate wave fields and currents, are derived by reflection mappings of the Lorentz-adjoint fields and currents (4.76),
$$\mathbf{e}^c(\mathbf{r}') = \mathsf{Q}_i\,\mathbf{e}^{(L)}(\mathbf{r}) = \mathsf{Q}_i\,\bar{\mathsf{I}}\,\bar{\mathbf{e}}(\mathbf{r}) = \overline{\mathsf{Q}}_i\bar{\mathbf{e}}(\mathbf{r}), \qquad \mathbf{j}^c(\mathbf{r}') = \overline{\mathsf{Q}}_i\,\bar{\mathbf{j}}(\mathbf{r}),$$
and obey the physical Maxwell equations in the reflected Lorentz-adjoint medium, $\mathsf{K}^c(\mathbf{r}') = \mathsf{Q}_i\,[\bar{\mathsf{I}}\;\mathsf{K}^T(\mathbf{r})\;\bar{\mathsf{I}}]\,\mathsf{Q}_i = \overline{\mathsf{Q}}_i\,\mathsf{K}^T(\mathbf{r})\,\overline{\mathsf{Q}}_i$, (4.37) and (4.75). If

the medium is not bianisotropic, the conjugate medium is then the reflected, transposed medium treated in previous chapters. All conjugate quantities—fields, currents and the constitutive tensor—are derived from the corresponding quantities in the given medium by a reflection-time-reversal transformation, $\mathrm{e}^c(\mathbf{r}')=\mathcal{RT}\,\mathrm{e}(\mathbf{r})$, $\mathsf{K}^c(\mathbf{r}')=\mathcal{RT}\,\mathsf{K}(\mathbf{r})$. The conjugate quantities are particularly useful when the medium is self-conjugate, i.e. when the medium has a plane of conjugation symmetry, so that we may then derive relations between currents and fields in the same medium. In Chap. 6 the concept of a conjugate medium, and of conjugate fields and currents, will be generalized to include any orthogonal mapping of the Lorentz-adjoint medium and of the Lorentz-adjoint fields and currents.

• $\bar{\mathrm{e}}'(\mathbf{r}')$ and $\bar{\mathrm{j}}'(\mathbf{r}')$ represent adjoint mappings of the given Maxwell fields and currents (4.78),

$$\bar{\mathrm{e}}'(\mathbf{r}')\equiv\bar{\mathrm{e}}^c(\mathbf{r}')=\overline{\mathbf{Q}}_y\,\mathrm{e}(\mathbf{r})=\bar{\mathsf{I}}\,\mathrm{e}'(\mathbf{r}') \qquad \bar{\mathrm{j}}'(\mathbf{r}')\equiv\bar{\mathrm{j}}^c(\mathbf{r}')=\overline{\mathbf{Q}}_y\mathrm{j}(\mathbf{r})=\bar{\mathsf{I}}\,\mathrm{j}'(\mathbf{r}')$$

They obey the adjoint Maxwell equations in the conjugate medium, $\widetilde{\mathsf{K}}^c(\mathbf{r}')=\mathsf{K}(-\mathrm{b}^c,\mathbf{r}')$ (4.75). The current $\bar{\mathrm{j}}'(\mathbf{r}')=\bar{\mathsf{I}}\,\mathrm{j}'(\mathbf{r}')=[\mathbf{J}'_{\mathrm{e}}(\mathbf{r}')\,,\,-\mathbf{J}'_{\mathrm{m}}(\mathbf{r}')]$ (4.78) is physically realizable, insofar as the constituent sources, $\mathbf{J}'_{\mathrm{e}}(\mathbf{r}')$ and $\mathbf{J}'_{\mathrm{m}}(\mathbf{r}')$, are independent. In the derivation of reciprocity relations in the next section, it will be useful to consider the *Maxwell fields* generated by the currents $\bar{\mathrm{j}}'$ in the conjugate medium.

4.4.3 Lorentz reciprocity in media with spatial conjugation symmetry

Suppose that $\mathrm{e}_1(\mathbf{r})$ and $\mathrm{j}_1(\mathbf{r})$ satisfy Maxwell's equations (4.28) in the given gyrotropic medium $\mathsf{K}(\mathrm{b},\mathbf{r})$, and specifically in the first quadrant in Fig. 4.2. Let $\mathrm{e}_2^c(\mathbf{r}')$ and $\mathrm{j}_2^c(\mathbf{r}')$ satisfy the equations in the conjugate medium which, as we recall, is a reflection mapping of the Lorentz-adjoint medium, (4.37) and (4.75),

$$\mathsf{K}^c(\mathrm{b}^c,\mathbf{r}')=\mathbf{Q}_y\left[\bar{\mathsf{I}}\,\mathsf{K}^T(\mathrm{b},\mathbf{r})\bar{\mathsf{I}}\right]\mathbf{Q}_y\equiv\mathbf{Q}_y\mathsf{K}^{(L)}(\mathbf{r})\mathbf{Q}_y \tag{4.79}$$

which in our case would be the given medium in the second quadrant of Fig. 4.2, with $\mathrm{b}^c\equiv\mathrm{b}(\mathbf{r}')=\mathbf{q}_y\mathrm{b}(\mathbf{r})$ and $\mathbf{r}'=\mathbf{q}_y\,\mathbf{r}$ (4.61):

$$\mathsf{L}(\mathbf{r})\,\mathrm{e}_1(\mathbf{r}):=\left[i\omega\mathsf{K}^T(\mathrm{b},\mathbf{r})+\mathsf{D}(\mathbf{r})\right]\mathrm{e}_1(\mathbf{r})=-\mathrm{j}_1(\mathbf{r}) \tag{4.80}$$

$$\mathsf{L}(\mathbf{r}')\,\mathrm{e}_2^c(\mathbf{r}'):=\left[i\omega\mathsf{K}^c(\mathrm{b}^c,\mathbf{r}')+\mathsf{D}(\mathbf{r})\right]\mathrm{e}_2^c(\mathbf{r}')=-\mathrm{j}_2^c(\mathbf{r}') \tag{4.81}$$

Now apply the reflection operator $\mathbf{Q}_y$ to (4.81), noting from (4.79) with $\mathbf{Q}_y=\mathbf{Q}_y^{-1}$ that $\mathsf{K}^{(L)}(\mathbf{r})=\mathbf{Q}_y\mathsf{K}^c(\mathbf{r}')\mathbf{Q}_y$, and hence

$$\mathbf{Q}_y\mathsf{L}(\mathbf{r}')\mathrm{e}_2^c(\mathbf{r}')=\left[\mathbf{Q}_y\mathsf{L}(\mathbf{r}')\mathbf{Q}_y\right]\mathbf{Q}_y\,\mathrm{e}_2^c(\mathbf{r}')=-\mathbf{Q}_y\,\mathrm{j}_2^c(\mathbf{r}')$$

or

$$[i\omega \mathsf{K}^{(L)}(\mathbf{r}) + \mathsf{D}(\mathbf{r})]\, \mathbf{e}_2^{c\prime}(\mathbf{r}) = -\mathbf{j}_2^{c\prime}(\mathbf{r}) \tag{4.82}$$

where, cf. (4.37),

$$\mathbf{e}_2^{c\prime}(\mathbf{r}) := \mathsf{Q}_y \mathbf{e}_2^c(\mathbf{r}') = \mathbf{e}^{(L)}(\mathbf{r}), \qquad \mathbf{j}_2^{c\prime}(\mathbf{r}) := \mathsf{Q}_y \mathbf{j}_2^c(\mathbf{r}') = \mathbf{j}^{(L)}(\mathbf{r}) \tag{4.83}$$

We have here identified the reflected conjugate fields and currents as the corresponding Lorentz-adjoint quantities, $\mathbf{e}^{(L)}(\mathbf{r})$ and $\mathbf{j}^{(L)}(\mathbf{r})$, but it will be convenient to retain the usage of (4.82) in order to emphasize that they are physical fields and currents, derived from the physical fields and currents in the conjugate medium by a reflection mapping. Now the Maxwell fields and currents in the given medium [in the present case $\mathbf{e}_1(\mathbf{r})$ and $\mathbf{j}_1(\mathbf{r})$], and the Maxwell fields in the Lorentz-adjoint medium [in the present case $\mathbf{e}_2^{c\prime} \equiv \mathbf{e}_2^{(L)}(\mathbf{r})$ and $\mathbf{j}_2^{c\prime} \equiv \mathbf{j}_2^{(L)}(\mathbf{r})$], are related in general by a Lorentz-reciprocity theorem (4.51)

$$\int (\tilde{\mathbf{e}}_1 \,\bar{\mathsf{I}}\, \mathbf{j}_2^{c\prime} - \tilde{\mathbf{e}}_2^{c\prime} \,\bar{\mathsf{I}}\, \mathbf{j}_1)\, d^3r = 0 \tag{4.84}$$

We now cast the the second term in the integrand, a scalar (invariant), into a different form, by mapping it with the aid of (4.14) and (4.67), into the conjugate space,

$$\tilde{\mathbf{e}}_2^{c\prime}(\mathbf{r}) \,\bar{\mathsf{I}}\, \mathbf{j}_1(\mathbf{r}) = \tilde{\mathbf{e}}_2^c(\mathbf{r}')\, \tilde{\mathsf{Q}}_y \,\bar{\mathsf{I}}\, \mathbf{j}_1(\mathbf{r}) = \tilde{\mathbf{e}}_2^c(\mathbf{r}') \,\bar{\mathsf{I}}\, \mathsf{Q}_y \mathbf{j}_1(\mathbf{r}) = \tilde{\mathbf{e}}_2^c(\mathbf{r}') \,\bar{\mathsf{I}}\, \mathbf{j}_1'(\mathbf{r}') \tag{4.85}$$

so that (4.84) becomes finally, in mixed vector-matrix notation,

$$\int \mathbf{e}_1 \cdot \bar{\mathsf{I}}\, \mathbf{j}_2^{c\prime}\, d^3r = \int \mathbf{e}_2^c \cdot \bar{\mathsf{I}}\, \mathbf{j}_1'\, d^3r' \tag{4.86}$$

This reciprocity relation may be expressed in compact form as an equality between the two inner products of the given wave fields, $\mathbf{e}_1$ and $\mathbf{e}_2^c$, and the adjoint reflected sources,

$$\bar{\mathsf{I}}\, \mathbf{j}_2^{c\prime}(\mathbf{r}) \equiv \overline{\mathsf{Q}}_y \mathbf{j}_2^c(\mathbf{r}') \qquad \text{and} \qquad \bar{\mathsf{I}}\, \mathbf{j}_1'(\mathbf{r}') \equiv \overline{\mathsf{Q}}_y \mathbf{j}_1(\mathbf{r}),$$

$$\langle \mathbf{e}_1\,,\, \bar{\mathsf{I}}\, \mathbf{j}_2^{c\prime} \rangle = \langle \mathbf{e}_2^c\,,\, \bar{\mathsf{I}}\, \mathbf{j}_1{}' \rangle \tag{4.87}$$

In terms of the constituent wave fields and currents, (4.86) becomes

$$\int (\mathbf{E}_1 \cdot \mathbf{J}_{e2}^c{}' - \mathbf{H}_1 \cdot \mathbf{J}_{m2}^c{}')\, d^3r = \int (\mathbf{E}_2^c \cdot \mathbf{J}_{e1}' - \mathbf{H}_2^c \cdot \mathbf{J}_{m1}')\, d^3r' \tag{4.88}$$

Using terminology due to Rumsey [106], we could describe the result (4.87) by stating that the *reaction* of the source $\mathbf{j}_1$ (through its field $\mathbf{e}_1$) on the source $\mathbf{j}_2^{c\prime}$, equals the reaction of the source $\mathbf{j}_2^c$ (through its field $\mathbf{e}_2^c$) on the source $\mathbf{j}_1'$.

If the two sources, $\mathbf{j}_1$ and $\mathbf{j}_2^{c\prime}$, are denoted abstractly as a and b, and the reflections of these sources, $\mathbf{j}_1'(\mathbf{r}')=\mathbf{Q}_y\mathbf{j}_1(\mathbf{r})$ and $\mathbf{j}_2^c(\mathbf{r}')=\mathbf{Q}_y\mathbf{j}_2^{c\prime}(\mathbf{r})$, as a' and b', then (4.87) may be written in the form

$$\langle a\,,\,b\rangle = \langle b'\,,\,a'\rangle \tag{4.89}$$

i.e. the reaction of (the field of) the source a on the source b, equals the reaction of (the field of) the reflected source b' on the reflected source a' in the conjugate medium. (The fact that $\mathbf{j}_2^{c\prime}(\mathbf{r}) \rightarrow b$ has been taken as the 'primary' or 'given' source, and $\mathbf{j}_2^c(\mathbf{r}') \rightarrow b'$ as the secondary, reflected source, is just a matter of convenience). With (4.89) expressed in terms of the currents

$$\mathbf{j}_a := \mathbf{j}_1(\mathbf{r}), \qquad \mathbf{j}_b := \mathbf{j}_2^{c\prime}(\mathbf{r}), \qquad \mathbf{j}_{a'} := \mathbf{j}_1'(\mathbf{r}'), \qquad \mathbf{j}_{b'} := \mathbf{j}_2^c(\mathbf{r}') \tag{4.90}$$

the Lorentz reciprocity theorem, in terms of inner products, becomes

$$\langle \mathbf{e}_a\,,\,\bar{\mathbf{I}}\,\mathbf{j}_b\rangle = \langle \mathbf{e}_{b'}\,,\,\bar{\mathbf{I}}\,\mathbf{j}_{a'}\rangle \tag{4.91}$$

where $\mathbf{e}_a$ and $\mathbf{e}_{b'}$ are the (physical) Maxwell fields generated by $\mathbf{j}_a$ and $\mathbf{j}_{b'}$ respectively. These results are illustrated schematically in Fig. 4.4. Their physical significance will be discussed in Sec. 4.5.

Equivalence of given and reflected systems

In the discussion of the reflection mapping of the Maxwell system, Sec. 4.4.1, it was noted that the reflected wave field, $\mathbf{e}'(\mathbf{r}')$, and the reflected current distribution, $\mathbf{j}'(\mathbf{r}')$, satisfy Maxwell's equations (4.70) in the reflected medium. In the symmetric magnetoplasma model of Fig. 4.4, the current distributions, $\mathbf{j}_a(\mathbf{r})$ and $\mathbf{j}_b(\mathbf{r})$, and the field, $\mathbf{e}_a(\mathbf{r})$, are mapped by reflection into $\mathbf{j}_a'(\mathbf{r}')$, $\mathbf{j}_b'(\mathbf{r}')$ and $\mathbf{e}_a'(\mathbf{r}')$, where typically (4.70)

$$\mathbf{j}_a'(\mathbf{r}') = \mathbf{Q}_z\,\mathbf{j}_a(\mathbf{r}), \qquad \mathbf{e}_a'(\mathbf{r}') = \mathbf{Q}_z\,\mathbf{e}_a(\mathbf{r}), \qquad \mathbf{r}' = \mathbf{q}_z\,\mathbf{r}$$

With the aid of these transformations, and with $\mathbf{Q}_z = {\mathbf{Q}_z}^T = {\mathbf{Q}_z}^{-1}$, we find the reaction between the two sources in the given and reflected media,

$$\begin{aligned}
\langle a\,,\,b\rangle &= \langle \mathbf{e}_a\,,\,\bar{\mathbf{I}}\,\mathbf{j}_b\rangle = \int \mathbf{e}_a^T(\mathbf{r})\,\bar{\mathbf{I}}\,\mathbf{j}_b(\mathbf{r})\,d^3r \\
&= \int \mathbf{e}_a'(\mathbf{r}')\,{\mathbf{Q}_z}^T\cdot\mathbf{Q}_z\,\bar{\mathbf{I}}\,\mathbf{j}_b'(\mathbf{r}')\,d^3r \\
&\equiv \langle \mathbf{e}_a'\,,\,\bar{\mathbf{I}}\,\mathbf{j}_b'\rangle
\end{aligned}$$

which is just the reaction of the (field of the) reflected source a_r on b_r,

$$\langle a\,,\,b\rangle = \langle a_r\,,\,b_r\rangle \tag{4.92}$$

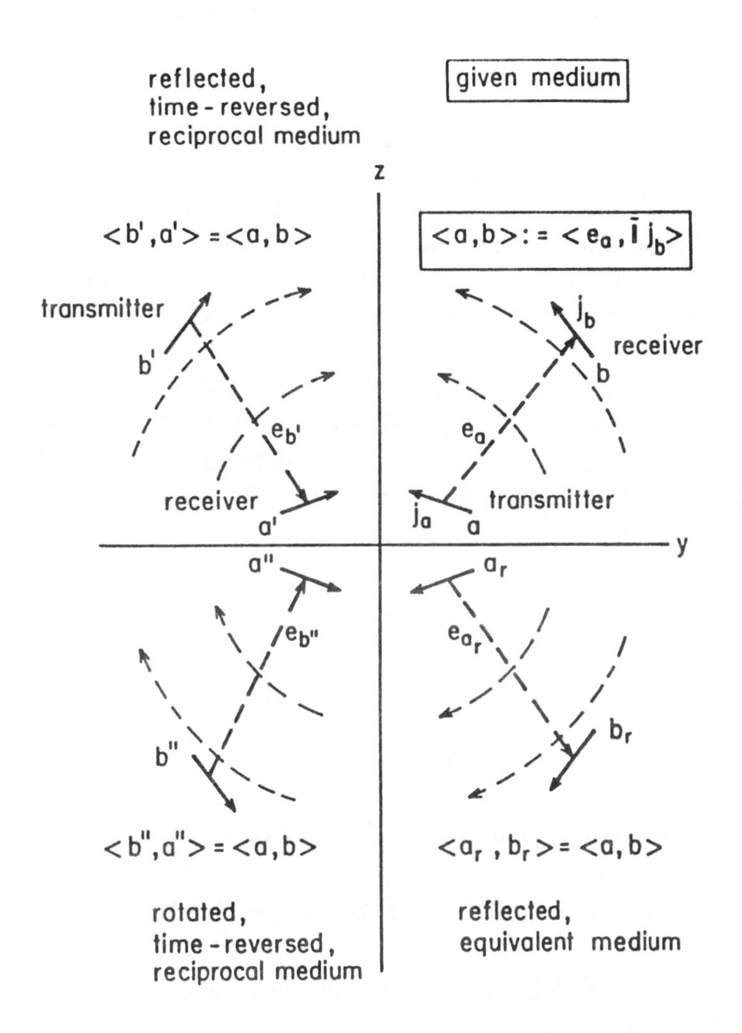

Figure 4.4: Reciprocity and equivalence between given and mapped currents and fields. $\mathbf{e}_{b'}$ and $\mathbf{e}_{b''}$ are the Maxwell fields generated by the sources b' and b''. Each current source, symbolized by a single arrow, is a 6-element column representing two independent current distributions of different types.

The given and reflected current distributions and fields are said to be *equivalent*. The *equivalence relation* (4.92) is illustrated in Fig. 4.4. We note also that the (Lorentz-)adjoint reflected system in quadrant 2 is equivalent to the adjoint-rotated (twice-reflected) system in quadrant 3. We consider next some of the consequences of the reciprocity and equivalence relations just derived.

4.5 Some consequences of Lorentz reciprocity

4.5.1 Media with symmetric constitutive tensors

In order to appreciate the features specific to Lorentz reciprocity in media possessing spatial symmetries, we shall first consider some of the consequences of the theorem in media with symmetric, but not bianisotropic, constitutive tensors, $\boldsymbol{\varepsilon}$ and $\boldsymbol{\mu}$, so that the Lorentz-adjoint media (4.37) and the given media coincide. We repeat the formalism of the previous section, although it may appear somewhat cumbersome in the present context, in order to highlight the common features.

Suppose that the current distribution $\mathbf{j}_a(\mathbf{r})$ and the field $\mathbf{e}_a(\mathbf{r})$ satisfy Maxwell's equations in the given medium, and that $\mathbf{j}_b(\mathbf{r})$ and $\mathbf{e}_b(\mathbf{r})$ satisfy the same (physical) equations in the Lorentz-adjoint medium, which in the present case coincides with the given medium. The Lorentz reciprocity theorem (4.91) takes the form

$$\int (\mathbf{e}_a \cdot \bar{\mathbf{I}}\, \mathbf{j}_b - \mathbf{e}_b \cdot \bar{\mathbf{I}}\, \mathbf{j}_a)\, d^3r = 0 \tag{4.93}$$

as in (4.86). This may be written, in terms of inner products, in the form

$$\langle \mathbf{e}_a \,,\, \bar{\mathbf{I}}\mathbf{j}_b \rangle = \langle \mathbf{e}_b \,,\, \bar{\mathbf{I}}\, \mathbf{j}_a \rangle \tag{4.94}$$

and in terms of the constituent wave fields and currents, as

$$\int (\mathbf{E}_a \cdot \mathbf{J}_{e,b} - \mathbf{H}_a \cdot \mathbf{J}_{m,b})\, d^3r = \int (\mathbf{E}_b \cdot \mathbf{J}_{e,a} - \mathbf{H}_b \cdot \mathbf{J}_{m,a})\, d^3r \tag{4.95}$$

With the sources, $\mathbf{j}_a$ and $\mathbf{j}_b$, denoted abstractly as a and b, (4.94) takes the form, cf. (4.89),

$$\langle a \,,\, b \rangle = \langle b \,,\, a \rangle \tag{4.96}$$

i.e. the reaction of (the field of) the source a on the source b, equals the reaction of (the field of) the source b on the source a.

4.5.2 Lorentz reciprocity for antennas

Suppose now, for simplicity, that the sources are electric currents, I_1 and I_2, flowing in isolated wires or antennas. The reciprocity theorem (4.95) reduces to

$$\int (\mathbf{E}_2 \cdot \mathbf{J}_{e1} - \mathbf{E}_1 \cdot \mathbf{J}_{e2})\, d^3r = 0 \tag{4.97}$$

Consider now an elementary length of wire $\mathbf{d}\boldsymbol{l}$, of cross-section $\Delta\boldsymbol{a}$ (whose normal is parallel to $\mathbf{d}\boldsymbol{l}$, which is taken in the direction of the current, I). Then

$$\mathbf{J}_e\, d^3r \to (\mathbf{J}_e \cdot \Delta a)\, dl = I\, dl$$

and (4.97) becomes

$$\int I_1\, \mathbf{E}_2 \cdot d\boldsymbol{l}_1 = \int I_2\, \mathbf{E}_1 \cdot d\boldsymbol{l}_2 \qquad (4.98)$$

where $\mathbf{d}\boldsymbol{l}_1$ and $\mathbf{d}\boldsymbol{l}_2$ are measured along the antenna wires, 1 and 2, respectively.

The simplest application is to *elementary dipole antennas* of lengths $\Delta\boldsymbol{l}_1$ and $\Delta\boldsymbol{l}_2$, situated at $\mathbf{r}_i$ (i=1 or 2), in which uniform currents I_1 or I_2 flow. The equivalent current density is then

$$\mathbf{J}_{e,i}(\mathbf{r}) = I_i\, \Delta\boldsymbol{l}_i\, \delta(\mathbf{r} - \mathbf{r}_i) \qquad (4.99)$$

in terms of the three-dimensional Dirac delta function. We could imagine the antennas to be terminated by two small spheres or discs on which charges $\pm q = \pm q_0 \exp(i\omega t)$ accumulate, from which it may be seen that our elementary antennas are equivalent to Hertzian dipoles of moments

$$\mathbf{p}_i := q\, \Delta\boldsymbol{l}_i \qquad \text{with} \qquad I\, \Delta\boldsymbol{l}_i = \frac{dq}{dt} \Delta\boldsymbol{l}_i = i\omega \mathbf{p}_i$$

If $I_1 = I_2 = I$, then (4.99) in (4.98) gives

$$I\, (\mathbf{E}_2 \cdot \Delta\boldsymbol{l}_1 - \mathbf{E}_1 \cdot \Delta\boldsymbol{l}_2) = 0$$

or

$$\mathcal{E}_{12} = \mathcal{E}_{21} \qquad (4.100)$$

which states that the EMF, $\mathcal{E}_{12}$, induced in $\Delta\boldsymbol{l}_1$ by the the current I in $\Delta\boldsymbol{l}_2$, equals the EMF, $\mathcal{E}_{21}$, induced in $\Delta\boldsymbol{l}_2$ by the same current I in $\Delta\boldsymbol{l}_1$; in other words, in a pair of elementary dipole antennas, one transmitting and one receiving, equal EMF's are induced by equal currents when the roles of transmitter and receiver are interchanged.

We consider next two representative types of antenna configurations: a straight wire transmitting (receiving) antenna with a pair of input (output) terminals, and an arbitrary closed-wire (loop) antenna, containing similarly a terminal pair (Fig. 4.5).

When the antenna is used as a transmitter, an input voltage $V(0)$ causes a current $I(s)$ to flow at a point s, measured along the antenna from the terminals at s=0. (An elementary length of arc along the antenna is denoted by the vector ds, which is tangential to the wire at that point. In the open

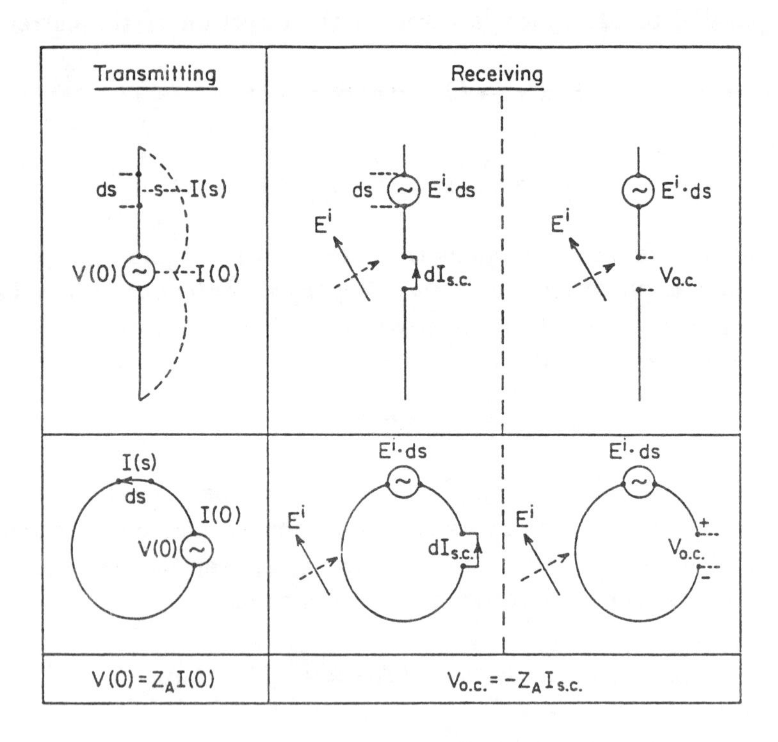

Figure 4.5: Voltages and currents in thin-wire transmitting and receiving antennas.

linear antenna, the values of s in the lower half are taken as negative.) The ratio of the applied voltage at the input terminals to the current at s, defines a transfer impedance, Z_{0s} [77, p.348],

$$Z_{0s} = V(0)/I(s) \tag{4.101}$$

and the current at the terminals is determined by the antenna input impedance, $Z_A \equiv Z_{00}$,

$$I(0) = V(0)/Z_A \tag{4.102}$$

When the antenna is used as a receiver, an incident wave field at s, $\mathbf{E}^i(s)$, induces an EMF, $\mathbf{E}^i(s) \cdot ds$, in an element ds of the antenna, and this causes a current $dI_{s.c.}$ to flow at the output terminals when they are short-circuited. It will be convenient in the discussion that follows to consider the element ds as a second 'port', which has been short-circuited in the transmitting antenna, and which has been connected to a zero-impedance voltage source, $\mathbf{E}^i(s) \cdot ds$, in the receiving antenna. The applied EMF and current are related by the transfer impedance, Z_{s0},

$$Z_{s0} = \frac{\mathbf{E}^i(s) \cdot ds}{dI_{s.c.}} \tag{4.103}$$

Now a fundamental reciprocity theorem concerning circuits composed of linear impedances [77, p.346-347], formulated originally by Rayleigh, states that

$$Z_{0s} = Z_{s0} \tag{4.104}$$

which gives, in the present context,

$$\frac{V(0)}{I(s)} = \frac{\mathbf{E}^i(s) \cdot ds}{dI_{s.c.}}$$

and the short-circuited current may be integrated to give

$$I_{s.c.} = \frac{1}{V(0)} \int I(s)\, \mathbf{E}^i(s) \cdot ds \tag{4.105}$$

Knowing the short-circuit current and the terminal impedance, we may apply Thévenin's equivalent network theorem [52, pp.46-48], [77, p.353], to obtain the open-circuit voltage at the terminals:

$$\begin{aligned} V_{o.c.} = -I_{s.c.}\, Z_A &= -\frac{Z_A}{V(0)} \int I(s)\, \mathbf{E}^i(s) \cdot ds \\ &= -\int \frac{I(s)}{I(0)} \mathbf{E}^i(s) \cdot ds \end{aligned} \tag{4.106}$$

which is a weighted sum of the induced elementary EMF's $\mathbf{E}^i \cdot ds$.

Consider now two arbitrary antennas, 1 and 2. We let one serve as a transmitter and the other as a receiver, and then reverse their roles. In the first instance a current distribution $I_1(s_1)$ in antenna 1 (with a value $I_1(0)$ at the terminals) generates a field $\mathbf{E}_1(s_2)$ at the second antenna. In the second instance a current distribution $I_2(s_2)$ in antenna 2, with $I_2(0)$ at the terminals, generates a field $\mathbf{E}_2(s_1)$ at 1.

The open-circuit voltages generated in each case at the receiving antenna terminals will be, by (4.106),

$$\begin{aligned} V_{o.c.}^{(2)} &= -\frac{1}{I_2(0)} \int I_2(s_2)\, \mathbf{E}_1(s_2) \cdot ds_2 \\ V_{o.c.}^{(1)} &= -\frac{1}{I_1(0)} \int I_1(s_1)\, \mathbf{E}_2(s_1) \cdot ds_1 \end{aligned} \tag{4.107}$$

Application of (4.98) to (4.107) then yields the result that the same input current in either antenna, $I_1(0)=I_2(0)$, induces the same open-circuit voltage in the other antenna:

$$V_{o.c.}^{(1)} = V_{o.c.}^{(2)} \quad \text{when} \quad I_2(0) = I_1(0) \tag{4.108}$$

It can similarly be shown by straightforward application of Thévenin's theorem that, if a voltage $V(0)$ applied to the terminals of antenna 1 produces a short-circuit current $I_{s.c.}$ in antenna 2, then an equal voltage applied to 2 will produce an equal short-circuit current in 1:

$$I_{s.c.}^{(2)} = I_{s.c.}^{(1)} \quad \text{when} \quad V_1(0) = V_2(0) \tag{4.109}$$

Eqs. (4.108) and (4.109) are simple and useful expressions of Lorentz reciprocity in isolated wire antennas.

Equality of directional patterns

A number of important receiving and transmitting characteristics of an antenna can be shown to be identical with the aid of the reciprocity theorem. Such relations are proved in many standard texts [28, Sec. 35],[77, Chap. 11],[43, Chap. 4], and will be mentioned only briefly here. However, attention will be drawn specifically to the directional properties of receiving and transmitting antennas, insofar as the relation between them will be modified in anisotropic media, and reciprocity between them will be restored only by some mapping transformation.

The directional pattern of a transmitting antenna indicates the relative strength of the radiated field at a large fixed distance in different directions in space. The directional pattern of a receiving antenna indicates the relative response of the antenna to an incident locally plane wave field of constant intensity from different directions. The directional pattern of an antenna in a homogeneous anisotropic medium, in contrast to the situation in free space, will depend on the orientation of the antenna in the medium, i.e. on its orientation with respect to the principal axes of the constitutive tensor.

The directional pattern of a transmitting antenna can be measured, in principle, by means of a short exploring dipole antenna moved about on, and tangential to, the surface of a large sphere centred at the antenna under test, and oriented in a direction parallel to the radiated electric field at the point of observation. (If the radiated field is elliptically polarized, two separate measurements would have to be made, with the dipole antenna oriented consecutively parallel and then perpendicular to the major axis of the polarization ellipse.) For a given voltage, $V(0)$, applied to the antenna being tested, the current $I_{s.c.}$ induced to flow in the short dipole antenna, will be a measure of the electric field radiated in that direction. If the same voltage, $V(0)$, is now applied to the dipole antenna, the same short-circuit current $I_{s.c.}$ will flow in the test antenna, in accord with (4.109), and this in turn will be a measure of its directional response as a receiving antenna. The equality of the short-circuit currents in the two cases thereby establishes the equality of the directional properties of the antenna as a receiver and as a transmitter.

The effective length

The effective length $\ell_{eff}^{(trans)}$ of a linear transmitting antenna, (assumed to lie parallel to the z-axis), is the length of an equivalent antenna in which a current $I(0)$ (the terminal current in the given antenna) flows at all points, z, along its length, and that radiates the same far field as the given antenna in a direction perpendicular to its length [77, p. 351]. Then

$$\ell_{eff}^{(trans)} = \frac{1}{I(0)} \int I(z)\, dz \tag{4.110}$$

where $I(z)$ is the actual current at z in the transmitting antenna.

The effective length of a receiving antenna, $\ell_{eff}^{(rec)}$, is such that an incident field $\mathbf{E}^i$, parallel to, and constant along the length of an equivalent linear antenna, induces an open-circuit voltage,

$$V_{o.c.} = -\mathbf{E}^i\, \ell_{eff}^{(rec)} \tag{4.111}$$

equal to that induced in the given antenna. But, as shown in (4.106), the open-circuit voltage induced in the receiving antenna by a constant electric field, $E^i(z) = E^i$=const., is

$$V_{o.c.} = -\frac{1}{I(0)} \int I(z)\, E^i(z)\, dz = -\frac{E^i}{I(0)} \int I(z)\, dz = -E^i\, \ell_{eff}^{(trans)} \tag{4.112}$$

using (4.110). Hence, comparing (4.111) and (4.112), we find

$$\ell_{eff}^{(rec)} = \ell_{eff}^{(trans)} \tag{4.113}$$

Directivity (gain) and effective area

The 'receiving cross section' or *effective area*, A_r, of a receiving antenna which is matched to its terminating load is defined [28, Sec. 37], [43, Sec. 4.4] as the ratio of the maximum power that can be absorbed by the receiver in its most favorable orientation to the incident power flux. (It is assumed that the polarization of the incoming radiation is that which gives maximum received power.) The directivity or *gain*, G, of a transmitting antenna is the ratio of the power flux radiated in the optimal direction to the mean radiated power flux averaged over all directions. For a given free-space wavelength, λ_0, it can be shown that the effective area of the antenna as a receiver is proportional to its gain as a transmitter [28, Sec. 37], [43, Sec. 4.4],

$$A_r = \frac{\lambda_0}{4\pi} G \tag{4.114}$$

4.5.3 Reciprocity relations for antennas in anisotropic media

Let a and b, in the first quadrant of Fig. 4.4, represent antennas in which electric currents, $I^{(a)}(s_a)$ and $I^{(b)}(s_b)$, flow. Then a', b' and a_r, b_r represent reflected antennas and current distributions in the reciprocal medium, quadrant 2, and in the equivalent medium, quadrant 4, respectively. a'' and b'' represent the rotated antennas and currents in the reciprocal medium in quadrant 3.

With the aid of (4.107) we may write down the induced open-circuit voltage, $V^{(b)}_{o.c.}$, induced in antenna b by the field $\mathbf{E}_a(s_b)$, which is generated by the current $I^{(a)}$ in antenna a:

$$V^{(b)}_{o.c.} = -\frac{1}{I^{(b)}(0)} \int I^{(b)}(s_b)\, \mathbf{E}_a(s_b) \cdot d\mathbf{s}_b \tag{4.115}$$

$I^{(b)}(s_b)$ is the current which flows in b when used as a transmitter, and the ratio $I^{(b)}(s_b)/I^{(b)}(0)$ serves as a weighting factor for the elementary induced voltages $\mathbf{E}_a(s_b) \cdot d\mathbf{s}_b$. Note that in the derivation of (4.115) we have assumed that the theorem of network reciprocity, and in particular the result, $Z_{0s}=Z_{s0}$ (4.104), is still valid when the antenna circuit is immersed in a non-reciprocal medium, such as a magnetoplasma.

The open-circuit voltage, $V^{(a')}_{o.c.}$, induced in a' by a current $I^{(b')}(s_{b'})$ in b', is given similarly by

$$V^{(a')}_{o.c.} = -\frac{1}{I^{(a')}(0)} \int I^{(a')}(s_{a'})\, \mathbf{E}_{b'}(s_{a'}) \cdot d\mathbf{s}_{a'} \tag{4.116}$$

Now the Lorentz reciprocity theorem (4.91),

$$\langle \mathbf{e}_a \,,\, \bar{\mathbf{I}}\,\mathbf{j}_b \rangle = \langle \mathbf{e}_{b'} \,,\, \bar{\mathbf{I}}\,\mathbf{j}_{a'} \rangle$$

when applied to electric currents in antennas, becomes

$$\int \left(\mathbf{E}_a \cdot \mathbf{J}_e^{(b)} - \mathbf{E}_{b'} \cdot \mathbf{J}_e^{(a')} \right) = 0 \tag{4.117}$$

or, as in (4.98),

$$\int I^{(b)}(s_b)\, \mathbf{E}_a(s_b) \cdot d\mathbf{s}_b = \int I^{(a')}(s_{a'}) \mathbf{E}_{b'}(s_{a'}) \cdot d\mathbf{s}_{a'} \tag{4.118}$$

Since the currents in a and in a', as well as in b and in b', have been assumed to be reflections of each other in the y=0 plane, i.e. $I^{(a)}(s_a)=I^{(a')}(s_{a'})$ etc., we may replace $I^{(b)}(0)$ by $I^{(b')}(0)$ in (4.115), and $I^{(a')}(0)$ by $I^{(a)}(0)$ in (4.116). Hence, comparing (4.115) with (4.118), we see that if an input current in a produces an open-circuit voltage in b, then the same input current in b' will produce the same open-circuit voltage in a'. This then is the expression of Lorentz reciprocity in an anisotropic (gyrotropic) medium, with the roles of receiver and transmitter interchanged in the mapped (reflected) antennas.

Equality of directional patterns

The results just obtained, when applied to the analysis of the directional properties of an antenna by means of a second, probing antenna, used alternatively as a transmitter and as a receiver (see the previous section), lead to the conclusion that the directional pattern of a transmitting antenna a in a gyrotropic medium, quadrant 1 in Fig. 4.6, is a reflection with respect to the y=0 plane of the directional pattern of the reflected antenna, a', in the conjugate (reciprocal) medium, quadrant 2, when used as a receiver. The directional characteristic of the reflected antenna, a_r, on the other hand, situated in quadrant 4 in Fig. 4.6, when used as a transmitter, is a reflection with respect to the z=0 plane of the pattern of the given transmitting antenna in quadrant 1.

Other reciprocal properties, mentioned in the previous section may be demonstrated analogously. The equivalent length of a linear transmitting antenna is equal to the equivalent length of the reflected receiving antenna in the conjugate medium. The receiving cross section or effective area of a receiving antenna equals a constant, $(\lambda_0^2/4\pi)$, times the directivity or gain of the reflected antenna in the conjugate medium, cf. (4.114).

These and other examples of reciprocity relations between sources and fields in non-reciprocal media, in which Lorentz-type reciprocity is recovered by orthogonal mappings of the sources and fields into a correspondingly mapped and time-reversed (conjugate) medium, have been discussed in a number of papers by Altman, Schatzberg and Suchy [8, 108, 9, 11, 121].

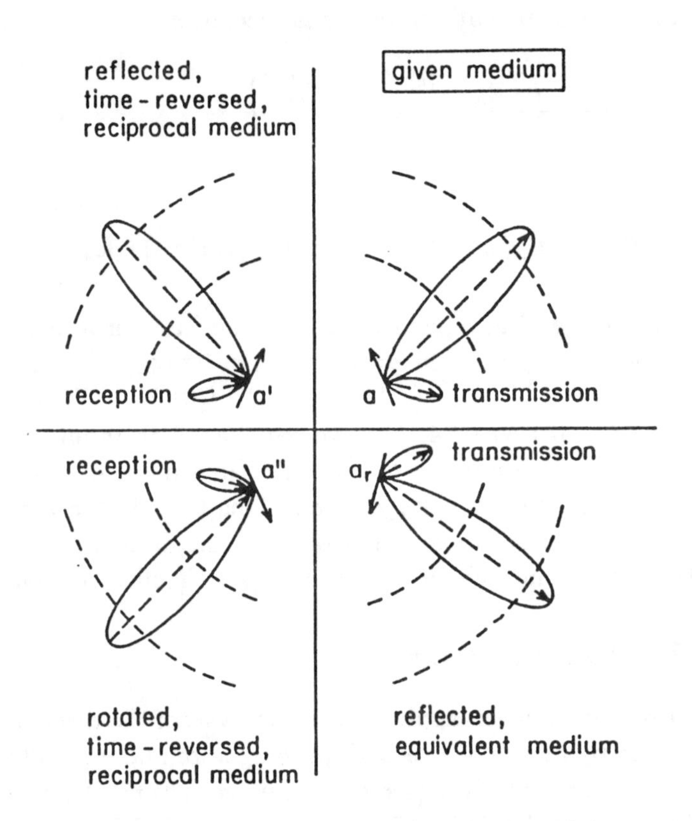

Figure 4.6: Directional patterns of given and mapped antennas.

4.5.4 Antennas in the magnetic meridian plane

Let us suppose the parameters of the medium (the plasma) and the external magnetic field to have conjugation symmetry with respect to the y=0 plane. Thus we could be considering the axisymmetric model discussed in the previous sections, in which any plane containing the symmetry axis (the magnetic dipole axis) could be taken as the conjugation symmetry plane. The dipole magnetic field would then be parallel to the plane at all points on it. We call such an arbitrarily chosen plane of reference, the 'magnetic meridian plane'. If this model is conceived as an idealization of the earth's ionosphere, then any vertical north-south plane could be considered to be the magnetic meridian plane, which would be the plane of conjugation symmetry. For propagation problems over distances which are short in comparison with the earth's radius, or the characteristic radius of curvature of the magnetic field lines, the medium could be regarded as plane stratified, and the local magnetic north-south vertical plane would be the magnetic meridian plane, as in Chaps. 2 and 3. We now consider the consequences of Lorentz reciprocity for a pair of antennas located on such a magnetic meridian plane, in a plane-stratified medium.

• If both antennas are parallel to the magnetic-meridian plane on which they are located, then each antenna is a reflection of itself with respect to this magnetic-meridian, conjugation plane. A reciprocity relation will then exist between the given pair of antennas and the reflected pair (Sec. 4.5.3), i.e. the same applied voltage in one antenna induces the same short-circuit current in the second, when the roles of transmitter and receiver are interchanged.

• If both antennas are perpendicular to the magnetic meridian plane on which they are located, then each antenna current is reversed in direction on reflection, i.e. the applied and induced currents are phase shifted by π, which is just the original situation with a half-period time shift. In other words, the two antennas are again reciprocal.

• If one antenna is parallel, and the other perpendicular, to the magnetic-meridian plane on which they are located, then reflection leaves one antenna current unchanged, but reverses the direction of the current in the other. The antenna pair then exhibit 'anti-reciprocity' [33, p.431], the received signal being reversed in sign when the roles of transmitter and receiver are interchanged.

The above reciprocity relations between antennas in the magnetic-meridian plane have been derived by Budden [32, Sec. 23.5], [33, p.578] by means of other considerations.

• As a final example, consider the above plane-stratified model, but with propagation between two antennas, each of which is parallel to the vertical magnetic-meridian north-south plane, but whose centres are located on an east-west line, equidistant from the magnetic-meridian plane, which lies in between. A given input current in the first antenna, let us say the antenna that lies to the east, produces some open-circuit voltage in the second antenna to the west. The reciprocal situation requires mirroring of the antennas with respect to the magnetic-meridian plane, which lies midway between them, and interchanging their roles, i.e. it is the second antenna which now transmits from east to west. The reciprocity theorem tells us that for east-west propagation, the same input current in the one antenna produces the same open-circuit voltage in the second, when their positions are interchanged. If the antennas do not lie in vertical, north-south planes, one must of course use mirrored antennas, which change their orientations, when their positions are interchanged [8].

Chapter 5

From scattering theorem to Lorentz reciprocity

5.1 Green's function in isotropic media

5.1.1 Statement of the problem

In Secs. 2.1.2 and 3.4.2 it was shown how Lorentz reciprocity in physical space led, via Kerns' formulation and its generaliztion to include eigenmodes within anisotropic media, to a scattering theorem in plane-stratified systems ('reciprocity in **k**-space'). In this chapter we consider the reverse derivation. Our starting point is the scattering theorem (2.112) for the plane-stratified medium (which for concreteness is taken to be a magnetoplasma), and from it we shall derive the Lorentz-reciprocity theorem, (4.89) and (4.91), relating currents and fields in a given medium and their mirror images, with respect to a magnetic meridian plane, in a conjugate medium. Since the plane-stratified magnetoplasma is self-conjugate, as we have seen in Chap. 2, all fields and currents, given and conjugate, are located in the same medium.

The basic idea, briefly outlined already in Sec. 2.1.4, is the following. In a plane-stratified medium, represented schematically in Fig. 2.4, with the z-axis normal to the stratification, the positive- and negative-going (with respect to z) propagation vectors $\mathbf{k}_{\pm}$ have transverse (to z) components $\mathbf{k}_t=\mathbf{k}_0(s_x, s_y)$, and the amplitudes $a_\alpha(\alpha=\pm1,\pm2)$ of the eigenmodes (cf. Sec. 2.3.2) are related by the 2×2 reflection and transmission matrices, $\mathbf{R}_{\pm}$ and $\mathbf{T}_{\pm}$ (2.104), which compose the 4×4 scattering matrix $\mathbf{S}$ (2.103). In a conjugate problem, in which the propagation vectors of the incident plane waves have transverse components $\mathbf{k}_t^c=\mathbf{k}_0(-s_x, s_y)$, the eigenmode amplitudes are a_α^c ($\alpha=\pm1,\pm2$), related by the reflection and transmission matrices $\mathbf{R}_{\pm}^c$ and $\mathbf{T}_{\pm}^c$which compose the scattering matrix $\mathbf{S}^c$. The scattering theorem relates the given and conjugate scattering matrices, (2.112) and (2.113),

$$\mathbf{S}^c = \tilde{\mathbf{S}}, \qquad \mathbf{R}^c_\pm = \tilde{\mathbf{R}}_\pm \qquad \mathbf{T}^c_\pm = \tilde{\mathbf{T}}_\mp$$

and these relations are illustrated in Fig. 2.4.

Now the propagation vectors $\mathbf{k}_\pm$ or $\mathbf{k}^c_\pm$ in Fig. 2.4 may be regarded as single components of an angular spectrum of plane waves or eigenmodes generated by a current element at A in the given medium, or at B' in the conjugate medium, and arriving at the corresponding current elements at B or at A' in the two respective media. The current elements at A' or B' are taken to be the (adjoint) reflection mappings (as in the passage from quadrant 1 to 2 in Fig. 4.3) of the current elements at A or B. In those sections of this chapter where only electric currents $\mathbf{J}_e$ are considered, it will not be necessary to distinguish between 'reflection' and 'adjoint-reflection' mappings of the current systems (see Sec. 4.4.1). The variation in amplitude of any one of the component eigenmodes in its passage from a level z' to z (the levels at which the 'transmitting' and 'receiving' points are located) is determined with the aid of the appropriate transmission or reflection matrices, $\mathbf{T}_\pm(\mathbf{k}_t, z, z')$, $\mathbf{R}_\pm(\mathbf{k}_t)$ or $\mathbf{T}^c_\pm(\mathbf{k}^c_t, z, z')$ and $\mathbf{R}^c_\pm(\mathbf{k}^c_t)$, which are presumed to be known, or which in principle can be calculated by numerical methods discussed in Chap. 1. The wave field at the receiving point, i.e. at the current element at B or A', would then be obtained by a Fourier synthesis of the angular spectrum of plane waves or eigenmodes reaching it. Integration over all elements in the current distributions a and b', or a' and b, of which the elements at A and B, or A' and B', form a part, would finally yield the 'reaction' $\langle a , b \rangle$ or $\langle a' , b' \rangle$ of the field of one source on the other in each of the two media. If the two reactions are shown to be equal, cf. (4.89) and (4.91), Lorentz reciprocity in physical space is thereby established.

To carry out this program it will be necessary initially to derive the angular spectrum of plane waves or eigenmodes generated by an arbitrary current distribution in free space, or in a plane-stratified gyrotropic medium. The angular spectrum, given explicitly as a function of the direction cosines, s_x and s_y, of the transverse propagation vector $\mathbf{k}_t$, which is a propagation constant (Snell's law), is then in a form permitting direct application of elements of the scattering matrix $\mathbf{S}(\mathbf{k}_t)$. The given and adjoint eigenmode spectrum leads to the construction of a dyadic (tensor) Green's function in transverse-**k** space which is applied to an elementary homogeneous slab in the stratification, and thence, with the aid of the elements of the scattering matrix, to the plane-stratified system as a whole. This approach, developed by Schatzberg and Altman [8,108] for determining the reaction of (the field of) one source on another, provides the framework for the analysis described in this chapter.

5.1.2 The plane-wave transverse-k spectrum of a current distribution in free space

We consider an electric current distribution, $\mathbf{J}(\mathbf{r}) \equiv \mathbf{J}_e(\mathbf{r})$, in free space outside a plane-stratified medium. The z-axis is normal to the stratification. We propose initially to determine the free-space wave fields, $\mathbf{E}(\mathbf{r})$ and $\mathbf{H}(\mathbf{r})$, as integrals over s_x and s_y, the direction cosines of $\mathbf{k}_t$. All currents and fields, $\mathbf{J}(\mathbf{r})$, $\mathbf{E}(\mathbf{r})$ and $\mathbf{H}(\mathbf{r})$, which are assumed to have $\exp(i\omega t)$ time dependence, with $k_0 := \omega/c$, are Fourier analyzed in k-space, so that typically

$$\begin{aligned} \mathbf{H}_k(k_x, k_y, k_z) &= \iiint \mathbf{H}(\mathbf{r}) \exp(i\mathbf{k} \cdot \mathbf{r})\, d^3r \\ &= \iiint \mathbf{H}(x, y, z) \exp[i(k_x x + k_y y + k_z z)]\, dx\, dy\, dz \end{aligned} \tag{5.1}$$

$$\mathbf{H}(\mathbf{r}) = \frac{1}{8\pi^2} \iiint \mathbf{H}_k(k_x, k_y, k_z) \exp[-i(k_x x + k_y y + k_z z)]\, dk_x\, dk_y\, dk_z \tag{5.2}$$

where $\mathbf{k} \equiv (k_x, k_y, k_z)$ is an arbitrary point in k-space, and all points in this space are included in the Fourier integration. Maxwell's equations in free space

$$\begin{aligned} \nabla \times \mathbf{E}(\mathbf{r}) &= -\frac{\partial \mathbf{B}(\mathbf{r})}{\partial t} = -i\omega\mu_0 \mathbf{H}(\mathbf{r}) \\ \nabla \times \mathbf{H}(\mathbf{r}) &= \frac{\partial \mathbf{D}(\mathbf{r})}{\partial t} + \mathbf{J}(\mathbf{r}) = i\omega\varepsilon_0 \mathbf{E}(\mathbf{r}) + \mathbf{J}(\mathbf{r}) \end{aligned} \tag{5.3}$$

when Fourier analyzed in k-space become, cf. (5.2), with $\mathbf{E}(\mathbf{r}) \rightarrow \mathbf{E}_k$, $\mathbf{H}(\mathbf{r}) \rightarrow \mathbf{H}_k$, $\nabla \rightarrow -i\mathbf{k}$,

$$\begin{aligned} \mathbf{k} \times \mathbf{E}_k &= k_0 Z_0 \mathbf{H}_k \\ \mathbf{k} \times \mathbf{H}_k &= -k_0 Y_0 \mathbf{E}_k + i\mathbf{J}_k \end{aligned} \tag{5.4}$$

where $Z_0 \equiv 1/Y_0 \equiv (\mu_0/\varepsilon_0)^{1/2}$. Eliminating $\mathbf{E}_k$ or $\mathbf{H}_k$ in (5.4), and using the relation $\mathbf{k} \cdot \mathbf{H}_k = 0$ (which follows from $\nabla \cdot \mathbf{H}(\mathbf{r}) = 0$), we obtain

$$\mathbf{H}_k = \frac{i\mathbf{k} \times \mathbf{J}_k}{k_0{}^2 - k^2} \tag{5.5}$$

$$\mathbf{E}_k = -\frac{iZ_0}{k_0} \left[\frac{\mathbf{k}(\mathbf{k} \cdot \mathbf{J}_k) - k_0{}^2 \mathbf{J}_k}{k_0{}^2 - k^2} \right] \tag{5.6}$$

Substitution of (5.5) in (5.2) yields

$$\mathbf{H}(\mathbf{r}) = \frac{1}{8\pi^3} \iiint \frac{\mathbf{k} \times \mathbf{J}_k}{k_0{}^2 - k_x{}^2 - k_y{}^2 - k_z{}^2} \exp[-i(k_x x + k_y y + k_z z)]\, dk_x\, dk_y\, dk_z \tag{5.7}$$

and substitution of (5.6) yields similarly

$$\mathbf{E}(\mathbf{r}) = -\frac{iZ_0}{8\pi^3}\iiint \frac{\left[\mathbf{k}\,(\frac{\mathbf{k}}{k_0}\cdot\mathbf{J}_k) - k_0\mathbf{J}_k\right]}{{k_0}^2 - {k_x}^2 - {k_y}^2 - {k_z}^2}\exp\left[-i(k_x x + k_y y + k_z z)\right] dk_x\, dk_y\, dk_z \tag{5.8}$$

Now we wish to get rid of the dependence on the longitudinal component of $\mathbf{k}$ by performing explicitly the integration over k_z. To be able to handle the poles of k_z in a physically meaningful way, we revert to a physically more transparent formulation by substituting for $\mathbf{J}_k$ its inverse Fourier transform, cf. (5.1). Thus (5.7) becomes

$$\mathbf{H}(\mathbf{r}) = \frac{i}{8\pi^3}\iiiint\!\!\iint \frac{\mathbf{k}\times\mathbf{J}(\mathbf{r}')}{{k_0}^2 - {k_x}^2 - {k_y}^2 - {k_z}^2} \cdot\exp\left[-i\left\{k_x(x-x') + k_y(y-y') + k_z(z-z')\right\}\right] d^3r'\, dk_x\, dk_y\, dk_z \tag{5.9}$$

and similarly for $\mathbf{E}(\mathbf{r})$. (Note that in the present chapter primed coordinates, z' or $\mathbf{r}'$, will be used for the source location, unprimed coordinates for the observation point.)

It will be convenient to express all three components of $\mathbf{k}$ in terms of k_0:

$$\mathbf{k} = k_0(s_x, s_y, n) = (\mathbf{k}_t, k_0 n) \qquad \mathbf{k}_t := k_0(s_x, s_y), \;\; k_z := k_0 n \tag{5.10}$$

so that (5.9) takes the form

$$\mathbf{H}(\mathbf{r}) = \frac{ik_0}{8\pi^3}\iiiint\!\!\iint \frac{\mathbf{k}\times\mathbf{J}(\mathbf{r}')}{1 - {s_x}^2 - {s_y}^2 - n^2} \cdot\exp\left[-ik_0\left\{s_x(x-x') + s_y(y-y') + n(z-z')\right\}\right] d^3r'\, ds_x\, ds_y\, dn \tag{5.11}$$

We now keep s_x and s_y fixed, i.e. $\mathbf{k}_t = k_0(s_x, s_y) = \text{const}$, and integrate over all n. If ${s_x}^2 + {s_y}^2 > 1$ then there are poles on the imaginary axis at $n = \pm q \equiv \pm i({s_x}^2 + {s_y}^2 - 1)^{1/2}$, and we get evanescent waves decaying away from the source. If ${s_x}^2 + {s_y}^2 - 1 < 1$, the poles are located at $n = \pm q \equiv \pm i(1 - {s_x}^2 - {s_y}^2)^{1/2}$. To facilitate the computation, we close the integration path along the $\mathcal{Re}(n)$ axis by a semi-circle at infinity in the upper or lower complex-n plane, see Fig. 5.1, according as $z > z'$ or $z < z'$ respectively, so that the additional integration path will contribute nothing to the integral [112, Sec. 7.15]. We may now indent around the poles, as in Fig. 5.1, which is equivalent to supposing that the positive (negative) pole has a small negative (positive) imaginary part. When $z > z'$, i.e. when the integration path is closed in the negative imaginary half plane, the pole at $n = q \equiv (1 - {s_x}^2 - {s_y}^2)^{1/2}$ is captured; when $z < z'$, i.e. when the integration path is closed in the positive imaginary half plane,

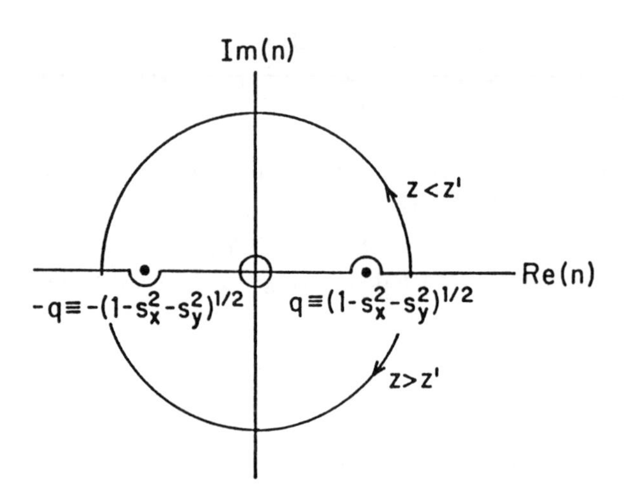

Figure 5.1: The complex-n plane integration contours.

the pole at $n = -q \equiv -(1 - s_x{}^2 - s_y{}^2)^{1/2}$ is captured. Hence, by the Cauchy residue theorem, (5.11) becomes

$$\mathbf{H}(\mathbf{r}) = (\mp 2\pi i)(\frac{-ik_0{}^2}{8\pi^3}) \iiiint\!\!\int \frac{(s_x, s_y, \pm q) \times \mathbf{J}(\mathbf{r}')}{\pm 2q} \cdot \exp\left[-ik_0 \left\{s_x(x - x') + s_y(y - y') \pm q(z - z')\right\}\right] d^3r' \, ds_x \, ds_y$$

in which the sign in $(\mp 2\pi i)$ is determined by the sense (clockwise or anticlockwise) in which the contour is traversed, and the sign of $(\pm q)$ in the denominator is determined by the sign of the pole, $\pm q$, which is captured. We have finally

$$\mathbf{H}(\mathbf{r}) = \frac{-k_0{}^2}{8\pi^2 q} \iiiint\!\!\int \hat{\mathbf{k}}_0^{\pm} \times \mathbf{J}(\mathbf{r}') \exp[-i\mathbf{k}_0^{\pm} \cdot (\mathbf{r} - \mathbf{r}')] \, d^3r' \, ds_x \, ds_y \tag{5.12}$$

where

$$\mathbf{k}_0^{\pm} := k_0(s_x, s_y, \pm q) \quad \text{and} \quad \hat{\mathbf{k}}_0^{\pm} := (s_x, s_y, \pm q) \tag{5.13}$$

and similarly for the electric field (5.8)

$$\mathbf{E}(\mathbf{r}) = \frac{k_0{}^2}{8\pi^2 Y_0 \, q} \iiiint\!\!\int \left[\hat{\mathbf{k}}_0^{\pm} \left\{\hat{\mathbf{k}}_0^{\pm} \cdot \mathbf{J}(\mathbf{r}')\right\} - \mathbf{J}(\mathbf{r}')\right] \cdot \exp[-i\mathbf{k}_0^{\pm} \cdot (\mathbf{r} - \mathbf{r}')] \, d^3r' \, ds_x \, ds_y \tag{5.14}$$

Eqs. (5.12) and (5.14) give the required plane-wave angular spectrum of an electric current distribution $\mathbf{J}(\mathbf{r}')$ in free space in terms of the direction cosines of $\mathbf{k}_t$ in transverse-$\mathbf{k}$ space. When the current distribution is confined to a plane (the 'transverse' plane), then (5.12) and (5.14) reduce to a form derived by Clemmow [38, Sec. 2.2.4].

Now there is more to (5.14) than meets the eye. For the present we note that

$$\hat{\mathbf{k}}_0^{\pm}\left\{\hat{\mathbf{k}}_0^{\pm}\cdot\mathbf{J}(\mathbf{r}')\right\}-\mathbf{J}(\mathbf{r}')=-\mathbf{J}_{\perp}^{\pm}(\mathbf{r}') \tag{5.15}$$

and the vector expression in the integrand is seen to be no more than the sign-reversed projection of the electric-currnt vector $\mathbf{J}(\mathbf{r}')$ on the plane perpendicular to the propagation vector $\mathbf{k}_0^{\pm}$, i.e. on the plane containing the field vectors $\mathbf{E}^{\pm}$ and $\mathbf{H}^{\pm}$, according as $z > z'$ or $z < z'$. Eq. (5.14) then simplifies to

$$\mathbf{E}(\mathbf{r})=\frac{-k_0^2}{8\pi^2 Y_0\, q}\iiiint\!\!\int \mathbf{J}_{\perp}^{\pm}(\mathbf{r}')\exp[-i\mathbf{k}_0^{\pm}\cdot(\mathbf{r}-\mathbf{r}')]\,d^3r'\,ds_x\,ds_y \tag{5.16}$$

5.1.3 The transverse-k eigenmode expansion in free space

Our next step is to replace $\mathbf{E}(\mathbf{r})$ and $\mathbf{J}_{\perp}^{\pm}(\mathbf{r}')$ in (5.16) by their Fourier transforms in $\mathbf{k}_t$-space. We note that

$$\iint \mathbf{J}_{\perp}^{\pm}(\mathbf{r}')\exp[ik_0(s_x x'+s_y y')]\,dx'\,dy'=\mathbf{J}_{\perp}^{\pm}(\mathbf{k}_t,z') \tag{5.17}$$

where $\mathbf{r}'=(x',y',z')$ and $\mathbf{k}_t=k_0(s_x,s_y)$, so that (5.16) with the aid of (5.17) becomes

$$\begin{aligned}\mathbf{E}(\mathbf{r}) &= \frac{-k_0^2}{8\pi^2 Y_0\, q}\iiint \mathbf{J}_{\perp}^{\pm}(\mathbf{r}')\exp\left[-ik_0\left\{s_x x+s_y y\pm q(z-z')\right\}\right]ds_x\,ds_y\,dz' \\ &= \frac{k_0^2}{4\pi^2}\iint \mathbf{E}(\mathbf{k}_t,z)\exp\left[-ik_0(s_x x+s_y y)\right]ds_x\,ds_y \end{aligned} \tag{5.18}$$

the second equality being just the Fourier transform of $\mathbf{E}(\mathbf{r})$ in $\mathbf{k}_t$-space. Equating the two integrands in (5.18), we get

$$\mathbf{E}(\mathbf{k}_t,z)=\frac{-1}{2Y_0\, q}\int \mathbf{J}_{\perp}^{\pm}(\mathbf{k}_t,z')\exp\left[\mp ik_0 q(z-z')\right]dz' \tag{5.19}$$

according as $z > z'$ or $z < z'$.

In order to facilitate comparison with results to be derived in anisotropic media, we would like to extract from (5.19) the eigenfields $\mathbf{e}_\alpha\equiv(\mathbf{E}_\alpha,\mathbf{H}_\alpha)$,

$\alpha=\pm1,\pm2$, generated by the current distribution $\mathbf{J}(\mathbf{k}_t,z')$. Consider the elliptically polarized eigenmodes $\mathbf{e}_\alpha$, whose electric-field components have the form

$$\mathbf{E}_\alpha = (1\,,\,\rho_\alpha\,,\,0)\,E_{\xi\alpha}\,, \qquad \alpha=\pm1,\pm2 \tag{5.20}$$

in the (ξ,η,ς) cartesian coordinate system (1.77) in which the ς-axis is along $\mathbf{k}_0^\pm$ and makes an angle θ with respect to the z-axis, with $\cos\theta=\pm q$ (5.13); the direction of the ξ-axis, which is in the plane of incidence, depends on the direction of the $\mathbf{k}_0^\pm$, and will be specified by the corresponding unit vectors $\hat{\xi}^\pm$; the direction of the η-axis is normal to the plane of incidence and does not depend on the direction of $\mathbf{k}_0^\pm$ in this plane. We recall that the direction of the z-axis, and through it the orientation of the 'plane of incidence', are determined by the stratified medium which bounds the free space in which our waves are propagating. To ensure eigenmode biorthogonality, the wave polarizations $\rho_\alpha{:=}E_{\eta\alpha}/E_{\xi\alpha}$, must satisfy the relation (1.79)

$$\rho_1\rho_2 = 1 = \rho_{-1}\rho_{-2} \tag{5.21}$$

In order to decompose the electric wave field $\mathbf{E}$ of a positive- or negative-going wave into the respective eigenmode wave fields $\mathbf{E}_{\pm1}$ and $\mathbf{E}_{\pm2}$, we construct a biorthogonal, rotating elliptic basis in (ξ,η) space, with base vectors and their adjoints given by

$$\hat{\epsilon}_\alpha = \frac{1}{(1-\rho_\alpha{}^2)^{1/2}}\left(\hat{\xi}^\pm + \rho_\alpha\hat{\eta}\right)\,, \quad \hat{\bar{\epsilon}}_\alpha = \frac{1}{(1-\rho_\alpha{}^2)^{1/2}}\left(\hat{\xi}^\pm - \rho_\alpha\hat{\eta}\right) \tag{5.22}$$

with $\alpha=\pm1,\pm2$. When $\rho_{\pm1}=i$ and $\rho_{\pm2}=-i$, these base vectors become rotating, circular base vectors [72, Sec. 7.2], and the adjoint base vectors reduce to the complex conjugates, $\hat{\bar{\epsilon}}_\alpha=\hat{\epsilon}_\alpha{}^*$. For linear modes, parallel and perpendicular to the plane of incidence, the given and the adjoint linear base modes coincide.

The base vectors (5.22) are biorthogonal,

$$\hat{\bar{\epsilon}}_\alpha\cdot\hat{\epsilon}_\beta = \delta_{\alpha\beta} \qquad (\alpha\,,\beta=1,2 \;\text{ or }\; -1,-2) \tag{5.23}$$

as can be verified by inspection, which permits the decomposition of any positive- or negative-going wave field, $\mathbf{E}(z){=}\mathbf{E}(\mathbf{k}_0^\pm,z)$, into the constituent eigenmodes. When $z=z'$, then

$$\begin{aligned}\mathbf{E}(\mathbf{k}_0^+,z') &= E_1(z')\hat{\epsilon}_1 + E_2(z')\hat{\epsilon}_2 = \left[\hat{\bar{\epsilon}}_1\cdot\mathbf{E}(z')\right]\hat{\epsilon}_1 + \left[\hat{\bar{\epsilon}}_2\cdot\mathbf{E}(z')\right]\hat{\epsilon}_2 \\ \mathbf{E}(\mathbf{k}_0^-,z') &= E_{-1}(z')\hat{\epsilon}_{-1} + E_{-2}(z')\hat{\epsilon}_{-2} = \left[\hat{\bar{\epsilon}}_{-1}\cdot\mathbf{E}(z')\right]\hat{\epsilon}_{-1} + \left[\hat{\bar{\epsilon}}_{-2}\cdot\mathbf{E}(z')\right]\hat{\epsilon}_{-2}\end{aligned} \tag{5.24}$$

Now the scalar product $\hat{\bar{\epsilon}}_\alpha\cdot\mathbf{E}(z')$, $\alpha=\pm1,\pm2$, represents the complex amplitude of the eigenmode α and carries the phase information of the mode (cf. Sec. 2.3.1). Thus at any level z, the complex modal amplitude becomes $\hat{\varepsilon}_\alpha \cdot \mathbf{E}(z') \exp[\mp i k_0 q(z-z')]$, according as $z > z'$ or $z < z'$, and so from (5.24) we obtain

$$\mathbf{E}(\mathbf{k}_0^\pm, z) = \sum_{\substack{\alpha=1,2\,(z>z') \\ \alpha=-1,-2\,(z<z')}} \left[\hat{\hat{\varepsilon}}_\alpha \cdot \mathbf{E}(\mathbf{k}_0^\pm, z')\right] \hat{\varepsilon}_\alpha \exp[\mp i k_0 q(z-z')] \tag{5.25}$$

The associated magnetic fields are given by, cf. (5.4),

$$\mathbf{H}(\mathbf{k}_0^\pm, z) = Y_0\, \hat{\mathbf{k}}_0^\pm \times \mathbf{E}(z), \quad \mathbf{H}_\alpha(z) = Y_0\, \hat{\mathbf{k}}_0^\pm \times \mathbf{E}_\alpha(z), \quad \alpha = \pm 1, \pm 2 \tag{5.26}$$

so that when only the electric wave vector $\mathbf{E}(\mathbf{k}_t,z)$ is specified, as in (5.19), the overall wave field, $\mathbf{e}(\mathbf{k}_t){:=}\{\mathbf{E}(\mathbf{k}_t),\mathbf{H}(\mathbf{k}_t)\}$, may be reconstructed from it with the aid of the normalized eigenvectors $\hat{\mathbf{e}}_\alpha$, (2.71) and (2.72). Thus, from (5.21) and (5.26),

$$\mathbf{e}(\mathbf{k}_t, z) = (2Y_0\, q)^{1/2} \left[\{\hat{\hat{\varepsilon}}_{\pm 1} \cdot \mathbf{E}(z')\}\hat{\mathbf{e}}_{\pm 1} + \{\hat{\hat{\varepsilon}}_{\pm 2} \cdot \mathbf{E}(z')\}\hat{\mathbf{e}}_{\pm 2}\right] \exp[\mp i k_0 q(z-z')] \tag{5.27}$$

where $\hat{\mathbf{e}}_\alpha$, the normalized eigenvectors in free space [with the refractive index $n_\alpha{=}1$, and the wave polarization ratios $\sigma_\alpha{=}0$, $\hat{\varsigma}\cdot\hat{\mathbf{z}}{=}q\,\mathrm{sgn}(\alpha)$ (5.13)], are given by

$$\hat{\mathbf{e}}_\alpha = (1\,, \rho_\alpha\,, 0\,;\, -Y_0\rho_\alpha\,, Y_0\,, 0)/[2Y_0\, q(1-{\rho_\alpha}^2)]^{1/2} \tag{5.28}$$

as in (2.72). Note that the term $\mathrm{sgn}(\alpha)$ in the denominator of (2.71) is cancelled when multiplied by the z-component of the vector $(0, 0, 1 - {\rho_\alpha}^2)$ in (ξ, η, ς) coordinates, which equals $(1-{\rho_\alpha}^2)\, q\, \mathrm{sgn}(\alpha)$.

The corresponding normalized adjoint eigenvectors $\hat{\hat{\mathbf{e}}}_\alpha$, (2.71) and (2.72), are obtained by means of $\rho_\alpha \to -\rho_\alpha$,

$$\hat{\hat{\mathbf{e}}}_\alpha = (1, -\rho_\alpha, 0\,;\, Y_0\rho_\alpha, Y_0, 0)/[2Y_0\, q(1-{\rho_\alpha}^2)]^{1/2} \tag{5.29}$$

It should be noted that the electric field components of $\hat{\mathbf{e}}_\alpha$ and $\hat{\hat{\mathbf{e}}}_\alpha$ are just the elliptic basis vectors $\hat{\varepsilon}_\alpha$ and $\hat{\hat{\varepsilon}}_\alpha$ (5.22) respectively, aside from a normalizing factor $(2Y_0\, q)^{1/2}$, which appears in (5.27).

The overall wave field $\mathbf{e}(\mathbf{k}_t,z)$ generated by the current distribution $\mathbf{J}(\mathbf{k}_t,z)$ (5.19) may now be obtained with the aid of (5.27):

$$\mathbf{e}(\mathbf{k}_t, z) = \frac{-1}{(2Y_0\, q)^{1/2}} \sum_{\substack{\alpha=1,2\,(z>z') \\ \alpha=-1,-2\,(z<z')}} \int \left[\hat{\hat{\varepsilon}}_\alpha \cdot \mathbf{J}_\perp^\pm(z')\right] \hat{\mathbf{e}}_\alpha \exp[\mp i k_0 q(z-z')]\, dz' \tag{5.30}$$

As a final step, we recall that $\hat{\boldsymbol{\epsilon}}_\alpha$ (α=1,2) and $\mathbf{J}_\perp^+$ both lie in a plane perpendicular to $\mathbf{k}_0^+$, just as $\hat{\boldsymbol{\epsilon}}_\alpha$ (α =−1,−2) lie in a plane perpendicular to $\mathbf{k}_0^-$. Hence the scalar product $\hat{\boldsymbol{\epsilon}}_\alpha \cdot \mathbf{J}_\perp^\pm$ in (5.30) can be replaced in both cases by $\hat{\boldsymbol{\epsilon}}_\alpha \cdot \mathbf{J}(\mathbf{k}_t)$, where $\mathbf{J}(\mathbf{k}_t)$ is the overall Fourier-transformed current rather than its projection. Furthermore, if $\mathbf{J} \equiv \mathbf{J}_e$ is replaced formally by the 6-current vector $\mathbf{j} := (\mathbf{J}_e, \mathbf{J}_m)$ with $\mathbf{J}_m = 0$, then the unit vector $\hat{\boldsymbol{\epsilon}}_\alpha$ in the scalar product can be replaced by $(2Y_0 q)^{1/2} \hat{\mathbf{e}}_\alpha$ [note the different normalizations of $\hat{\boldsymbol{\epsilon}}_\alpha$ and $\hat{\mathbf{e}}_\alpha$ in (5.22) and (5.29)]:

$$\hat{\boldsymbol{\epsilon}}_\alpha \cdot \mathbf{J}_\perp^\pm = \hat{\boldsymbol{\epsilon}}_\alpha \cdot \mathbf{J}(\mathbf{k}_t) = (2Y_0 q)^{1/2} \hat{\mathbf{e}}_\alpha \cdot \mathbf{j}(\mathbf{k}_t) \tag{5.31}$$

and we obtain finally, with (5.31) in (5.30),

$$\mathbf{e}(\mathbf{k}_t, z) = -\sum_{\substack{\alpha = 1,2\ (z > z') \\ \alpha = -1,-2\ (z < z')}} \int [\hat{\mathbf{e}}_\alpha \cdot \mathbf{j}(\mathbf{k}_t, z')]\ \hat{\mathbf{e}}_\alpha \exp[\mp i k_0 q (z - z')]\, dz' \tag{5.32}$$

The reader who is still with us at this point may well ask what we have achieved by converting a relatively straightforward expression such as (5.16), involving 'tangible' physical quantities such as electric fields and currents, into a form which, albeit having a certain formal symmetry and simplicity, involves abstract quantities such as normalized 6-component eigenfields and their adjoints. Well, we have in fact arrived at a formulation which is quite general, and will be shown to apply also to anisotropic media where the eigenmode expansion must of necessity replace the plane-wave spectrum. The detailed derivation for the free-space currents and fields should then, hopefully, provide some insight into the eigenmode expansion in the more general case.

Eq. (5.32) gives us the field $\mathbf{e}(\mathbf{k}_t, z)$ as the integrated sum of eigenmodes a_α, each of amplitude $-\hat{\mathbf{e}}_\alpha \cdot \mathbf{j}(z')\, dz'$, generated by the current elements $\mathbf{j}(\mathbf{k}_t, z')\, dz'$. It is interesting to note that a result of the same form is obtained in waveguides [72, eq. (8.140)], where the amplitude A_λ^+ of a positive-going eigenmode λ, generated by a current $\mathbf{J}(\mathbf{r}')$ is given by

$$A_\lambda^+ = -\int \hat{\mathbf{E}}_\lambda^- \cdot \mathbf{J}(\mathbf{r}')\, d^3 r'$$

(aside from a normalizing factor), where $\hat{\mathbf{E}}_\lambda^-$ is the normalized wavefield of the negative-going (adjoint) eigenmode.

5.1.4 From eigenmode expansion to Green's functions

The current distribution $\mathbf{j}(\mathbf{k}_t,z')$ and the field it generates, $\mathbf{e}(\mathbf{k}_t,z')$, are related by a 6×6 dyadic (tensor) Green's function $\mathsf{G}(\mathbf{k}_t, z, z')$:

$$\mathbf{e}(\mathbf{k}_t, z) = \int \mathsf{G}(\mathbf{k}_t, z, z')\mathbf{j}(\mathbf{k}_t, z')\, dz' \tag{5.33}$$

and comparison with (5.32) yields an expression for the Green's function in transverse-**k** space:

$$\mathsf{G}(\mathbf{k}_t, z, z') = - \sum_{\substack{\alpha = 1,2\ (z>z') \\ \alpha=-1,-2\ (z<z')}} \hat{\mathbf{e}}_\alpha \hat{\bar{\mathbf{e}}}_\alpha^T \exp[-ik_0\, q_\alpha(z-z')] \tag{5.34}$$

where $q_\alpha := q\,\mathrm{sgn}(\alpha)$.

The Green's function in physical space is defined by

$$\mathbf{e}(\mathbf{r}) = \int \mathsf{G}(\mathbf{r},\mathbf{r}')\mathbf{j}(\mathbf{r}')\, d^3r' \tag{5.35}$$

and may be derived from (5.33) by expressing $\mathbf{e}(\mathbf{r})$ and $\mathbf{j}(\mathbf{k}_t,z')$ in terms of their transforms:

$$\mathbf{e}(\mathbf{r}) = \frac{k_0^2}{4\pi^2} \iint \mathbf{e}(\mathbf{k}_t, z) \exp[-ik_0(s_x x + s_y y)]\, ds_x\, ds_y$$

$$\mathbf{j}(\mathbf{k}_t, z') = \iint \mathbf{j}(\mathbf{r}') \exp[ik_0(s_x x' + s_y y')]\, dx'\, dy'$$

These, in conjunction with (5.33), give

$$\mathbf{e}(\mathbf{r}) = \frac{k_0^2}{4\pi^2} \iint \mathsf{G}(\mathbf{k}_t, z, z')\mathbf{j}(\mathbf{r}') \exp\left[-ik_0 \left\{s_x(x-x') + s_y(y-y')\right\}\right] d^3r'\, d^2s \tag{5.36}$$

with $\iint \cdots ds_x\, ds_y \to \int \cdots d^2s$. Substitution of (5.34) in (5.36), and comparison with (5.35), give finally the Green's function in physical space,

$$\begin{aligned} \mathsf{G}(\mathbf{r},\mathbf{r}') &= \frac{k_0^2}{4\pi^2} \iint \mathsf{G}(\mathbf{k}_t, z, z') \exp\left[-ik_0 \left\{s_x(x-x') + s_y(y-y')\right\}\right] ds_x\, ds_y \\ &= - \sum_{\substack{\alpha = 1,2\ (z>z') \\ \alpha=-1,-2\ (z<z')}} \frac{k_0^2}{4\pi^2} \int \hat{\mathbf{e}}_\alpha\, \hat{\bar{\mathbf{e}}}_\alpha^T \exp[-i\mathbf{k}_\alpha \cdot (\mathbf{r}-\mathbf{r}')]\, d^2s \end{aligned} \tag{5.37}$$

with $\mathbf{k}_\alpha := k_0(s_x, s_y, q_\alpha)$. A similar result is given by Felsen and Marcuvitz [53, Sec. 1.4a, eq. (17)].

5.2 Green's function in anisotropic media

5.2.1 Green's function outside the source region

Before attempting to construct a Green's function in a multilayer anisotropic medium, we first address the problem of finding one in transverse-**k** space in an infinite homogeneous medium. An elementary layer or slab imbedded in the multilayer system may be considered as such a homogeneous medium, with the influence of the two bounding interfaces of each layer being treated separately by a multiple-reflection analysis or, equivalently, by wave-field matching at the interfaces.

In order to formulate the Green's function compactly, it will be convenient to include the phase factors $\exp(-ik_0\, q_\alpha z)$ and $\exp(ik_0\, q_\alpha z')$, cf. (5.34), in the given and adjoint normalized eigenvectors, so that we define

$$\hat{\mathbf{e}}_\alpha(z) := \hat{\mathbf{e}}^0_\alpha \exp(-ik_0\, q_\alpha z)\,, \qquad \hat{\bar{\mathbf{e}}}_\alpha(z') := \hat{\bar{\mathbf{e}}}^0_\alpha \exp(ik_0\, q_\alpha z') \tag{5.38}$$

with $\hat{\mathbf{e}}^0_\alpha \equiv \hat{\mathbf{e}}_\alpha(0)$ and $\hat{\bar{\mathbf{e}}}^0_\alpha \equiv \hat{\bar{\mathbf{e}}}_\alpha(0)$ corresponding to what until now were denoted $\hat{\mathbf{e}}_\alpha$ and $\hat{\bar{\mathbf{e}}}_\alpha$, (2.71) and (2.72). Then $\hat{\mathbf{e}}_\alpha(z)$ and $\hat{\bar{\mathbf{e}}}_\alpha(z)$ represent plane-wave ansätze for the given and adjoint eigenmodes, cf. (2.31) and (2.43), whose amplitudes satisfy the biorthogonality normalization. The eigenvectors $\hat{\mathbf{e}}_\alpha(z)$ and $\hat{\bar{\mathbf{e}}}_\alpha(z)$ themselves satisfy the normalization condition when both are at the same level z,

$$\hat{\bar{\mathbf{e}}}_\alpha^{\,T}(z)\, \mathbf{U}_z\, \hat{\mathbf{e}}_\beta(z) = \delta_{\alpha\beta}\, \mathrm{sgn}(\alpha) \tag{5.39}$$

and they obey the homogeneous Maxwell's equations, (2.30) and (2.32), and the adjoint equations, (2.34) and (2.44), respectively:

$$\mathbf{L}\, \hat{\mathbf{e}}_\alpha(z) := ik_0 \left[\mathbf{C} - \frac{i}{k_0}\mathbf{U}_z \frac{d}{dz}\right] \hat{\mathbf{e}}_\alpha(z) = ik_0 \left[\mathbf{C} - q_\alpha \mathbf{U}_z\right] \hat{\mathbf{e}}_\alpha(z) = 0 \tag{5.40}$$

$$\bar{\mathbf{L}}\, \hat{\bar{\mathbf{e}}}_\alpha(z) := ik_0 \left[\mathbf{C}^T + \frac{i}{k_0}\mathbf{U}_z \frac{d}{dz}\right] \hat{\bar{\mathbf{e}}}_\alpha(z) = ik_0 \left[\mathbf{C}^T - \bar{q}_\alpha \mathbf{U}_z\right] \hat{\bar{\mathbf{e}}}_\alpha(z) = 0 \tag{5.41}$$

with $\bar{q}_\alpha = q_\alpha$ (2.46). The amplitude-normalized eigenmodes are finally grouped together into modal matrices, $\mathbf{E}(z)$ and $\bar{\mathbf{E}}(z)$, by analogy with (2.100) and (2.101),

$$\mathbf{E}(z) := [\hat{\mathbf{e}}_1(z)\;\; \hat{\mathbf{e}}_2(z)\;\; \hat{\mathbf{e}}_{-1}(z)\;\; \hat{\mathbf{e}}_{-2}(z)]\,, \quad \bar{\mathbf{E}}(z) := [\hat{\bar{\mathbf{e}}}_1(z)\;\; \hat{\bar{\mathbf{e}}}_2(z)\;\; \hat{\bar{\mathbf{e}}}_{-1}(z)\;\; \hat{\bar{\mathbf{e}}}_{-2}(z)] \tag{5.42}$$

and eigenmode biorthogonality (5.39) may be expressed with the aid of (2.81) as

$$\tilde{\bar{\mathbf{E}}}(z)\, \mathbf{U}_z\, \mathbf{E}(z) = \bar{\mathbf{I}}^{(6)} \tag{5.43}$$

Let us return to the Maxwell system (2.29),

$$\mathsf{L}\,\mathbf{e}(\mathbf{k}_t, z) := ik_0 \left[\mathsf{C} - \frac{i}{k_0}\mathsf{U}_z\frac{d}{dz}\right] \mathbf{e}(\mathbf{k}_t, z) = -\mathbf{j}(\mathbf{k}_t, z) \tag{5.44}$$

If $\mathbf{e}(\mathbf{k}_t,z)$ is related to $\mathbf{j}(\mathbf{k}_t,z')$ by a dyadic Green's function $\mathsf{G}(\mathbf{k}_t, z, z')$, as in (5.33), then (5.44) yields

$$ik_0 \left[\mathsf{C} - \frac{i}{k_0}\mathsf{U}_z\frac{d}{dz}\right] \mathsf{G}(\mathbf{k}_t, z, z') = -\mathsf{I}^{(6)}\,\delta(z - z') \tag{5.45}$$

as can be verified by multiplying both sides from the right by $\mathbf{j}(\mathbf{k}_t,z')$ and integrating over z'. Our aim in this section is to find the functional form or structure of the 6×6 matrix $\mathsf{G}(\mathbf{k}_t, z, z')$ that will satisfy (5.45).

Now $\mathsf{G}(\mathbf{k}_t, z, z')$ should satisfy the following requirements:

- When $z \neq z'$, i.e. outside the source region, the columns of G must satisfy the homogeneous system (5.40), and hence should consist of linear combinations of the eigenvectors $\hat{\mathbf{e}}_\alpha(z)$ (5.38).

- If the medium is isotropic, then G should reduce to the form (5.34), as derived in Secs. 5.1.3 and 5.1.4.

- G depends on z and z' only through their difference $(z - z')$, [see the governing equation (5.45)], and we also demand that G vanish for $(z - z') \rightarrow \pm\infty$, at least for slightly absorbing media.

All these requirements are satisfied by the Green's function given in (5.34), with q_α now representing one of the roots of the Booker quartic (2.33):

$$\begin{aligned} \mathsf{G}(\mathbf{k}_t, z, z') &= -\sum_{\substack{\alpha = 1,2\,(z > z') \\ \alpha = -1,-2\,(z < z')}} \hat{\mathbf{e}}^0_\alpha \hat{\mathbf{\hat{e}}}^{0\,T}_\alpha \exp[-ik_0\, q_\alpha(z - z')] \\ &= -\sum_{\substack{\alpha = 1,2\,(z > z') \\ \alpha = -1,-2\,(z < z')}} \hat{\mathbf{e}}_\alpha(z)\, \hat{\mathbf{\hat{e}}}^T_\alpha(z') \end{aligned} \tag{5.46}$$

An equivalent result has been given by Felsen and Marcuvitz [53, Sec. 1.4, eqs. (14) and (16)].

Substituting (5.46) in (5.33) in order to relate fields to currents, we recover (5.32) with $\hat{\mathbf{e}}^0_\alpha$ and $\hat{\mathbf{\hat{e}}}^0_\alpha$ replacing $\hat{\mathbf{e}}_\alpha$ and $\hat{\mathbf{\hat{e}}}_\alpha$, and q_α replacing $\pm q$:

$$\begin{aligned} \mathbf{e}(\mathbf{k}_t, z) &= -\sum_{\substack{\alpha = 1,2\,(z > z') \\ \alpha = -1,-2\,(z < z')}} \int [\hat{\mathbf{\hat{e}}}^0_\alpha \cdot \mathbf{j}(\mathbf{k}_t, z')]\, \hat{\mathbf{e}}^0_\alpha \exp[-ik_0\, q_\alpha(z - z')]\, dz' \\ &= -\sum_\alpha a_\alpha(\mathbf{k}_t, z)\, \hat{\mathbf{e}}^0_\alpha(\mathbf{k}_t, z) \end{aligned} \tag{5.47}$$

in which we have decomposed the field $\mathbf{e}(\mathbf{k}_t,z)$ into its constituent eigenmodes (2.98). We may therefore consider the elementary amplitude $da_\alpha(\mathbf{k}_t, z')$, generated by the current element $\mathbf{j}(\mathbf{k}_t, z')dz'$ to be given by

$$da_\alpha(\mathbf{k}_t, z') = \hat{\tilde{\mathbf{e}}}^0_\alpha \cdot \mathbf{j}(\mathbf{k}_t, z')\, dz' \tag{5.48}$$

with the term $\exp[-ik_0\, q_\alpha(z - z')]$ giving the phase change in $da_\alpha(z')$ at any other level z.

Although (5.46) clearly satisfies (5.45) when $z \neq z'$, i.e. outside of the source plane, it remains to be seen to what extent the solution is complete when the source plane is included. In order to facilitate the mathematical manipulation of the discontinuities at $z{=}z'$, implied by the summation conditions in (5.46), we introduce the unit step function $h(z - z')$, defined by

$$h(z - z') := \begin{cases} 0, & z < z' \\ 1, & z > z' \end{cases} \tag{5.49}$$

which can be conceived as an integrated Dirac delta function,

$$h(z - z') = \int \delta(z - z')\, dz\,, \qquad \frac{d}{dz} h(z - z') = \delta(z - z') \tag{5.50}$$

We now incorporate the unit step function into a 4×4 unit step matrix,

$$\mathsf{H}(z - z') := \begin{bmatrix} h(z - z') & 0 & 0 & 0 \\ 0 & h(z - z') & 0 & 0 \\ 0 & 0 & 1 - h(z - z') & 0 \\ 0 & 0 & 0 & 1 - h(z - z') \end{bmatrix} \tag{5.51}$$

and by means of it we may express the Green's function (5.46), using the modal matrices (5.42), as

$$\mathsf{G}(\mathbf{k}_t, z, z') = -\mathsf{E}(\mathbf{k}_t, z)\, \mathsf{H}(z - z')\, \bar{\tilde{\mathsf{E}}}(\mathbf{k}_t, z') \tag{5.52}$$

We note, for later use, that in view of (5.50) we may write

$$\frac{d}{dz} \mathsf{H}(z - z') = \bar{\mathsf{I}}^{(4)}\, \delta(z - z') \tag{5.53}$$

with $\bar{\mathsf{I}}^{(4)}$ as in (2.81).

5.2.2 The need for a Green's function in the source region

The Green's function, (5.46) or (5.52), which relates a 6-component wave field to an arbitrary 6-component current distribution should span a 6-dimensional space. But, in effect, it is constructed from only four independent eigenvectors $\mathbf{e}_\alpha$, which span a 4-dimensional space. That something is lacking can be seen when (5.52) is substituted in (5.45). The differential operator $\mathbf{U}_z\, d/dz$, when operating on the unit step matrix $\mathsf{H}(z-z')$, see (5.53), produces four of the six delta functions of (5.45), (details are given in Sec 5.2.3), but since the third and sixth rows of $\mathbf{U}_z$ are null (2.25), the delta functions appearing on the right-hand side in the third and sixth rows are unaccounted for. The delta functions must therefore appear explicitly in the third and sixth rows of $\mathbf{G}(\mathbf{k}_t, z, z')$.

Now we could have defined an adjoint Green's function $\overline{\mathbf{G}}(\mathbf{k}_t, z, z')$ relating adjoint fields $\bar{\mathbf{e}}(\mathbf{r})$ and adjoint currents $\bar{\mathbf{j}}(\mathbf{r}')$, by analogy with (5.33), and governed by the adjoint Maxwell equations:

$$ik_0 \left[\mathbf{C}^T + \frac{i}{k_0}\mathbf{U}_z \frac{d}{dz}\right] \overline{\mathbf{G}}(\mathbf{k}_t, z, z') = -\mathbf{I}^{(6)}\, \delta(z-z') \tag{5.54}$$

By reasoning similar to the above we find, after interchanging z and z', that

$$\begin{aligned}\overline{\mathbf{G}}(\mathbf{k}_t, z', z) &= \overline{\mathbf{E}}(\mathbf{k}_t, z')\, \mathsf{H}(z'-z)\, \tilde{\mathbf{E}}(\mathbf{k}_t, z) \\ &= \mathbf{G}^T(\mathbf{k}_t, z, z')\end{aligned} \tag{5.55}$$

(remembering that $\mathsf{H}(z-z')$ is odd in its argument). To account for the delta functions on the right-hand side, in the third and sixth rows, we require that the delta functions appear explicitly in the third and sixth rows of $\overline{\mathbf{G}}$, i.e. in the third and sixth columns of $\mathbf{G}$. The upshot then is that the Green's matrix has delta functions symmetrically placed in the third and sixth rows and columns, to account for the source region. In Sec. 5.2.3 it is shown in fact that for media which are not bianisotropic, the Maxwell system (5.45) is satisfied, in the source region too, by adding a pair of delta functions to the (3,3) and (6,6) positions of the dyadic Green's function.

To put the present analysis into perspective we note that Tai [123]and Collin [41] had long drawn attention to the incompleteness of the eigenmode expansion of the Green's function in waveguides, and various approaches have since been adopted [123, 41, 103, 136, 97] to obtain the complete dyadic Green's function, including explicit delta-function terms in the source region. Weiglhofer [132, 133] has used rather simple algebraic matrix identities to invert the tensor differential operators governing the field equations, and has thereby obtained the complete dyadic Green's functions for anisotropic and chiral media. In general, the form of the Green's function, i.e. of the eigenmode expansion and the source term, is determined by the geometry or the boundary conditions of the problem considered, and the source term derived in the next section is specific to the plane-stratified anisotropic media considered in this chapter.

5.2.3 The complete Green's function in homogeneous anisotropic media

In order to determine the exact form of the complete Green's function, and to verify that it satisfies the governing differential equations (5.45), it will be useful to separate fields, currents, tensors and matrix operators into terms which are transverse (t) and normal (z) to the stratification. For instance,

$$\mathbf{e} \rightarrow \begin{bmatrix} \mathbf{e}_t \\ \mathbf{e}_z \end{bmatrix}, \quad \tilde{\mathbf{e}}_t = [E_x,\ E_y,\ H_x,\ H_y], \quad \tilde{\mathbf{e}}_z = [E_z,\ H_z], \quad \mathsf{E}(z) \rightarrow \begin{bmatrix} \mathsf{E}_t(z) \\ \mathsf{E}_z(z) \end{bmatrix} \tag{5.56}$$

so that (5.43) may be written in the form

$$\left[\tilde{\tilde{\mathsf{E}}}_t(z),\ \tilde{\tilde{\mathsf{E}}}_z(z)\right] \begin{bmatrix} \mathsf{U}_t & 0 \\ 0 & 0 \end{bmatrix} \begin{bmatrix} \mathsf{E}_t(z) \\ \mathsf{E}_z(z) \end{bmatrix} = \begin{bmatrix} \bar{\mathsf{I}}^{(4)} & 0 \\ 0 & 0 \end{bmatrix} \tag{5.57}$$

with $\mathsf{U}_t \equiv \mathsf{U}_z^{(4)}$, defined in (3.2). Hence, multiplying from the left with $\bar{\mathsf{I}}^{(4)} = \left(\bar{\mathsf{I}}^{(4)}\right)^{-1}$, we get

$$\begin{aligned} \left[\bar{\mathsf{I}}^{(4)} \tilde{\tilde{\mathsf{E}}}_t(z)\right] \left[\mathsf{U}_t\, \mathsf{E}_t(z)\right] &= \mathsf{I}^{(4)} \\ &= \mathsf{U}_t\, \mathsf{E}_t(z)\, \bar{\mathsf{I}}^{(4)} \tilde{\tilde{\mathsf{E}}}_t(z) \end{aligned} \tag{5.58}$$

in which we have interchanged the order of $\left[\bar{\mathsf{I}}^{(4)} \tilde{\tilde{\mathsf{E}}}_t(z)\right]$ and $\left[\mathsf{U}_t\, \mathsf{E}_t(z)\right]$, since they are reciprocal. We shall need this relation presently.

The homogeneous Maxwell system (5.40) and its adjoint (5.41) become

$$\left\{ \begin{bmatrix} \mathsf{C}_t & \mathsf{C}_1 \\ \mathsf{C}_2 & \mathsf{C}_z \end{bmatrix} - \frac{i}{k_0} \begin{bmatrix} \mathsf{U}_t & 0 \\ 0 & 0 \end{bmatrix} \frac{d}{dz} \right\} \begin{bmatrix} \mathsf{E}_t(z) \\ \mathsf{E}_z(z) \end{bmatrix} = 0 \tag{5.59}$$

and

$$\left\{ \begin{bmatrix} \tilde{\mathsf{C}}_t & \tilde{\mathsf{C}}_1 \\ \tilde{\mathsf{C}}_2 & \tilde{\mathsf{C}}_z \end{bmatrix} + \frac{i}{k_0} \begin{bmatrix} \mathsf{U}_t & 0 \\ 0 & 0 \end{bmatrix} \frac{d}{dz} \right\} \begin{bmatrix} \bar{\mathsf{E}}_t(z) \\ \bar{\mathsf{E}}_z(z) \end{bmatrix} = 0 \tag{5.60}$$

The elements of C and $\tilde{\mathsf{C}}$ (2.29) and (2.20), have been regrouped in (5.59) and (5.60) so that typically, for media which are not bianisotropic,

$$\mathsf{C}_z = \tilde{\mathsf{C}}_z = c \begin{bmatrix} \varepsilon_{33} & 0 \\ 0 & \mu_{33} \end{bmatrix} \tag{5.61}$$

The transposed z-component of (5.60) gives

$$\tilde{\tilde{\mathbf{E}}}_t(z)\,\mathbf{C}_1 + \tilde{\tilde{\mathbf{E}}}_z(z)\,\mathbf{C}_z = 0$$

or

$$\left(\tilde{\tilde{\mathbf{E}}}_t(z)\right)^{-1}\tilde{\tilde{\mathbf{E}}}_z(z) + \mathbf{C}_1\,\mathbf{C}_z^{-1} = 0 \tag{5.62}$$

In view of the discussion in Sec. 5.2.2, the proposed Green's function, after regrouping the matrix elements is, with (5.52),

$$\mathbf{G}(\mathbf{k}_t, z, z') = -\begin{bmatrix} \mathbf{E}_t(z) \\ \mathbf{E}_z(z) \end{bmatrix} \mathbf{H}(z-z')\,\left[\tilde{\tilde{\mathbf{E}}}_t(z'),\ \tilde{\tilde{\mathbf{E}}}_z(z')\right] + \mathbf{G}^0(\mathbf{k}_t, z, z') \tag{5.63}$$

where $\mathbf{G}^0$, the additional source-term dyadic, is assumed to have the form

$$\mathbf{G}^0 = \begin{bmatrix} 0 & 0 \\ 0 & \mathbf{G}_z^0 \end{bmatrix}, \qquad \mathbf{G}_z^0 = \frac{i}{k_0}\mathbf{A}\,\delta(z-z') \tag{5.64}$$

where the 2×2 matrix $\mathbf{A}$ is still to be determined. [That the form of the dyadic source term $\mathbf{G}^0$ is indeed given by (5.64) can be demonstrated by direct manipulation of the governing equations [107], without the need to make *a priori* assumptions.] We wish to confirm that $\mathbf{G}$, with the appropriate value of $\mathbf{A}$ inserted, satisfies (5.45) which becomes

$$ik_0\left\{\begin{bmatrix} \mathbf{C}_t & \mathbf{C}_1 \\ \mathbf{C}_2 & \mathbf{C}_z \end{bmatrix} - \frac{i}{k_0}\begin{bmatrix} \mathbf{U}_t & 0 \\ 0 & 0 \end{bmatrix}\frac{d}{dz}\right\}\mathbf{G}(\mathbf{k}_t, z, z') = -\mathbf{I}^{(6)}\,\delta(z-z') \tag{5.65}$$

Note that the differential operator d/dz in (5.65) operates successively on the matrices $\mathbf{E}(z) \equiv \{\mathbf{E}_t(z)\,,\,\mathbf{E}_z(z)\}$ and $\mathbf{H}(z-z')$ composing $\mathbf{G}$ (5.63). Thus, with the aid of (5.59) and (5.53), the left-hand side of (5.65) becomes

$$-\delta(z-z')\left\{\begin{bmatrix} \mathbf{U}_t & 0 \\ 0 & 0 \end{bmatrix}\begin{bmatrix} \mathbf{E}_t(z) \\ \mathbf{E}_z(z) \end{bmatrix}\bar{\mathbf{I}}^{(4)}\,\left[\tilde{\tilde{\mathbf{E}}}_t(z)\,,\tilde{\tilde{\mathbf{E}}}_z(z)\right] + \begin{bmatrix} 0 & \mathbf{C}_1\mathbf{A} \\ 0 & \mathbf{C}_z\mathbf{A} \end{bmatrix}\right\}$$

$$= -\delta(z-z')\begin{bmatrix} \mathbf{U}_t\mathbf{E}_t(z)\,\bar{\mathbf{I}}^{(4)}\tilde{\tilde{\mathbf{E}}}_t(z) & \mathbf{U}_t\mathbf{E}_t(z)\,\bar{\mathbf{I}}^{(4)}\tilde{\tilde{\mathbf{E}}}_z(z) + \mathbf{C}_1\mathbf{A} \\ 0 & \mathbf{C}_z\mathbf{A} \end{bmatrix}$$

$$= -\delta(z-z')\begin{bmatrix} \mathbf{I}^{(4)} & \left[\tilde{\tilde{\mathbf{E}}}_t(z)\right]^{-1}\tilde{\tilde{\mathbf{E}}}_z(z) + \mathbf{C}_1\mathbf{A} \\ 0 & \mathbf{C}_z\mathbf{A} \end{bmatrix} \tag{5.66}$$

the last step stemming from (5.58). With the aid of (5.62) the above expression is seen to be equal to

$$-\delta(z-z')\begin{bmatrix} \mathbf{I}^{(4)} & 0 \\ 0 & \mathbf{I}^{(2)} \end{bmatrix} \equiv -\delta(z-z')\,\mathbf{I}^{(6)}$$

provided that $\mathsf{A}=\mathsf{C}_z^{-1}$, and the Maxwell system (5.65) for the complete Green's function is thereby satisfied. For a medium that is not bianisotropic we have, with C_z given by (5.61),

$$\mathsf{A} = \mathsf{C}_z^{-1} = \frac{1}{c}\begin{bmatrix} \varepsilon_{33}^{-1} & 0 \\ 0 & \mu_{33}^{-1} \end{bmatrix}$$

and from (5.64) the source term in the Green's dyadic becomes

$$\mathbf{G}_z^0(\mathbf{k}_t, z, z') = \frac{i}{\omega}\begin{bmatrix} \varepsilon_{33}^{-1} & 0 \\ 0 & \mu_{33}^{-1} \end{bmatrix}\delta(z-z') \tag{5.67}$$

with $\omega = k_0 c$.

Just what is the significance of this source term? If we start with the inhomogeneous Maxwell system (2.24) and solve it for the longitudinal field components (E_z, H_z) in terms of the other field components and currents, the result can be shown to be, cf. Felsen and Marcuvitz [53, Sec. 8.2, eq.(3)],

$$\begin{bmatrix} E_z \\ H_z \end{bmatrix} = \begin{bmatrix} \dfrac{-\tilde{\varepsilon}_{zt}}{\varepsilon_{zz}} & \dfrac{\nabla_t \times \mathsf{I}^{(2)}}{i\omega\varepsilon_{zz}} \\ \dfrac{-\nabla_t \times \mathsf{I}^{(2)}}{i\omega\mu_{zz}} & \dfrac{-\tilde{\mu}_{zt}}{\mu_{zz}} \end{bmatrix}\begin{bmatrix} \mathbf{E}_t \\ \mathbf{H}_t \end{bmatrix} - \begin{bmatrix} \dfrac{J_{ez}}{i\omega\varepsilon_{zz}} \\ \dfrac{J_{mz}}{i\omega\mu_{zz}} \end{bmatrix} \tag{5.68}$$

where

$$\tilde{\varepsilon}_{zt} := [\varepsilon_{zx}\,,\, \varepsilon_{zy}], \qquad \tilde{\mu}_{zt} := [\mu_{zx}\,,\, \mu_{zy}], \qquad \nabla_t \times \mathsf{I}^{(2)} := \left[-\frac{\partial}{\partial y}, \frac{\partial}{\partial x}\right]$$

The eigenmode expansion of the Green's function, (5.46) or (5.52), with its built-in discontinuity at the current source, satisfies (5.68) when $(J_{ez}\,,\, J_{mz})=0$. Thus the first row of (5.68) with $J_{ez}=0$ becomes

$$\nabla_t \times \mathbf{H}_t - i\omega(\varepsilon_{zt}\cdot\mathbf{E}_t + \varepsilon_{zz}E_z) = (\nabla\times\mathbf{H})_z - \frac{\partial D_z}{\partial t} = 0 \tag{5.69}$$

which solves one of Maxwell's equations outside the source region, in which there are only outgoing wave fields. In the source region itself the additional contribution, $(-J_{ez}/i\omega\varepsilon_{zz}\,,\, -J_{mz}/i\omega\mu_{zz})$, to E_z and H_z must be taken into account to satisfy the inhomogeneous Maxwell system, and this additional point relation is expressed by means of the two delta functions we have found in the dyadic Green's function, (5.63) and (5.67).

These source terms in the Green's function will have no influence on the propagation analysis, and will therefore be ignored henceforth. Their importance from our point of view is that they have enabled us to resolve the problem of completeness of the Green's function.

5.3 Green's function in a multilayer medium

5.3.1 Transfer matrices in the multilayer medium

We have seen in Secs. 5.1.4 and 5.2.1, eqs. (5.33) and (5.48), how to determine the amplitude $da_\alpha(z')$ of eigenmodes radiated by a current element $\mathbf{j}(\mathbf{k}_t, z')\,dz'$ in an elementary layer of thickness dz'. To determine the amplitude $da_\beta(z)$ of the eigenmodes reaching an elementary layer at z (supposing $z > z'$), we could simply multiply the initial amplitude $da_\alpha(z')$ by the transmission matrix element $T_{\beta\alpha}$ (2.13), i.e. $da_\beta(z) = T_{\beta\alpha}(z, z')\,da_\alpha(z')$, on condition that z' and z were located in infinite homogeneous half-spaces bounding the multilayer system at both ends. If however z' and z are located in elementary layers embedded in the multilayer system, then two modifications must be made. First, the eigenmodes generated at z' are multiply reflected with accompanying mode conversion within the layer, so that the net positive-going wave amplitudes differ from the primary radiated amplitudes. Second, the eigenmodes arriving at z from z' will be multiply reflected with mode conversion within that elementary layer too, with consequent modification of amplitudes. Thus a multiple reflection analysis must be performed at both ends of the propagation path.

We have seen, cf. (5.48), that the 'primary' amplitudes da_α of the eigenmodes generated by the current element $\mathbf{j}(\mathbf{k}_t, z')\,dz'$ in an elementary layer of thickness dz', are given by

$$da_\alpha = -\hat{\bar{\mathbf{e}}}^0_\alpha(\mathbf{k}_t, z') \cdot \mathbf{j}(\mathbf{k}_t, z')\,dz', \qquad \alpha = \pm 1, \pm 2 \tag{5.70}$$

(We note that since the parameters of the medium are functions of z', so too is $\hat{\bar{\mathbf{e}}}^0_\alpha$ (5.38) which is constant for homogeneous media.) Grouping the four elementary amplitudes into a column matrix, as in (2.101), we may write (5.70) in matrix form,

$$da = -\widetilde{\overline{\mathsf{E}}}^0(\mathbf{k}_t, z')\,\mathbf{j}(\mathbf{k}_t, z')\,dz' \tag{5.71}$$

with the notation

$$da := \begin{bmatrix} da_+ \\ da_- \end{bmatrix}, \qquad da_\pm := \begin{bmatrix} da_{\pm 1} \\ da_{\pm 2} \end{bmatrix} \tag{5.72}$$

$$\widetilde{\mathsf{E}}^0(z) := [\hat{\mathbf{e}}^0_1(z) \;\; \hat{\mathbf{e}}^0_2(z) \;\; \hat{\mathbf{e}}^0_{-1}(z) \;\; \hat{\mathbf{e}}^0_{-2}(z)]^T \tag{5.73}$$

and $\widetilde{\overline{\mathsf{E}}}^0(z)$ similarly defined. Note that $\mathsf{E}^0(z)$ and $\overline{\mathsf{E}}^0(z)$ do not carry the phase factors $\exp(\mp i k_0 q_\alpha z)$, as do the corresponding modal matrices $\mathsf{E}(z)$ and $\overline{\mathsf{E}}(z)$ (5.42).

The eigenmodes $\mathbf{e}_\alpha$ undergo multiple reflections and multiple mode conversions within the elementary layer dz', and these may be summed as an infinite geometric series in the matrix products [4], or calculated directly as follows, to yield the net positive- or negative-going wave amplitudes $d\mathbf{A}_\pm$:

$$\begin{aligned} d\mathbf{A}_+ &= d\mathbf{a}_+ + \mathbf{R}_-\, d\mathbf{A}_- \\ d\mathbf{A}_- &= d\mathbf{a}_- + \mathbf{R}_+\, d\mathbf{A}_+ \end{aligned} \tag{5.74}$$

with $d\mathbf{A}_\pm$ defined in analogy with $d\mathbf{a}_\pm$ (5.72), and $\mathbf{R}_\pm \equiv \mathbf{R}_\pm(\mathbf{k}_t)$ is the reflection matrix (2.13). All quantities in (5.74) are measured at z'. We may solve (5.74) for $d\mathbf{A}_+$ and $d\mathbf{A}_-$:

$$[\mathbf{I}^{(2)} - \mathbf{R}_-\mathbf{R}_+]\, d\mathbf{A}_+ = d\mathbf{a}_+ + \mathbf{R}_-\, d\mathbf{a}_- = \left[\mathbf{I}^{(2)}, \mathbf{R}_-\right] \begin{bmatrix} d\mathbf{a}_+ \\ d\mathbf{a}_- \end{bmatrix}$$

or

$$d\mathbf{A}_+(z') = \mathbf{N}_+(z') \left[\mathbf{I}^{(2)}, \mathbf{R}_-(z')\right] d\mathbf{a}(z'), \qquad \mathbf{N}_+ := \left[\mathbf{I}^{(2)} - \mathbf{R}_-\mathbf{R}_+\right]^{-1} \tag{5.75}$$

and similarly

$$d\mathbf{A}_-(z') = \mathbf{N}_-(z') \left[\mathbf{R}_+(z'), \mathbf{I}^{(2)}\right] d\mathbf{a}(z'), \qquad \mathbf{N}_- := \left[\mathbf{I}^{(2)} - \mathbf{R}_+\mathbf{R}_-\right]^{-1} \tag{5.76}$$

(Had we used the alternative multiple-reflection procedure [4], as we presently do, the matrices $\mathbf{N}_\pm \equiv [\mathbf{I}^{(2)} - \mathbf{R}_\mp\mathbf{R}_\pm]^{-1}$ would have appeared as a consequence of summing an infinite geometric series of multiply-reflected waves.) We note, for later use, that

$$\mathbf{R}_\pm[\mathbf{I}^{(2)} - \mathbf{R}_\mp\mathbf{R}_\pm]^{-1} = [\mathbf{I}^{(2)} - \mathbf{R}_\pm\mathbf{R}_\mp]^{-1}\,\mathbf{R}_\pm$$

and hence $\mathbf{N}_+$ and $\mathbf{N}_-$, (5.75) and (5.76), are related by

$$\mathbf{N}_\mp\mathbf{R}_\pm = \mathbf{R}_\pm\mathbf{N}_\pm \tag{5.77}$$

The initial, or primary, amplitudes $d\mathbf{a}_\pm(z)$ of the eigenmodes reaching an elementary layer at z will be

$$d\mathbf{a}_\pm(z) = \mathbf{T}_\pm(\mathbf{k}_t, z, z')\, d\mathbf{A}_\pm(z'), \qquad z \gtrless z' \tag{5.78}$$

where $\mathbf{T}_\pm(z,z')$ is the transmission matrix (2.13) from z' to z, the plus/minus sign corresponding to $z \gtrless z'$. We now perform a multiple-reflection analysis

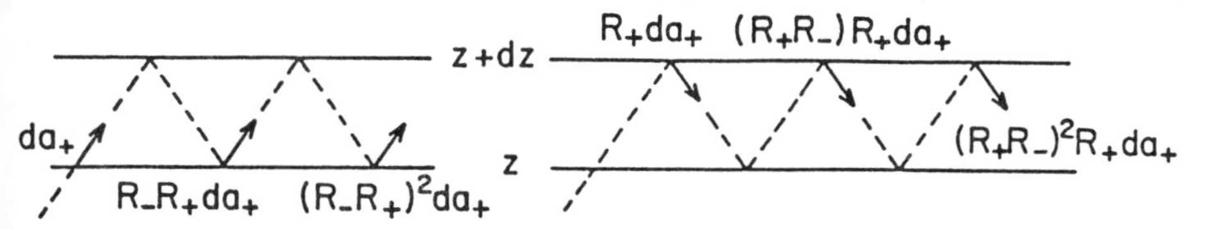

Figure 5.2: Multiple reflections in an elementary layer.

(see Fig. 5.2) in the layer dz. The resultant positive-going wave amplitude, with $da_+(z)$ given by (5.78), is

$$\begin{aligned} d\mathbf{A}_+(z) &= \left[\mathbf{I}^{(2)} + \mathbf{R}_-\mathbf{R}_+ + (\mathbf{R}_-\mathbf{R}_+)^2 + \cdots\right] da_+(z) \\ &= \left[\mathbf{I}^{(2)} - \mathbf{R}_-\mathbf{R}_+\right]^{-1} da_+ \equiv \mathbf{N}_+(z)\, da_+(z), \qquad z > z' \end{aligned} \tag{5.79}$$

in which we have summed the infinite geometric series. [As could be expected this is just the result given by (5.75), with z replacing z', when the 'primary' input wave is upgoing only, i.e. when $da_-=0$, $da=(da_+\,,0)$]. Summing all elementary waves reflected back from the upper interface we obtain

$$\begin{aligned} d\mathbf{A}_-(z) &= \left[\mathbf{I}^{(2)} + \mathbf{R}_+\mathbf{R}_- + (\mathbf{R}_+\mathbf{R}_-)^2 + \cdots\right] \mathbf{R}_+\, da_+ \\ &= \left[\mathbf{I}^{(2)} - \mathbf{R}_+\mathbf{R}_-\right]^{-1} \mathbf{R}_+\, da_+ = \mathbf{N}_-(z)\,\mathbf{R}_+(z)\, da_+(z), \quad z > z' \end{aligned} \tag{5.80}$$

which again is just (5.76), with $z \leftrightarrow z'$ and $da_-=0$.

Similarly, if $z < z'$ we get with $da_-(z)$ given by (5.78),

$$d\mathbf{A}_+(z) = \mathbf{N}_+(z)\,\mathbf{R}_-(z)\, da_-(z), \qquad z < z' \tag{5.81}$$

$$d\mathbf{A}_-(z) = \mathbf{N}_-(z)\, da_-(z), \qquad z < z' \tag{5.82}$$

which are also derivable from (5.75) and (5.76) with $z \leftrightarrow z'$ and $da_+=0$.

With the aid of (5.77), equations (5.79) – (5.82) may be written compactly in the form:

$$d\mathbf{A}(z) := \begin{bmatrix} d\mathbf{A}_+(z) \\ d\mathbf{A}_-(z) \end{bmatrix} = \begin{cases} \begin{bmatrix} \mathbf{I}^{(2)} \\ \mathbf{R}_+(z) \end{bmatrix} \mathbf{N}_+(z)\, da_+(z)\,, & z > z' \\[2ex] \begin{bmatrix} \mathbf{R}_-(z) \\ \mathbf{I}^{(2)} \end{bmatrix} \mathbf{N}_-(z)\, da_-(z)\,, & z < z' \end{cases} \tag{5.83}$$

Substituting (5.75) and (5.76) in (5.78), and the result in (5.83), we obtain finally the *transfer matrices*, $\mathsf{F}_{\pm}(\mathbf{k}_t, z, z')$, $z \gtrless z'$, relating primary modal amplitudes at z' to resultant amplitudes at z:

$$d\mathsf{A}(z) = \mathsf{F}_{\pm}(\mathbf{k}_t, z, z')\, da(z'), \qquad z \gtrless z' \tag{5.84}$$

with

$$\mathsf{F}_{+}(\mathbf{k}_t, z, z') = \begin{bmatrix} \mathsf{I}^{(2)} \\ \mathsf{R}_{+}(z) \end{bmatrix} \mathsf{N}_{+}(z)\, \mathsf{T}_{+}(z, z')\, \mathsf{N}_{+}(z') \left[\mathsf{I}^{(2)}\,,\ \mathsf{R}_{-}(z')\right], \qquad z > z' \tag{5.85}$$

$$\mathsf{F}_{-}(\mathbf{k}_t, z, z') = \begin{bmatrix} \mathsf{R}_{-}(z) \\ \mathsf{I}^{(2)} \end{bmatrix} \mathsf{N}_{-}(z)\, \mathsf{T}_{-}(z, z')\, \mathsf{N}_{-}(z') \left[\mathsf{R}_{+}(z')\,,\ \mathsf{I}^{(2)}\right], \qquad z < z' \tag{5.86}$$

in which $\mathsf{R}_{\pm}$, $\mathsf{T}_{\pm}$ and $\mathsf{N}_{\pm}$ are all functions of $\mathbf{k}_t$. We note that these transfer matrices are not directly comparable with the transfer matrices $\mathsf{P}(\mathbf{k}_t, z, z')$ defined in Sec. 3.3. The latter related four eigenmode amplitudes at one level to four at another level (3.102) in a source-free medium, and the analysis was based on the homogeneous Maxwell equations. In the present analysis on the other hand, the source of the eigenmodes at z is the current element in the layer at z' (5.70), and the transfer matrix F relates explicitly to the elementary modal amplitudes in that layer radiated by the current element.

5.3.2 From transfer matrix to Green's function

The elementary amplitudes $d\mathsf{A}(z)$ determined in (5.84) correspond to an elementary wave field, cf. (2.76), (5.72), (5.73) and (5.74),

$$d\mathrm{e}(\mathbf{k}_t, z) = \sum_{\alpha} d\mathsf{A}_{\alpha}(z)\, \hat{\mathrm{e}}^{0}_{\alpha}(\mathbf{k}_t, z) = \mathsf{E}^{0}(\mathbf{k}_t, z)\, d\mathsf{A} \tag{5.87}$$

Substitution of (5.84) in (5.87) gives

$$d\mathrm{e}(\mathbf{k}_t, z) = \mathsf{E}^{0}(\mathbf{k}_t, z)\, \mathsf{F}_{\pm}(\mathbf{k}_t, z, z')\, da(z'), \qquad z \gtrless z' \tag{5.88}$$

Now substitute the value of da given by (5.71) and integrate over all z' to obtain

$$\mathrm{e}(\mathbf{k}_t, z) = -\int \mathsf{E}^{0}(\mathbf{k}_t, z)\, \mathsf{F}_{\pm}(\mathbf{k}_t, z, z')\, \tilde{\tilde{\mathsf{E}}}^{0}(\mathbf{k}_t, z')\, \mathrm{j}(\mathbf{k}_t, z')\, dz' \tag{5.89}$$

Comparing this with our definition of the Green's function (5.33), yields finally

$$\mathsf{G}(\mathbf{k}_t, z, z') = -\mathsf{E}^{0}(\mathbf{k}_t, z)\, \mathsf{F}_{\pm}(\mathbf{k}_t, z, z')\, \tilde{\tilde{\mathsf{E}}}^{0}(\mathbf{k}_t, z'), \quad z \gtrless z' \tag{5.90}$$

which is the dyadic Green's function expressed in terms of the given and adjoint eigenmodes, and a transfer matrix which is a function of the reflection and transmission matrices, (5.85) and (5.86), at the two levels of the multilayer system.

5.4 Conjugate medium and Lorentz reciprocity

5.4.1 Green's function in the conjugate medium

The quantities $\mathsf{R}_{\pm}$, $\mathsf{T}_{\pm}$ and $\mathsf{N}_{\pm}$, appearing in the expressions for the transfer matrices $\mathsf{F}_{\pm}$ in the previous section, are functions of $\mathbf{k}_t \equiv k_0(s_x, s_y)$. The corresponding quantities in the conjugate system, in which $\mathbf{k}_t^c := k_0(-s_x, s_y)$, are related to those in the given system by the scattering theorem (2.113), in which

$$\mathsf{R}^c_{\pm}(\mathbf{k}^c_t, z) = \tilde{\mathsf{R}}_{\pm}(\mathbf{k}_t, z), \qquad \mathsf{T}^c_{\pm}(\mathbf{k}^c_t, z, z') = \tilde{\mathsf{T}}_{\mp}(\mathbf{k}_t, z', z) \tag{5.91}$$

In the conjugate system $\mathsf{N}^c_{\pm}(\mathbf{k}_t, z)$, (5.75) and (5.76), becomes with the aid of (5.91),

$$\begin{aligned} \mathsf{N}^c_{\pm}(\mathbf{k}^c_t, z) &:= \left[\mathsf{I}^{(2)} - \mathsf{R}^c_{\mp}\,\mathsf{R}^c_{\pm}\right]^{-1} = \left[\mathsf{I}^{(2)} - \tilde{\mathsf{R}}_{\mp}\tilde{\mathsf{R}}_{\pm}\right]^{-1} \\ &= \left\{\left[\mathsf{I}^{(2)} - \mathsf{R}_{\pm}\,\mathsf{R}_{\mp}\right]^{-1}\right\}^T = \tilde{\mathsf{N}}_{\mp}(\mathbf{k}_t, z) \end{aligned} \tag{5.92}$$

The conjugate transfer matrix $\mathsf{F}^c_{+}(\mathbf{k}^c_t, z', z)$, cf. (5.85), becomes with the aid of (5.91), (5.92) and (5.86)

$$\begin{aligned} \mathsf{F}^c_{+}(\mathbf{k}^c_t, z', z) &= \begin{bmatrix} \mathsf{I}^{(2)} \\ \mathsf{R}^c_{+}(z') \end{bmatrix} \mathsf{N}^c_{+}(z')\,\mathsf{T}^c_{+}(z', z)\,\mathsf{N}^c_{+}(z)\left[\mathsf{I}^{(2)}\,,\ \mathsf{R}^c_{-}(z)\right] \\ &= \mathsf{J}\begin{bmatrix} \tilde{\mathsf{R}}_{+}(z') \\ \mathsf{I}^{(2)} \end{bmatrix} \tilde{\mathsf{N}}_{-}(z')\,\tilde{\mathsf{T}}_{-}(z, z')\,\tilde{\mathsf{N}}_{-}(z)\left[\tilde{\mathsf{R}}_{-}(z)\,,\ \mathsf{I}^{(2)}\right]\mathsf{J} \\ &= \text{trans}\left\{\mathsf{J}\begin{bmatrix} \mathsf{R}_{-}(z) \\ \mathsf{I}^{(2)} \end{bmatrix} \mathsf{N}_{-}(z)\,\mathsf{T}_{-}(z, z')\,\mathsf{N}_{-}(z')\left[\mathsf{R}_{+}(z')\,,\ \mathsf{I}^{(2)}\right]\mathsf{J}\right\} \\ &= \mathsf{J}\,\tilde{\mathsf{F}}_{-}(\mathbf{k}_t, z, z')\,\mathsf{J} \end{aligned} \tag{5.93}$$

where 'trans' means 'the transpose of', and J (3.17), given by

$$\mathsf{J} := \begin{bmatrix} 0 & \mathsf{I}^{(2)} \\ \mathsf{I}^{(2)} & 0 \end{bmatrix} = \tilde{\mathsf{J}} = \mathsf{J}^{-1} \tag{5.94}$$

has been used to interchange the order of columns (rows) when used as a pre- (post-) multiplier. In a similar fashion

$$\mathsf{F}^c_-(\mathbf{k}_t, z', z) = \mathsf{J}\,\tilde{\mathsf{F}}_+(\mathbf{k}_t, z, z')\,\mathsf{J} \tag{5.95}$$

As a final step towards relating the Green's function in the given medium (5.90) to that in the conjugate medium, we need to map the modal matrices, $\mathsf{E}^0(\mathbf{k}_t, z)$ and $\overline{\mathsf{E}}^0(\mathbf{k}_t, z')$, into the conjugate medium. With the aid of (2.92) and (2.93), viz.

$$\mathbf{e}^c_{-\alpha}(\mathbf{k}^c_t, z) = \overline{\mathsf{Q}}_y\,\bar{\mathbf{e}}_\alpha(\mathbf{k}_t, z)$$

with

$$\overline{\mathsf{Q}}_y \equiv \mathsf{Q}^c_y := \begin{bmatrix} \mathsf{q}_y & 0 \\ 0 & \mathsf{q}_y \end{bmatrix} = \tilde{\overline{\mathsf{Q}}}_y\,, \qquad \mathbf{k}^c_t := k_0(-s_x, s_y)$$

we find

$$\mathsf{E}^0_c(\mathbf{k}^c_t, z) = \overline{\mathsf{Q}}_y\,[\hat{\bar{\mathbf{e}}}^0_{-1}\,,\,\bar{\mathbf{e}}^0_{-2}\,,\,\bar{\mathbf{e}}^0_1\,,\,\bar{\mathbf{e}}^0_2] = \overline{\mathsf{Q}}_y\,\overline{\mathsf{E}}^0(\mathbf{k}_t, z)\,\mathsf{J} \tag{5.96}$$

and similarly

$$\overline{\mathsf{E}}^0_c(\mathbf{k}^c_t, z) = \overline{\mathsf{Q}}_y\,\mathsf{E}^0(\mathbf{k}_t, z)\,\mathsf{J} \tag{5.97}$$

Green's function in the conjugate system, $\mathsf{G}^c(\mathbf{k}^c_t, z', z)$, by analogy with (5.90), then becomes with the aid of (5.93) – (5.97),

$$\begin{aligned} \mathsf{G}^c(\mathbf{k}^c_t, z', z) &= -\mathsf{E}^0_c(\mathbf{k}^c_t, z')\,\mathsf{F}^c_\pm(\mathbf{k}^c_t, z', z)\,\tilde{\overline{\mathsf{E}}}^0_c(\mathbf{k}^c_t, z) \\ &= -\overline{\mathsf{Q}}_y\,\overline{\mathsf{E}}^0(\mathbf{k}_t, z')\,\tilde{\mathsf{F}}_\mp(\mathbf{k}_t, z, z')\,\tilde{\mathsf{E}}^0(\mathbf{k}_t, z)\,\overline{\mathsf{Q}}_y \\ &= \overline{\mathsf{Q}}_y\,\tilde{\mathsf{G}}(\mathbf{k}_t, z, z')\,\overline{\mathsf{Q}}_y \end{aligned} \tag{5.98}$$

or, equivalently,

$$\mathsf{G}(\mathbf{k}_t, z, z') \rightarrow \mathsf{G}(s_x, s_y; z, z') = \overline{\mathsf{Q}}_y\,\tilde{\mathsf{G}}^c(-s_x, s_y; z', z)\,\overline{\mathsf{Q}}_y \tag{5.99}$$

As a final step towards relating the Green's functions in physical rather than transverse-k space, we substitute (5.99) in (5.37):

$$\mathsf{G}(\mathbf{r}, \mathbf{r}') = \frac{k_0^2}{4\pi^2}\iint \mathsf{G}(s_x, s_y; z, z')\,\exp\left[-ik_0\left\{s_x(x - x') + s_y(y - y')\right\}\right]\,d^2s$$

$$= \frac{k_0^2}{4\pi^2}\overline{\mathsf{Q}}_y\left\{\iint \tilde{\mathsf{G}}^c(-s_x, s_y; z', z)\exp\left[-ik_0\left\{-s_x(x' - x) - s_y(y' - y)\right\}\right]d^2s\right\}\overline{\mathsf{Q}}_y$$

Finally, changing $s_x \rightarrow -s_x$ as an integration variable, and comparing the second with the first line in the above equation, we find

$$\mathsf{G}(\mathbf{r}, \mathbf{r}') = \overline{\mathsf{Q}}_y\,\tilde{\mathsf{G}}^c(\mathbf{r}'_c, \mathbf{r}_c)\,\overline{\mathsf{Q}}_y \tag{5.100}$$

with

$$\mathbf{r}_c' := \mathbf{q}_y \mathbf{r}' = (x', -y', z'), \qquad \mathbf{r}_c := \mathbf{q}_y \mathbf{r} = (x, -y, z)$$

Eq. (5.100), which is an expression of Lorentz reciprocity, as will now be seen, has been reached after a lengthy proof requiring, inter alia, application of the scattering theorem (2.112) and the relation between eigenmodes in the adjoint and conjugate media (2.92). A much shorter and straightforward proof [9, 121], based on a reflection mapping of the Lorentz-adjoint system in a medium possessing a plane of conjugation symmetry, is given in Chap. 6.

5.4.2 From conjugate Green's function to Lorentz reciprocity

Suppose that the field $\mathbf{e}_1(\mathbf{r})$ and the current distribution $\mathbf{j}_1(\mathbf{r})$ satisfy Maxwell's equations in the given medium. We shall calculate the reaction $\langle \mathbf{e}_1, \bar{\mathbf{I}}\,\mathbf{j}_2 \rangle$ of $\mathbf{e}_1$ on a second, independent current distribution $\mathbf{j}_2(\mathbf{r})$. First, we define the reflection mappings, cf. (2.85), of $\mathbf{j}_1(\mathbf{r})$ and $\mathbf{j}_2(\mathbf{r})$ with respect to a plane of conjugation symmetry, y=0,

$$\mathbf{j}_i^c(\mathbf{r}_c) = \mathbf{Q}_y \mathbf{j}_i(\mathbf{r}), \qquad i = 1\,, 2, \qquad \mathbf{r}_c = \mathbf{q}_y \mathbf{r} \tag{5.101}$$

and suppose that the current $\mathbf{j}_2^c(\mathbf{r}_c)$ and the field $\mathbf{e}_2^c(\mathbf{r}_c)$ satisfy Maxwell's equations in the conjugate medium.

With the help of (4.86), (4.87), (5.35), (5.100) and (5.101) we have

$$\begin{aligned}
\langle \mathbf{e}_1, \bar{\mathbf{I}}\,\mathbf{j}_2 \rangle &:= \int \tilde{\mathbf{e}}_1(\mathbf{r})\, \bar{\mathbf{I}}\, \mathbf{j}_2(\mathbf{r})\, d^3r \\
&= \iint \tilde{\mathbf{j}}_1(\mathbf{r}')\, \tilde{\mathbf{G}}(\mathbf{r}, \mathbf{r}')\, \bar{\mathbf{I}}\, \mathbf{j}_2(\mathbf{r})\, d^3r'\, d^3r \\
&= \iint \left\{ \tilde{\mathbf{j}}_1^c(\mathbf{r}_c')\, \tilde{\mathbf{Q}}_y \right\} \left\{ \overline{\mathbf{Q}}_y\, \mathbf{G}^c(\mathbf{r}_c', \mathbf{r}_c)\, \overline{\mathbf{Q}}_y \right\} \bar{\mathbf{I}}\, \mathbf{Q}_y\, \mathbf{j}_2^c(\mathbf{r}_c)\, d^3r_c\, d^3r_c'
\end{aligned}$$

in which the currents and Green's function have been mapped into the conjugate medium in the last line. With $\overline{\mathbf{Q}}_y := \bar{\mathbf{I}}\, \mathbf{Q}_y$ and $\mathbf{Q}_y = \tilde{\mathbf{Q}}_y = \mathbf{Q}_y{}^{-1}$ (2.93), this becomes

$$\begin{aligned}
\langle \mathbf{e}_1, \bar{\mathbf{I}}\,\mathbf{j}_2 \rangle &= \iint \left\{ \tilde{\mathbf{j}}_1^c(\mathbf{r}_c')\, \tilde{\bar{\mathbf{I}}} \right\} \mathbf{G}^c(\mathbf{r}_c', \mathbf{r}_c)\, \mathbf{j}_2^c(\mathbf{r}_c)\, d^3r_c\, d^3r_c' \\
&= \int \mathbf{e}_2^c(\mathbf{r}_c') \cdot \bar{\mathbf{I}}\, \mathbf{j}_1^c(\mathbf{r}_c')\, d^3r_c' \\
&= \langle \mathbf{e}_2^c, \bar{\mathbf{I}}\, \mathbf{j}_1^c \rangle
\end{aligned} \tag{5.102}$$

This is just the Lorentz reciprocity theorem (4.91) found in Sec. 4.4.3, which states that reciprocity between two current distributions or antennas, in a non-reciprocal medium possessing a plane of conjugation symmetry, may be achieved by going over to a reflected pair of current distributions and the fields they generate in a symmetrically disposed conjugate medium.

We have thereby achieved our aim of showing how the scattering theorem (2.112), in plane-stratified source-free media, leads to Lorentz reciprocity when the relation between currents and fields is expressed by means of an eigenmode expansion (Green's function) in the given and conjugate media.

Chapter 6

Orthogonal mappings of fields and sources

In previous chapters we considered the mapping of a region into itself if the medium possessed a plane of reflection or conjugation symmetry. In the one case the mapping yielded an equivalence relation, in the other a reciprocity relation. In this chapter we shall generalize the previous procedures in two essential ways. First, we shall no longer confine ourselves to media possessing a spatial symmetry structure, but shall compare two different spatial regions in which the constitutive tensor in the one can be mapped from the given, or from the Lorentz-adjoint constitutive tensor in the other, by means of an orthogonal spatial mapping (rotation, reflection or inversion). Second, the orthogonal spatial mappings will no longer be restricted to reflections or to rotations through an angle π, but will be extended to include the 'full rotation group' [61, Sec. 2-7] which comprises rotation through arbitrary angles, with or without reflection or inversion. Conceptually there will be little new in these generalizations, but they will permit the formal systematization of the ideas developed till now.

Much of the analysis will be limited to rotations through an angle π about the three coordinate axes, to reflections with respect to the three coordinate planes or to inversion with respect to the origin. These, together with the identity operator, constitute a subgroup of the full rotation group. The results that will be derived in this restricted framework will be valid also for the full rotation group, as will be shown, but the simpler mathematical representation allows a rather simple derivation of the results and highlights their main features.

6.1 Mapping of the vector fields

6.1.1 Orthogonal transformations in a cartesian basis

Consider two spatial regions V_1 and V_2 which are symmetrically related, so that for every point $\mathbf{r}=(x_1,x_2,x_3)$ in V_1, there is a corresponding point $\mathbf{r}'=(x_1',x_2',x_3')$ in V_2, such that

$$\mathbf{r}' = \boldsymbol{\lambda}\mathbf{r} \tag{6.1}$$

where $\boldsymbol{\lambda}$ is an orthonormal matrix which preserves the length of the vectors on which it operates,

$$\boldsymbol{\lambda} := \begin{bmatrix} \tilde{\boldsymbol{\ell}}_1 \\ \tilde{\boldsymbol{\ell}}_2 \\ \tilde{\boldsymbol{\ell}}_3 \end{bmatrix}, \qquad \boldsymbol{\lambda}\tilde{\boldsymbol{\lambda}} = \tilde{\boldsymbol{\lambda}}\boldsymbol{\lambda} = \mathsf{I}^{(3)} \quad \text{i.e.} \quad \tilde{\boldsymbol{\ell}}_i\boldsymbol{\ell}_j = \delta_{ij} \tag{6.2}$$

We have here expressed $\boldsymbol{\lambda}$ in terms of its rows $\tilde{\boldsymbol{\ell}}_i$, each of which is a unit vector orthogonal to the other two. The transformation (6.1) is often used in physics to represent a transformation of the coordinate system while the point $\mathbf{r}$ is fixed in space. This is sometimes called a *passive* transformation [59, end of Sec. 4-2]. In the present discussion it is the coordinate system that is considered fixed, and we then have an *active* transformation or *mapping* of $\mathbf{r}$ from one region of space into another. From (6.2) we obtain

$$\det\left[\tilde{\boldsymbol{\lambda}}\boldsymbol{\lambda}\right] = (\det\boldsymbol{\lambda})^2 = 1\,, \qquad \det\boldsymbol{\lambda} = \pm 1 \tag{6.3}$$

Hence, with (6.2),

$$\det\boldsymbol{\lambda} = \boldsymbol{\ell}_1\cdot(\boldsymbol{\ell}_2\times\boldsymbol{\ell}_3) = \pm 1$$

or, in general, in terms of the third-rank antisymmetric Levi-Civita tensor ϵ_{ijl} [87, Sec. 6],

$$\boldsymbol{\ell}_i\cdot(\boldsymbol{\ell}_j\times\boldsymbol{\ell}_l) = \epsilon_{ijl}\det\boldsymbol{\lambda} \tag{6.4}$$

Since $\boldsymbol{\ell}_1$, $\boldsymbol{\ell}_2$ and $\boldsymbol{\ell}_3$ are orthonormal, eq. (6.4) may be written in the form

$$\boldsymbol{\ell}_i = (\boldsymbol{\ell}_j\times\boldsymbol{\ell}_l)\,\epsilon_{ijl}\det\boldsymbol{\lambda} \tag{6.5}$$

(without summation over repeated indices).

If the mapping transformation is one of rotation then $\det\boldsymbol{\lambda}=1$; if it is one of reflection or inversion then $\det\boldsymbol{\lambda}=-1$. These statements become self-evident when the transformation matrix $\boldsymbol{\lambda}$ is diagonal, and the symbol $\mathbf{q}$ will then be used instead of $\boldsymbol{\lambda}$, as in the previous chapters.

6.1.2 The D_{2h} point-symmetry group

With a diagonal transformation matrix, $\mathbf{q}$, the mapping of the position vector $\mathbf{r}$ is given by

$$\mathbf{r}' = \mathbf{q}\,\mathbf{r} \equiv \pm\mathbf{q}_n\,\mathbf{r} \qquad n = 0, 1, 2, 3 \tag{6.6}$$

The matrices, $\mathbf{q}{=}{+}\mathbf{q}_n$, $n{=}1,2,3$,

$$\mathbf{q}_1 := \begin{bmatrix} 1 & 0 & 0 \\ 0 & -1 & 0 \\ 0 & 0 & -1 \end{bmatrix} \quad \mathbf{q}_2 := \begin{bmatrix} -1 & 0 & 0 \\ 0 & 1 & 0 \\ 0 & 0 & -1 \end{bmatrix} \quad \mathbf{q}_3 := \begin{bmatrix} -1 & 0 & 0 \\ 0 & -1 & 0 \\ 0 & 0 & 1 \end{bmatrix} \tag{6.7}$$

with $\det\mathbf{q}{=}1$, produce rotations through an angle π about the coordinate axes x_1, x_2 and x_3, respectively, while the matrices $\mathbf{q}{=}{-}\mathbf{q}_n$ $(n = 1,2,3)$, with $\det\mathbf{q}{=}{-}1$, produce reflections with respect to the coordinate planes, $x_n{=}0$. Two successive rotations (reflections) with respect to the same axis (plane) generate the identity operation:

$$(\pm\mathbf{q}_n)(\pm\mathbf{q}_n)\,\mathbf{r} = \mathbf{q}_0\,\mathbf{r} \equiv \mathbf{I}^{(3)}\mathbf{r} = \mathbf{r} \tag{6.8}$$

Successive rotations with respect to two perpendicular axes are equivalent to a rotation about the third axis; successive reflections with respect to two perpendicular planes are equivalent to a rotation about their line of intersection:

$$(\pm\mathbf{q}_1)(\pm\mathbf{q}_2) = \mathbf{q}_3 \tag{6.9}$$

The inversion operation,

$$\mathbf{r}' = -\mathbf{q}_0\,\mathbf{r} = -\mathbf{r} \tag{6.10}$$

is produced by successive reflection and rotation about the $x_n{=}0$ plane and the x_n axis respectively, or equivalently, by successive reflections with respect to three perpendicular planes:

$$-\,\mathbf{q}_0 = (-\mathbf{q}_n)(+\mathbf{q}_n) = (-\mathbf{q}_1)(-\mathbf{q}_2)(-\mathbf{q}_3) \tag{6.11}$$

The eight transformations generated by $\mathbf{q}{=}{\pm}\mathbf{q}_n$, constitute the D_{2h} point-symmetry group [61, Sec. 2-7], which is a subgroup of the full rotation group. In general

$$\mathbf{q} = \tilde{\mathbf{q}} = \mathbf{q}^{-1} \tag{6.12}$$

6.1.3 Mapping of polar vector fields

In mapping a vector field from a region V_1 to another region V_2, we must distinguish between polar (proper) and axial (pseudo) vectors. Polar vectors, such as the electric field $\mathbf{E}(\mathbf{r})$, the electric current density $\mathbf{J}_e(\mathbf{r})$, the propagation vector $\mathbf{k}(\mathbf{r})$ and the velocity vector $\boldsymbol{v}(\mathbf{r})$ (of a moving fluid or plasma medium, for instance), are mapped like the position vector $\mathbf{r}$. If $\mathbf{E}'(\mathbf{r}')$ and $\mathbf{k}'(\mathbf{r}')$, for instance, denote the mapped vector fields at the position $\mathbf{r}'$, then

$$\mathbf{E}'(\mathbf{r}') = \boldsymbol{\lambda}\mathbf{E}(\mathbf{r}) , \quad \mathbf{k}'(\mathbf{r}') = \boldsymbol{\lambda}\mathbf{k}(\mathbf{r}) ; \quad \mathbf{r}' = \boldsymbol{\lambda}\mathbf{r} , \quad \mathbf{r} \in V_1 , \mathbf{r}' \in V_2 \tag{6.13}$$

and expressing the matrix $\boldsymbol{\lambda}$ in terms of its rows (6.2), we have typically

$$k_i'(\mathbf{r}') = \boldsymbol{\ell}_i \cdot \mathbf{k}(\mathbf{r}) , \quad i = 1, 2, 3 \quad \text{with} \quad \mathbf{r}' = \boldsymbol{\lambda}\mathbf{r} \tag{6.14}$$

We note that if instead of mapping both $\mathbf{E}$ and its argument $\mathbf{r}$ separately, we had performed a 'passive' coordinate transformation with the aid of $\boldsymbol{\lambda}$ (in which 'reflection', for instance, would imply going over from a right-handed to a left-handed system), the result mathematically would have been the same, but the procedure is conceptually different. The relation between 'active' (mapping) and 'passive' transformations will become apparent in the discussion of the full rotation group in Sec. 6.1.5. The mapping procedure is the better suited to our needs.

The differential operator $\boldsymbol{\nabla} \equiv \partial/\partial\mathbf{r}$ must also be mapped. With $\boldsymbol{\lambda}$ orthonormal (6.2), i.e. $\boldsymbol{\lambda}^{-1} = \tilde{\boldsymbol{\lambda}}$, we have

$$\mathbf{r}' = \boldsymbol{\lambda}\mathbf{r} , \quad \mathbf{r} = \tilde{\boldsymbol{\lambda}}\mathbf{r}' ; \quad \mathbf{r} := (x_1, x_2, x_3) , \ \mathbf{r}' := (x_1', x_2', x_3')$$

or in indicial notation, with summation over repeated indices assumed,

$$x_i' = \lambda_{ij}\, x_j , \quad x_j = (\tilde{\lambda})_{ji}\, x_i' = \lambda_{ij}\, x_i' , \quad \frac{\partial x_j}{\partial x_i'} = \lambda_{ij} \tag{6.15}$$

Hence, with the aid of the last equation in (6.15),

$$\frac{\partial}{\partial x_i'} = \frac{\partial x_j}{\partial x_i'}\frac{\partial}{\partial x_j} = \lambda_{ij}\frac{\partial}{\partial x_j} \tag{6.16}$$

and $\boldsymbol{\nabla}$ is seen to transform like the position vector:

$$\boldsymbol{\nabla}' \equiv \frac{\partial}{\partial \mathbf{r}'} = \boldsymbol{\lambda}\frac{\partial}{\partial \mathbf{r}} \equiv \boldsymbol{\lambda}\boldsymbol{\nabla} \tag{6.17}$$

i.e. like a polar vector.

6.1.4 Mapping of axial vector fields

We consider next the mapping of axial vector fields such as $\mathbf{H}(\mathbf{r}), \mathbf{B}(\mathbf{r}), \mathbf{J}_{\mathrm{m}}(\mathbf{r})$. Now the different behaviour of these vectors under reflection and inversion mappings, as compared with that of the polar vectors, is a ***physical*** property, expressing the physical nature of the entities which are mapped. However, the physical nature of axial vectors is expressed in the mathematical formulae, such as Maxwell's equations, which describe the physical laws. Consequently we can deduce the behaviour of these vectors under mapping transformations from the structure of the mathematical formulae in which they appear.

We examine first the transformation properties of the antisymmetric matrix operators, $\nabla\times\mathsf{I}^{(3)}$ or $\mathbf{k}(\mathbf{r})\times\mathsf{I}^{(3)}$, (2.22) or (4.42), which appear in Maxwell's equations. These operators are defined by the relations

$$\left[\nabla\times\mathsf{I}^{(3)}\right]\boldsymbol{v}:=\nabla\times\boldsymbol{v}\,,\qquad\left[\mathbf{k}(\mathbf{r})\times\mathsf{I}^{(3)}\right]\boldsymbol{v}(\mathbf{r}):=\mathbf{k}(\mathbf{r})\times\boldsymbol{v}(\mathbf{r})$$

($\boldsymbol{v}(\mathbf{r})$ represents any vector field), and may be expressed in matrix form or in terms of the antisymmetric Levi-Civita tensor ϵ_{ijl},

$$\left[\mathbf{k}(\mathbf{r})\times\mathsf{I}^{(3)}\right]:=\begin{bmatrix}0&-k_3&k_2\\k_3&0&-k_1\\-k_2&k_1&0\end{bmatrix}=\left[\epsilon_{ijl}k_l(\mathbf{r})\right]\tag{6.18}$$

Consider the transformation of $\mathbf{k}(\mathbf{r})\times\mathsf{I}^{(3)}$ by means of the orthonormal matrix $\boldsymbol{\lambda}$ (6.2):

$$\boldsymbol{\lambda}\left[\mathbf{k}(\mathbf{r})\times\mathsf{I}^{(3)}\right]\tilde{\boldsymbol{\lambda}}=\begin{bmatrix}\tilde{\boldsymbol{\ell}}_1\\\tilde{\boldsymbol{\ell}}_2\\\tilde{\boldsymbol{\ell}}_3\end{bmatrix}\mathbf{k}(\mathbf{r})\times\left[\boldsymbol{\ell}_1\,,\,\boldsymbol{\ell}_2\,,\,\boldsymbol{\ell}_3\right]\tag{6.19}$$

The (i,j)th component of (6.19) may be written with the aid of (6.5) and (6.14) as

$$\begin{aligned}\boldsymbol{\ell}_i\cdot\{\mathbf{k}(\mathbf{r})\times\boldsymbol{\ell}_j\}=-\mathbf{k}(\mathbf{r})\cdot(\boldsymbol{\ell}_i\times\boldsymbol{\ell}_j)&=-\mathbf{k}(\mathbf{r})\cdot\boldsymbol{\ell}_l\,\epsilon_{ijl}\det\boldsymbol{\lambda}\\&=-k_l'(\mathbf{r}')\,\epsilon_{ijl}\det\boldsymbol{\lambda}\end{aligned}\tag{6.20}$$

Comparing this with (6.18) we conclude this is just the (i,j)th component of $(\det\boldsymbol{\lambda})[\mathbf{k}'(\mathbf{r}')\times\mathsf{I}^{(3)}]$, and hence, with $\mathbf{r}'=\boldsymbol{\lambda}\,\mathbf{r}$, we get

$$\left[\mathbf{k}'(\mathbf{r}')\times\mathsf{I}^{(3)}\right]=(\det\boldsymbol{\lambda})\,\boldsymbol{\lambda}\left[\mathbf{k}(\mathbf{r})\times\mathsf{I}^{(3)}\right]\tilde{\boldsymbol{\lambda}}\tag{6.21}$$

Similarly, with $\nabla':=\partial/\partial\mathbf{r}'$ (6.17), and $\mathbf{r}'=\boldsymbol{\lambda}\mathbf{r}$, we obtain

$$\left[\nabla'\times\mathsf{I}^{(3)}\right]=(\det\boldsymbol{\lambda})\,\boldsymbol{\lambda}\left[\nabla\times\mathsf{I}^{(3)}\right]\tilde{\boldsymbol{\lambda}}\tag{6.22}$$

Chen [36, eq. (1.118)] has derived a relation equivalent to (6.21) or (6.22).

If we now postmultiply both sides of (6.22) with the polar vector $\mathbf{E}'(\mathbf{r}') = \boldsymbol{\lambda}\mathbf{E}(\mathbf{r})$ we get, with $\tilde{\boldsymbol{\lambda}}\boldsymbol{\lambda} = \mathbf{I}^{(3)}$ (6.2),

$$\nabla' \times \mathbf{E}'(\mathbf{r}') = (\det \boldsymbol{\lambda})\, \boldsymbol{\lambda} \left[\nabla \times \mathbf{E}(\mathbf{r})\right] \tag{6.23}$$

and since, from Maxwell's equations,

$$\nabla \times \mathbf{E} = -\frac{\partial \mathbf{B}}{\partial t} = -i\omega \mathbf{B}$$

with ω a scalar, we infer that

$$\mathbf{B}'(\mathbf{r}') = (\det \boldsymbol{\lambda})\, \boldsymbol{\lambda}\mathbf{B}(\mathbf{r}) \tag{6.24}$$

Vectors which are mapped like $\mathbf{B}(\mathbf{r})$ in (6.24) are called *axial* or *pseudo vectors*. For rotations, when $\det \boldsymbol{\lambda}=1$, they are mapped like polar vectors, but under reflection or inversion when $\det \boldsymbol{\lambda} = -1$, they are reversed in sign (i.e. in direction) as compared with the correspondingly mapped polar-vector fields. Axial vectors (such as $\mathbf{B}$) are sometimes represented as antisymmetric second-rank tensors, (B_{ij}), which are said to be 'dual' to the axial vectors [87, Sec. 6]. Such representations are very useful in relativity theory, since there is no 4-dimensional analogue of the axial vectors in 3-dimensional space.

The cross product of two polar vectors, in analogy with (6.23), transforms like and therefore is an axial vector, whereas the cross product of a polar and an axial vector transforms as a polar vector. Thus the Poynting vector, $\mathbf{S}(\mathbf{r}) := \mathbf{E}(\mathbf{r}) \times \mathbf{H}(\mathbf{r})$, is polar:

$$\mathbf{S}'(\mathbf{r}') := \mathbf{E}'(\mathbf{r}') \times \mathbf{H}'(\mathbf{r}') = \boldsymbol{\lambda}\,\mathbf{S}(\mathbf{r}) = \boldsymbol{\lambda}\left[\mathbf{E}(\mathbf{r}) \times \mathbf{H}(\mathbf{r})\right], \qquad \mathbf{r}' = \boldsymbol{\lambda}\mathbf{r} \tag{6.25}$$

as was implied in the discussion on the conjugate wave fields in Sec. 2.4.

The axial-vector nature of vector fields like $\mathbf{H}(\mathbf{r})$ can be understood if one considers the field $\mathbf{H}(\mathbf{r})$ to have been generated by solenoidal currents, or more specifically by circulating electric charges. The behaviour of the circulating charges under reflection then determines the behaviour of $\mathbf{H}(\mathbf{r})$. This is illustrated in Fig. 6.1.

We consider finally the transformation (mapping) of the mixed 6-element polar-axial fields and currents used in the previous chapters. In general each constituent field will be transformed differently, so that

$$\mathbf{e}'(\mathbf{r}') := \begin{bmatrix} \mathbf{E}'(\mathbf{r}') \\ \mathbf{H}'(\mathbf{r}') \end{bmatrix} = \begin{bmatrix} \boldsymbol{\lambda} & 0 \\ 0 & (\det \boldsymbol{\lambda})\boldsymbol{\lambda} \end{bmatrix} \begin{bmatrix} \mathbf{E}(\mathbf{r}) \\ \mathbf{H}(\mathbf{r}) \end{bmatrix} \equiv \boldsymbol{\Lambda}\,\mathbf{e}(\mathbf{r}), \qquad \mathbf{r}' = \boldsymbol{\lambda}\mathbf{r} \tag{6.26}$$

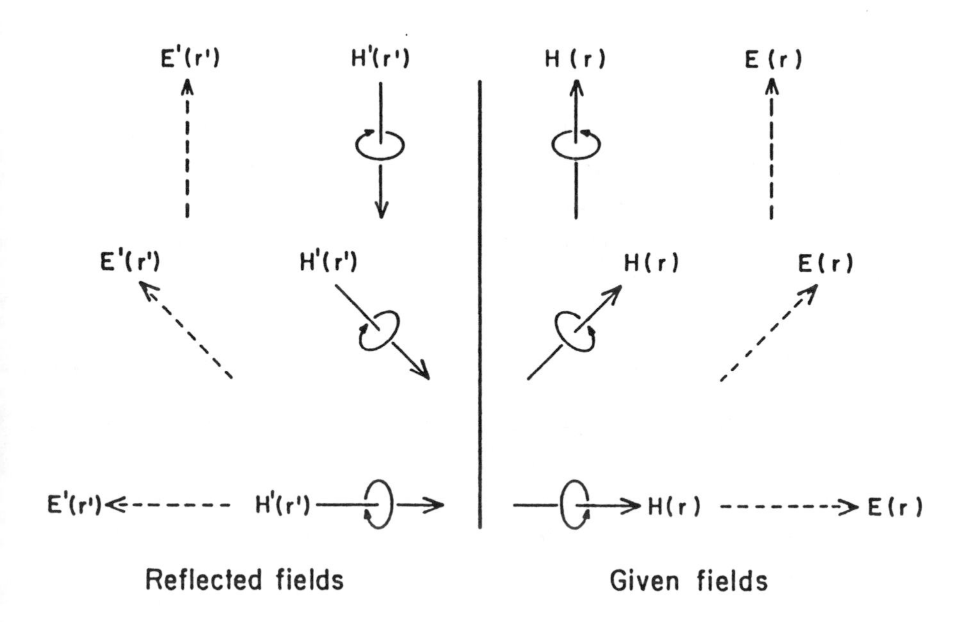

Figure 6.1: Reflection of an axial-vector field $\mathbf{H}(\mathbf{r})$ considered as the reflection of rotating electric charges. The reflected electric vectors are shown for comparison.

with

$$\Lambda := \begin{bmatrix} \boldsymbol{\lambda} & \mathbf{0} \\ \mathbf{0} & (\det \boldsymbol{\lambda})\, \boldsymbol{\lambda} \end{bmatrix}$$

and similarly for $\mathbf{j}' := \{\mathbf{J}'_e(\mathbf{r}'), \mathbf{J}'_m(\mathbf{r}')\}$. In the D_{2h} symmetry group when $\boldsymbol{\lambda}$ is diagonal, cf. (6.7), these become

$$\mathbf{e}'(\mathbf{r}') = \begin{bmatrix} \mathbf{q} & 0 \\ 0 & (\det \mathbf{q})\, \mathbf{q} \end{bmatrix} \begin{bmatrix} \mathbf{E}(\mathbf{r}) \\ \mathbf{H}(\mathbf{r}) \end{bmatrix} \equiv \mathbf{Q}\, \mathbf{e}(\mathbf{r}) \tag{6.27}$$

with $\mathbf{q} = \pm \mathbf{q}_n$ (6.7), and

$$\mathbf{Q} := \begin{bmatrix} \mathbf{q} & 0 \\ 0 & (\det \mathbf{q})\, \mathbf{q} \end{bmatrix} = \mathbf{Q}^T = \mathbf{Q}^{-1}$$

For reflection mappings, when $\boldsymbol{\lambda} \rightarrow \mathbf{q} = -\mathbf{q}_n$ in the D_{2h} symmetry group, with $\det \mathbf{q}{=}-1$ (Sec. 6.1.2), the mapping of an axial-vector field such as $\mathbf{B}(\mathbf{r})$ (6.24) is given by

$$\mathbf{B}'(\mathbf{r}') = -\mathbf{q}\,\mathbf{B}(\mathbf{r}) = \mathbf{q}_n\,\mathbf{B}(\mathbf{r}) \,, \qquad \mathbf{r}' = \mathbf{q}\,\mathbf{r} = -\mathbf{q}_n\,\mathbf{r} \tag{6.28}$$

Thus it might seem that the $\mathbf{B}$ field is rotated by $+\mathbf{q}_n$. However, although the vector $\mathbf{B}$ itself is rotated, the overall $\mathbf{B}(\mathbf{r})$ field, with $\mathbf{r}'{=}-\mathbf{q}_n\mathbf{r}$, is reflected—as can be seen if the field is time-harmonic. In that case the configuration of the reflected $\mathbf{B}'(\mathbf{r}')$ field a half period later, when $\mathbf{B}' \rightarrow -\mathbf{B}'$, is exactly the same as it would have been if it were a reflected polar-vector field.

6.1.5 The full rotation group: coordinate-free representation

In Sec. 6.1.1 we considered the full rotation group in a cartesian coordinate basis. Insofar as the transformations were linked to the coordinate system chosen, the formalism was best suited for describing 'passive' transformations of the coordinate system, in which the vector (or vector field) was fixed in space. In the present section we consider 'coordinate-free' mapping transformations, in which the 'active' transformation of the vector field is performed without reference to a coordinate system. Although the coordinate-linked mapping operators $\boldsymbol{\lambda}$ (or $\boldsymbol{\Lambda}$) used in the cartesian coordinate basis in Sec. 6.1.1, and the mapping operator $\boldsymbol{\gamma}$ (or $\boldsymbol{\Gamma}$) that will be used in the coordinate-free representation in this section, are mathematically equivalent, it will be convenient to differentiate between them by using different symbols in the two cases.

In three-dimensional space an orthonormal tensor is represented by three parameters and the value of its determinant (± 1). The original nine tensor elements must satisfy three orthogonality conditions and three normalization conditions, thus leaving three independent tensor elements. Since these six conditions are second degree in the tensor elements, the signs of these elements are left undetermined, and hence also the sign of the determinant.

Proper rotations of vectors

The three independent parameters of an orthonormal tensor can be taken as the three components of a vector $\boldsymbol{\phi}$, or preferably as the two independent components of the unit vector $\widehat{\boldsymbol{\phi}}$, and the modulus φ, i.e $\boldsymbol{\phi}{=}\varphi\widehat{\boldsymbol{\phi}}$. We shall now show that when the tensor

$$\boldsymbol{\gamma}_+(\boldsymbol{\phi}) := \exp[\boldsymbol{\phi} \times \mathbf{I}] \,, \qquad \mathbf{I} \equiv \mathbf{I}^{(3)} \tag{6.29}$$

operates on a vector $\boldsymbol{v}$, it conserves the longitudinal component $\boldsymbol{v}_{\parallel} = \widehat{\boldsymbol{\phi}}\,\widehat{\boldsymbol{\phi}}^T\boldsymbol{v}$ of $\boldsymbol{v}$ along $\widehat{\boldsymbol{\phi}}$ but rotates the transverse (to $\widehat{\boldsymbol{\phi}}$) component $\boldsymbol{v}_{\perp}=(\boldsymbol{v} - \widehat{\boldsymbol{\phi}}\,\widehat{\boldsymbol{\phi}}^T\boldsymbol{v})$ through an angle φ about $\widehat{\boldsymbol{\phi}}$.

The exponential of the antisymmetric tensor $\boldsymbol{\phi}\times\mathbf{I}$ is defined by its usual Taylor expansion, where the powers of $\boldsymbol{\phi}\times\mathbf{I}$ are defined as repeated matrix products of $\boldsymbol{\phi}\times\mathbf{I}$. Hence

$$\gamma_+(\boldsymbol{\phi}) := \exp[\boldsymbol{\phi}\times\mathbf{I}] = \mathbf{I} + [\boldsymbol{\phi}\times\mathbf{I}] + \frac{[\boldsymbol{\phi}\times\mathbf{I}]^2}{2!} + \frac{[\boldsymbol{\phi}\times\mathbf{I}]^3}{3!} + \cdots \tag{6.30}$$

We thus have

$$[\boldsymbol{\phi}\times\mathbf{I}]\,\boldsymbol{v} = \boldsymbol{\phi}\times\boldsymbol{v}$$

$$[\boldsymbol{\phi}\times\mathbf{I}]^2\,\boldsymbol{v} = \boldsymbol{\phi}\times(\boldsymbol{\phi}\times\boldsymbol{v}) = \boldsymbol{\phi}(\boldsymbol{\phi}\cdot\boldsymbol{v}) - \varphi^2\boldsymbol{v} = -\varphi^2\left[\mathbf{I} - \widehat{\boldsymbol{\phi}}\,\widehat{\boldsymbol{\phi}}^T\right]\boldsymbol{v}$$

$$[\boldsymbol{\phi}\times\mathbf{I}]^3\,\boldsymbol{v} = -\varphi^3\left[\widehat{\boldsymbol{\phi}}\times\boldsymbol{v}\right] = -\varphi^3\left[\widehat{\boldsymbol{\phi}}\times\mathbf{I}\right]\boldsymbol{v}$$

$$[\boldsymbol{\phi}\times\mathbf{I}]^4\,\boldsymbol{v} = -\varphi^4\left[\widehat{\boldsymbol{\phi}}\times\mathbf{I}\right]^2 = \varphi^4\left[\mathbf{I} - \widehat{\boldsymbol{\phi}}\,\widehat{\boldsymbol{\phi}}^T\right]\boldsymbol{v}$$

with $\widehat{\boldsymbol{\phi}}\times\widehat{\boldsymbol{\phi}}\,\widehat{\boldsymbol{\phi}}^T=0$. Hence, by iteration,

$$[\boldsymbol{\phi}\times\mathbf{I}]^{2n} = (-1)^n\,\varphi^{2n}\left[\mathbf{I} - \widehat{\boldsymbol{\phi}}\,\widehat{\boldsymbol{\phi}}^T\right]$$

$$[\boldsymbol{\phi}\times\mathbf{I}]^{2n+1} = (-1)^n\,\varphi^{2n+1}\left[\widehat{\boldsymbol{\phi}}\times\mathbf{I}\right]$$

so that

$$\begin{aligned}\gamma_+(\boldsymbol{\phi}) &:= \exp[\boldsymbol{\phi}\times\mathbf{I}] \\ &= \widehat{\boldsymbol{\phi}}\,\widehat{\boldsymbol{\phi}}^T + \left(1 - \frac{\varphi^2}{2!} + \frac{\varphi^4}{4!}\cdots\right)\left[\mathbf{I} - \widehat{\boldsymbol{\phi}}\,\widehat{\boldsymbol{\phi}}^T\right] + \left(\varphi - \frac{\varphi^3}{3!} + \frac{\varphi^5}{5!}\cdots\right)\left[\widehat{\boldsymbol{\phi}}\times\mathbf{I}\right] \\ &= \widehat{\boldsymbol{\phi}}\,\widehat{\boldsymbol{\phi}}^T + \cos\varphi\left[\mathbf{I} - \widehat{\boldsymbol{\phi}}\,\widehat{\boldsymbol{\phi}}^T\right] + \sin\varphi\left[\widehat{\boldsymbol{\phi}}\times\mathbf{I}\right]\end{aligned} \tag{6.31}$$

Now let $\gamma_+(\boldsymbol{\phi})$ operate on a vector $\boldsymbol{v}$:

$$\begin{aligned}\gamma_+(\boldsymbol{\phi})\boldsymbol{v} &= \widehat{\boldsymbol{\phi}}\,\widehat{\boldsymbol{\phi}}^T\boldsymbol{v} + \cos\varphi\left[\mathbf{I} - \widehat{\boldsymbol{\phi}}\,\widehat{\boldsymbol{\phi}}^T\right]\boldsymbol{v} + \sin\varphi\left[\widehat{\boldsymbol{\phi}}\times\mathbf{I}\right]\boldsymbol{v} \\ &= \boldsymbol{v}_{\parallel} + \boldsymbol{v}_{\perp}\cos\varphi + (\widehat{\boldsymbol{\phi}}\times\boldsymbol{v}_{\perp})\sin\varphi \\ &= \boldsymbol{v}_{\parallel} + \boldsymbol{v}'_{\perp} = \boldsymbol{v}'\end{aligned} \tag{6.32}$$

and we have thereby obtained a transformed vector $\boldsymbol{v}'$, whose projection $\boldsymbol{v}'_{\perp}$ on the plane transverse to $\widehat{\boldsymbol{\phi}}$ has been rotated by an angle φ with respect to $\boldsymbol{v}_{\perp}$. In the second line of (6.32), the terms $\boldsymbol{v}_{\perp}\cos\varphi$ and $(\widehat{\boldsymbol{\phi}}\times\boldsymbol{v}_{\perp})\sin\varphi$ represent respectively the projections of $\boldsymbol{v}'_{\perp}$ on two unit vectors, one along $\boldsymbol{v}_{\perp}$ and the other perpendicular to $\boldsymbol{v}_{\perp}$ in the transverse plane.

To express $\gamma_+(\phi)$ in terms of cartesian components, we let $\widehat{\phi} \to \hat{\mathbf{z}}$, so that

$$\begin{gathered} \boldsymbol{v}_\parallel = \boldsymbol{v}_z\,, \qquad \boldsymbol{v}_\perp = \boldsymbol{v}_x + \boldsymbol{v}_y = [\hat{\mathbf{x}}\hat{\mathbf{x}}^T + \hat{\mathbf{y}}\hat{\mathbf{y}}^T]\,\boldsymbol{v} \\ \widehat{\phi} \times \boldsymbol{v}_\perp = \hat{\mathbf{z}} \times [\hat{\mathbf{x}}\hat{\mathbf{x}}^T + \hat{\mathbf{y}}\hat{\mathbf{y}}^T]\,\boldsymbol{v}_\perp = [\hat{\mathbf{y}}\hat{\mathbf{x}}^T - \hat{\mathbf{x}}\hat{\mathbf{y}}^T]\,\boldsymbol{v}_\perp \end{gathered} \tag{6.33}$$

and hence, going back to the column-matrix representation of a vector, we have

$$\begin{bmatrix} v'_x \\ v'_y \\ v'_z \end{bmatrix} = \begin{bmatrix} \hat{\mathbf{x}}^T \\ \hat{\mathbf{y}}^T \\ \hat{\mathbf{z}}^T \end{bmatrix} \boldsymbol{v}' = \begin{bmatrix} \cos\varphi & -\sin\varphi & 0 \\ \sin\varphi & \cos\varphi & 0 \\ 0 & 0 & 1 \end{bmatrix} \begin{bmatrix} v_x \\ v_y \\ v_z \end{bmatrix}$$

with

$$\gamma_+(\varphi\hat{\mathbf{z}}) = \begin{bmatrix} \cos\varphi & -\sin\varphi & 0 \\ \sin\varphi & \cos\varphi & 0 \\ 0 & 0 & 1 \end{bmatrix} \tag{6.34}$$

We may now better appreciate the difference between a 'passive' coordinate transformation, and an 'active' mapping transformation of a (polar) vector. In both cases the mathematical relation between the original and the transformed vector is the same: $\boldsymbol{v}'=\gamma(\phi)\boldsymbol{v}$. However, in the mapping transformation (6.32), $\boldsymbol{v}'$ has been rotated through an angle φ about $\widehat{\phi}$, whereas the equivalent transformation of a vector generated by (6.34) is easily recognized as representing a rotation of the (x',y',z') system through an angle $-\varphi$ about the z-axis, while $\boldsymbol{v}$ is kept fixed in space—a result that is not unexpected.

Since $[\phi \times \mathsf{I}]$ which appears in the definition (6.29) of $\gamma_+(\phi)$ is an antisymmetric tensor, (6.18), we have with (6.30),

$$[\phi \times \mathsf{I}]^T = -[\phi \times \mathsf{I}]\,, \quad \text{and} \quad {\gamma_+}^T(\phi) = \exp\,[\phi \times \mathsf{I}]^T = \exp\,[-\phi \times \mathsf{I}] = \gamma_+(-\phi)$$

and so

$${\gamma_+}^T(\phi)\,\gamma_+(\phi) = \mathsf{I}\,, \qquad {\gamma_+}^T(\phi) = (\gamma_+)^{-1}(\phi) \tag{6.35}$$

which proves the orthonormality of the tensor $\gamma_+(\phi)$. Furthermore, if $\gamma_+(\phi)$ is written as a matrix as in (6.34), we see that

$$\det \gamma_+(\phi) = 1 \tag{6.36}$$

Note finally that when $\phi=\pi\hat{\mathbf{z}}$ we obtain, with the aid of (6.34) and (6.7),

$$\gamma_+(\pi\hat{\mathbf{z}}) = \begin{bmatrix} -1 & 0 & 0 \\ 0 & -1 & 0 \\ 0 & 0 & 1 \end{bmatrix} \equiv \mathsf{q}_3 \tag{6.37}$$

giving, as expected, a rotation of π about the z-axis.

Rotation with inversion or reflection of a vector

Consider next the transformation generated by the tensor

$$\gamma_-(\boldsymbol{\phi}) := -\exp[\boldsymbol{\phi}\times\mathsf{I}] , \quad \text{with} \quad {\gamma_-}^T = (\gamma_-)^{-1} , \quad \det\gamma_-(\boldsymbol{\phi}) = -1 \quad (6.38)$$

by analogy with (6.33) and (6.34). This tensor effects a rotation of φ about $\widehat{\boldsymbol{\phi}}$ generated by the tensor $\gamma_+(\boldsymbol{\phi})$, followed by an inversion generated by $-\mathsf{I}$. The net effect is thus equivalent to a rotation of $(\varphi\pm\pi)\widehat{\boldsymbol{\phi}}$ followed by reflection with respect to a plane perpendicular to $\widehat{\boldsymbol{\phi}}$. This can be seen in (6.34) with $\widehat{\boldsymbol{\phi}}=\hat{\mathbf{z}}$: the rotation through $(\varphi\pm\pi)$ changes the signs of the trigonometric functions in (6.34), and the reflection with respect to the z=0 plane changes the sign of the (z,z) component from +1 to −1. When $\varphi=\pi\hat{\mathbf{z}}$ we have, with (6.7)

$$\gamma_-(\pi\hat{\mathbf{z}}) = \begin{bmatrix} 1 & 0 & 0 \\ 0 & 1 & 0 \\ 0 & 0 & -1 \end{bmatrix} \equiv -\mathbf{q}_3 \quad (6.39)$$

which generates a pure reflection with respect to the z=0 plane. When φ=0 we have pure inversion:

$$\gamma_-(0) = \begin{bmatrix} -1 & 0 & 0 \\ 0 & -1 & 0 \\ 0 & 0 & -1 \end{bmatrix} \equiv -\mathbf{q}_0 \quad (6.40)$$

It is only in the D_{2h} subgroup (Sec. 6.1.2), with $\mathbf{q}=\pm\mathbf{q}_n$ (6.7), that we have pure rotation only (through π), pure reflection or inversion.

The transformations generated by $\gamma:=\gamma_\pm$ or equivalently by the matrix operator $\boldsymbol{\lambda}$ discussed in Sec. 6.1.1, constitute the full rotation group. The use of $\gamma_\pm$ in systematic discussion as in this chapter, has the advantage that the operations performed are transparent, with the axis and angle of rotation being given explicitly as arguments of the operator, without reference to an irrelevant coordinate system.

6.1.6 Mapping of mixed vector and tensor fields

Polar-vector fields such as $\mathbf{E}(\mathbf{r})$ are mapped from a region V_1 to another region V_2, just as in other representations when the matrix operators $\boldsymbol{\lambda}$ (6.13) or $\mathbf{q}$ (Sec. 6.1.2) were used, by means of

$$\mathbf{E}'(\mathbf{r}') = \gamma\mathbf{E}(\mathbf{r}) , \quad \mathbf{r}' = \gamma\mathbf{r} , \quad \gamma := \gamma_\pm , \quad \mathbf{r}\in V_1 , \quad \mathbf{r}'\in V_2 \quad (6.41)$$

The operators which generate vector products, such as $[\mathbf{k} \times \mathbf{I}^{(3)}]$ or $[\nabla \times \mathbf{I}^{(3)}]$ in (6.21) and (6.22), transform as:

$$\left[\mathbf{k}' \times \mathbf{I}^{(3)}\right] = (\det \gamma)\, \gamma\, [\mathbf{k} \times \mathbf{I}^{(3)}]\, \gamma^T , \qquad \mathbf{I} \equiv \mathbf{I}^{(3)} \tag{6.42}$$

$$\left[\nabla' \times \mathbf{I}^{(3)}\right] = (\det \gamma)\, \gamma\, [\nabla \times \mathbf{I}^{(3)}]\, \gamma^T \tag{6.43}$$

in which, with (6.33), (6.34) and (6.38),

$$\gamma := \gamma_{\pm}(\boldsymbol{\phi}) = \pm \exp\left[\boldsymbol{\phi} \times \mathbf{I}\right], \quad \gamma^T = \gamma^{-1}, \quad \det \gamma = \pm 1 \tag{6.44}$$

From these we infer that axial vectors, such as $\mathbf{B}(\mathbf{r}) \sim [\nabla \times \mathbf{I}^{(3)}]\, \mathbf{E}(\mathbf{r})$, are mapped according to

$$\mathbf{B}'(\mathbf{r}') = (\det \gamma)\, \gamma\, \mathbf{B}(\mathbf{r}); \quad \mathbf{r}' = \gamma \mathbf{r}, \quad \mathbf{r} \in V_1 \quad \mathbf{r}' \in V_2 \tag{6.45}$$

whereas the electric field, $\mathbf{E}(\mathbf{r}) \sim \nabla \times \mathbf{H}(\mathbf{r})$, or the Poynting vector, $\mathbf{S}(\mathbf{r}) = \mathbf{E}(\mathbf{r}) \times \mathbf{H}(\mathbf{r})$, are polar vectors.

In two-constituent polar-axial vector fields or current distributions such as

$$\mathrm{e}(\mathbf{r}) := \begin{bmatrix} \mathbf{E}(\mathbf{r}) \\ \mathbf{H}(\mathbf{r}) \end{bmatrix}, \qquad \mathrm{j}(\mathbf{r}) := \begin{bmatrix} \mathbf{J}_{\mathrm{e}}(\mathbf{r}) \\ \mathbf{J}_{\mathrm{m}}(\mathbf{r}) \end{bmatrix}$$

each constituent is mapped separately, so that

$$\begin{aligned} \mathrm{e}'(\mathbf{r}') := \begin{bmatrix} \mathbf{E}'(\mathbf{r}') \\ \mathbf{H}'(\mathbf{r}') \end{bmatrix} = \begin{bmatrix} \gamma & 0 \\ 0 & (\det \gamma)\gamma \end{bmatrix} \begin{bmatrix} \mathbf{E}(\mathbf{r}) \\ \mathbf{H}(\mathbf{r}) \end{bmatrix} =: \boldsymbol{\Gamma} \mathrm{e}(\mathbf{r}) \\ \mathrm{j}'(\mathbf{r}') = \boldsymbol{\Gamma} \mathrm{j}(\mathbf{r}), \qquad \mathbf{r}' = \gamma \mathbf{r} \end{aligned} \tag{6.46}$$

in which $\boldsymbol{\Gamma}$ is the analogue of $\boldsymbol{\Lambda}$ (6.26) in the cartesian representation, or of $\mathbf{Q}$ in the $\mathrm{D_{2h}}$ symmetry group (6.27). From (6.44) we deduce that $\boldsymbol{\Gamma}$ is also orthonormal:

$$\boldsymbol{\Gamma}^T = \boldsymbol{\Gamma}^{-1}, \qquad \det \boldsymbol{\Gamma} = \det \gamma = \pm 1 \tag{6.47}$$

Consider next the mapping of the constitutive tensor $\mathsf{K}(\mathbf{r})$, which relates two mixed vector fields (2.20),

$$\mathrm{d}(\mathbf{r}) := \begin{bmatrix} \mathbf{D}(\mathbf{r}) \\ \mathbf{B}(\mathbf{r}) \end{bmatrix} = \begin{bmatrix} \boldsymbol{\varepsilon}(\mathbf{r}) & \boldsymbol{\xi}(\mathbf{r}) \\ \boldsymbol{\eta}(\mathbf{r}) & \boldsymbol{\mu}(\mathbf{r}) \end{bmatrix} \begin{bmatrix} \mathbf{E}(\mathbf{r}) \\ \mathbf{H}(\mathbf{r}) \end{bmatrix} \equiv \mathsf{K}(\mathbf{r})\, \mathrm{e}(\mathbf{r}) \tag{6.48}$$

With the aid of (6.46) and (6.47), eq. (6.48) becomes

$$\mathrm{d}'(\mathbf{r}') = \boldsymbol{\Gamma} \mathsf{K}(\mathbf{r}) \boldsymbol{\Gamma}^T \mathrm{e}'(\mathbf{r}') = \mathsf{K}'(\mathbf{r}')\mathrm{e}'(\mathbf{r}')$$

so that

$$\mathsf{K}'(\mathbf{r}') = \boldsymbol{\Gamma}\,\mathsf{K}(\mathbf{r})\,\boldsymbol{\Gamma}^T\,, \qquad \mathbf{r}' = \gamma\mathbf{r} \tag{6.49}$$

and in the D_{2h} point-symmetry group, with $\mathbf{Q}=\mathbf{Q}^T$ (6.27),

$$\mathsf{K}'(\mathbf{r}') = \mathbf{Q}\mathsf{K}(\mathbf{r})\mathbf{Q}\,, \qquad \mathbf{r}' = \mathbf{q}\,\mathbf{r} \tag{6.50}$$

We note, from (6.46), (6.48) and (6.49), that the constitutive tensors that link a pair of polar or a pair of axial vectors, are mapped as

$$\boldsymbol{\varepsilon}'(\mathbf{r}') = \gamma\boldsymbol{\varepsilon}(\mathbf{r})\gamma^T\,, \qquad \boldsymbol{\mu}'(\mathbf{r}') = \gamma\boldsymbol{\mu}(\mathbf{r})\gamma^T\,, \qquad \mathbf{r}' = \gamma\mathbf{r} \tag{6.51}$$

Those which link a polar to an axial vector are mapped as

$$\boldsymbol{\xi}'(\mathbf{r}') = (\det\gamma)\,\gamma\boldsymbol{\xi}(\mathbf{r})\,\gamma^T\,, \qquad \boldsymbol{\eta}'(\mathbf{r}') = (\det\gamma)\,\gamma\,\boldsymbol{\eta}(\mathbf{r})\,\gamma^T\,, \qquad \mathbf{r}' = \gamma\mathbf{r} \tag{6.52}$$

$\boldsymbol{\varepsilon}$ and $\boldsymbol{\mu}$ are proper tensors, $\boldsymbol{\xi}$ and $\boldsymbol{\eta}$ are *pseudotensors* which change sign in reflection or inversion mappings ($\det\gamma = -1$). The pseudotensor character of $\boldsymbol{\xi}$ and $\boldsymbol{\eta}$ is evident in the constitutive relations for moving media. There we had, cf. (4.4a),

$$\boldsymbol{\xi} = \xi(\boldsymbol{v} \times \mathsf{I}^{(3)})\,, \qquad \xi := \frac{v}{c^2}\,\frac{c^2\varepsilon\mu - 1}{1 - v^2\varepsilon\mu}$$

which, as we have seen from its mapping (6.42), is indeed a pseudotensor. In Chap. 7, eq. (7.19), we shall encounter an example of a *biisotropic* constitutive tensor [83, end of Sec. 1.2c] in isotropic chiral media, which has the general structure

$$\mathsf{K} = \begin{bmatrix} \varepsilon\mathsf{I}^{(3)} & -i\alpha\mathsf{I}^{(3)} \\ i\alpha\mathsf{I}^{(3)} & \mu\mathsf{I}^{(3)} \end{bmatrix}$$

Since $\pm i\alpha\mathsf{I}^{(3)}$ are pseudotensors, it is evident that the scalar multiplier α is in fact a *pseudoscalar* that changes sign in reflection or inversion mappings [72, Sec. 6.11].

The antisymmetric tensor $[\nabla \times \mathsf{I}^{(3)}]$, is also mapped like a pseudotensor (6.43), and consequently the mapping of the differential operator $\mathsf{D}(\mathbf{r})$ (2.22) has the form

$$\begin{aligned} \mathsf{D}(\mathbf{r}') &:= \begin{bmatrix} 0 & -\nabla' \times \mathsf{I}^{(3)} \\ \nabla' \times \mathsf{I}^{(3)} & 0 \end{bmatrix} = \boldsymbol{\Gamma}\begin{bmatrix} 0 & -\nabla \times \mathsf{I}^{(3)} \\ \nabla \times \mathsf{I}^{(3)} & 0 \end{bmatrix}\boldsymbol{\Gamma}^T \\ &\equiv \boldsymbol{\Gamma}\mathsf{D}(\mathbf{r})\boldsymbol{\Gamma}^T \end{aligned} \tag{6.53}$$

with $\nabla := \partial/\partial\mathbf{r}$, $\nabla' := \partial/\partial\mathbf{r}'$. This was proved in Sec. 4.4.1, in the case of a reflection mapping (4.68), with the aid of the antisymmetric cartesian tensors U_x, U_y and U_z.

Consider finally the mapping of surface impedances in closed systems, cf. Sec. 4.3. The tensor (dyadic) surface impedance $\mathsf{Z}_s(\mathbf{r}_s)$ links the tangential electric and magnetic wave fields at a point $\mathbf{r}_s$ on the bounding surface (4.52):

$$\hat{\mathbf{n}}(\mathbf{r}_s) \times \mathbf{E}(\mathbf{r}_s) = \mathsf{Z}_s(\mathbf{r}_s)\mathbf{H}(\mathbf{r}_s) \quad \text{with} \quad \mathsf{Z}_s\hat{\mathbf{n}} = 0 \tag{6.54}$$

where $\hat{\mathbf{n}}(\mathbf{r}_s)$ is an outward unit normal vector on the surface S. The stipulation that $\hat{\mathbf{n}}(\mathbf{r}_s)$ and its mapping $\hat{\mathbf{n}}'(\mathbf{r}'_s)$ are both *outward*, determines that it is a polar vector. (If $\hat{\mathbf{n}}$ were taken in the direction of the elementary surface element $d\mathbf{r}_1 \times d\mathbf{r}_2$, it would be an axial vector, so that an outward normal would map under reflection into an inward normal.) Thus the vector $\hat{\mathbf{n}} \times \mathbf{E}$ is axial, and Z_s which links two axial vectors is a tensor (not a pseudo-tensor) cf. (6.51), with

$$\mathsf{Z}_s{}'(\mathbf{r}'_s) = \gamma \mathsf{Z}_s(\mathbf{r}_s)\gamma^T \,, \qquad \mathbf{r}'_s = \gamma \mathbf{r} \tag{6.55}$$

6.2 Mapping the Maxwell system

6.2.1 Invariance of Maxwell's equations under orthogonal mapping

Suppose that two symmetrically related regions V_1 and V_2 are filled with media having constitutive tensors $\mathsf{K}_1(\mathbf{r})$ and $\mathsf{K}_2(\mathbf{r}')$, $\mathbf{r} \in V_1$ and $\mathbf{r}' \in V_2$, and possibly enclosed by bounding surfaces S_1 and S_2 whose surface impedances are $\mathsf{Z}_1(\mathbf{r})$ and $\mathsf{Z}_2(\mathbf{r}')$ respectively. With $\exp(i\omega t)$ time dependence, Maxwell's equations in V_1 (2.21), as we have seen, may be written in the form

$$\mathsf{L}\,\mathrm{e} := \left[i\omega \mathsf{K}_1(\mathbf{r}) + \mathsf{D}(\mathbf{r})\right] \mathrm{e}(\mathbf{r}) = -\mathrm{j}(\mathbf{r}) \,, \qquad \mathbf{r} \in V_1 \tag{6.56}$$

If V_1 is enclosed by impedance walls, then the relation between the tangential components of $\mathbf{E}$ and $\mathbf{H}$ on the bounding surface S_1 is given by (6.54):

$$\hat{\mathbf{n}}(\mathbf{r}) \times \mathbf{E}(\mathbf{r}) = \mathsf{Z}_1(\mathbf{r})\mathbf{H}(\mathbf{r}) \,, \qquad \mathsf{Z}_1\hat{\mathbf{n}} = 0 \,, \qquad \mathbf{r} \in S_1 \tag{6.57}$$

Eqs. (6.56) and (6.57) may be mapped into the region V_2 which is symmetrically related to the region V_1, viz. for every point $\mathbf{r}$ in V_1 there is a corresponding point $\mathbf{r}'$ in V_2 such that $\mathbf{r}'=\gamma\mathbf{r}$. We premultiply (6.56) by $\varGamma$ and (6.57) by $(\det\gamma)\,\gamma$, noting that both $\varGamma$ and γ are orthonormal: $\varGamma^T\varGamma = \mathsf{I}^{(6)}$ (6.47) and $\gamma^T\gamma = \mathsf{I}^{(3)}$ (6.44). Thus

$$\varGamma \left[i\omega \mathsf{K}_1(\mathbf{r}) + \mathsf{D}(\mathbf{r})\right] \varGamma^T \cdot \varGamma \mathrm{e}(\mathbf{r}) = -\varGamma \mathrm{j}(\mathbf{r}) \tag{6.58}$$

and

$$(\det\gamma)\,\gamma \left[\hat{\mathbf{n}}(\mathbf{r}) \times \mathbf{E}(\mathbf{r})\right] = \gamma \mathsf{Z}(\mathbf{r})\,\gamma^T \cdot \gamma\,(\det\gamma)\,\mathbf{H}(\mathbf{r}) \tag{6.59}$$

With the aid of (6.46), (6.49), (6.53) and (6.55), which specify the behaviour e, j, K, D and Z under orthogonal transformations, these become

$$\left[i\omega \mathsf{K}'(\mathbf{r}') + \mathsf{D}(\mathbf{r}')\right] \mathbf{e}'(\mathbf{r}') = -\mathbf{j}'(\mathbf{r}'), \qquad \mathbf{r}' \in V_2 \tag{6.60}$$

$$\hat{\mathbf{n}}'(\mathbf{r}') \times \mathbf{E}'(\mathbf{r}') = \mathsf{Z}_1'(\mathbf{r}')\,\mathbf{H}'(\mathbf{r}'), \qquad \mathbf{r}' \in S_2 \tag{6.61}$$

and we have retrieved Maxwell's equations, which are thereby seen to be invariant under orthogonal mappings, (6.58) and (6.59).

If the constitutive tensor $\mathsf{K}_2(\mathbf{r}')$ of the medium occupying the region V_2 is just the mapped tensor $\mathsf{K}_1'(\mathbf{r}')$, and the surface impedance $\mathsf{Z}_2(\mathbf{r}')$ is the mapped impedance $\mathsf{Z}_1'(\mathbf{r}')$, i.e. if

$$\left.\begin{aligned} \mathsf{K}_2(\mathbf{r}') &\equiv \mathsf{K}_1'(\mathbf{r}') = \Gamma \mathsf{K}_1(\mathbf{r}) \Gamma^T \\ \mathsf{Z}_2(\mathbf{r}') &\equiv \mathsf{Z}_1'(\mathbf{r}') = \gamma \mathsf{Z}_1(\mathbf{r})\, \gamma^T \end{aligned}\right\} \mathbf{r} \in V_1, \quad \mathbf{r}' \in V_2, \quad \mathbf{r}' = \gamma \mathbf{r} \tag{6.62}$$

then the two media are said to be *equivalent*; a system of currents and fields which satisfies Maxwell's equations in one medium, will satisfy them also in the second medium (cf. Sec. 4.4.1 and Fig. 4.2).

6.2.2 The adjoint mapping

Examples of adjoint mappings were first encountered in Chap. 2, in which we used the adjoint reflection operator (2.93)

$$\overline{\mathsf{Q}}_y \equiv \mathsf{Q}_y^c := \mathsf{Q}_y \bar{\mathsf{I}} = \bar{\mathsf{I}} \mathsf{Q}_y \tag{6.63}$$

to transform unphysical, adjoint eigenmodes into physical, conjugate eigenmodes:

$$\mathbf{e}_{-\alpha}^c(-s_x, s_y) = \mathsf{Q}_y^c\, \bar{\mathbf{e}}_\alpha(s_x, s_y) \tag{6.64}$$

with the matrix $\bar{\mathsf{I}}$ (2.81) in the product $\mathsf{Q}_y^c = \mathsf{Q}_y \bar{\mathsf{I}}$, serving to reverse the direction of the Poynting vector by changing the sign of the axial-vector constituent $\mathbf{H}$ when operating on a wave field $\mathbf{e}' := (\mathbf{E}', \mathbf{H}')$.

We encountered a similar adjoint reflection mapping in Sec. 4.4.1 in which, in addition to mapping the wave fields (4.76),

$$\mathbf{e}^c(\mathbf{r}') = \overline{\mathsf{Q}}_y\, \bar{\mathbf{e}}(\mathbf{r}) \quad \text{or} \quad \bar{\mathbf{e}}^c(\mathbf{r}') = \overline{\mathsf{Q}}_y\, \mathbf{e}(\mathbf{r}), \qquad \mathbf{r}' = \mathsf{q}_y \mathbf{r} \tag{6.65}$$

the adjoint operator $\overline{\mathsf{Q}}_y$ also served to change the sign of the differential operator $\mathsf{D}(\mathbf{r})$ (4.69),

$$\overline{\mathsf{Q}}_y\, \mathsf{D}(\mathbf{r})\, \overline{\mathsf{Q}}_y = -\mathsf{D}(\mathbf{r}'), \qquad \mathbf{r}' = \mathsf{q}_y \mathbf{r} \tag{6.66}$$

These results can be generalized immediately by defining the adjoint mapping operator

$$\bar{\Gamma} := \Gamma \bar{\mathsf{I}} = \bar{\mathsf{I}} \Gamma \,, \quad \text{with} \quad \bar{\Gamma}^T = \bar{\Gamma}^{-1} \tag{6.67}$$

the last equality stemming from (6.47), with $\bar{\mathsf{I}} = \bar{\mathsf{I}}^T = \bar{\mathsf{I}}^{-1}$ (2.81).

By analogy with (6.66), application of $\bar{\Gamma}$ to the differential operator $\mathsf{D}(\mathbf{r})$ yields, with the aid of (6.53) and (4.69),

$$\bar{\Gamma}\, \mathsf{D}(\mathbf{r})\, \bar{\Gamma}^T = \bar{\mathsf{I}} \left[\Gamma \mathsf{D}(\mathbf{r})\, \Gamma^T \right] \bar{\mathsf{I}} = \bar{\mathsf{I}}\, \mathsf{D}(\mathbf{r}')\, \bar{\mathsf{I}} = -\mathsf{D}(\mathbf{r}') \tag{6.68}$$

with $\mathbf{r}' = \gamma \mathbf{r}$.

Suppose that the mapped constitutive tensor $\bar{\Gamma}\, \mathsf{K}_1(\mathbf{r})\, \bar{\Gamma}^T$ equals the *transpose* of the constitutive tensor $\mathsf{K}_2(\mathbf{r}')$ in the second medium:

$$\bar{\Gamma}\, \mathsf{K}_1(\mathbf{r})\, \bar{\Gamma}^T = \Gamma \left[\bar{\mathsf{I}}\, \mathsf{K}_1(\mathbf{r})\, \bar{\mathsf{I}} \right] \Gamma^T = {\mathsf{K}_2}^T(\mathbf{r}') \,, \quad \mathbf{r}' = \gamma \mathbf{r} \tag{6.69}$$

or, on transposition and use of the orthonormality of Γ (6.47),

$$\mathsf{K}_2(\mathbf{r}') = \Gamma \left[\bar{\mathsf{I}}\, {\mathsf{K}_1}^T(\mathbf{r})\, \bar{\mathsf{I}} \right] \Gamma^T \,, \qquad \mathbf{r} \in V_1, \quad \mathbf{r}' \in V_2 \tag{6.70}$$

Thus $\mathsf{K}_2(\mathbf{r}')$ represents an orthogonal mapping of the Lorentz-adjoint constitutive tensor $\mathsf{K}^{(L)} := \bar{\mathsf{I}}\, \mathsf{K}^T\, \bar{\mathsf{I}}$ (4.37) of the medium in V_1, and we call such a medium a *conjugate* medium, or a medium which is conjugate to the given medium. We thereby generalize our earlier usage, cf. Sec. 2.4 and eqs. (4.75) and (4.79), in which the concept of a conjugate medium was restricted to a reflection mapping of the Lorentz-adjoint medium.

Let us now map Maxwell's equations (2.21) from V_1 to V_2 by means of the adjoint operator $\bar{\Gamma}$, (6.67),

$$\bar{\Gamma}\, \mathsf{L}(\mathbf{r})\, \mathrm{e}(\mathbf{r}) := \bar{\Gamma} \left[i\omega \mathsf{K}_1(\mathbf{r}) + \mathsf{D}(\mathbf{r}) \right] \bar{\Gamma}^T \bar{\Gamma} \mathrm{e}(\mathbf{r}) = -\bar{\Gamma}\, \mathrm{j}(\mathbf{r})$$

yielding, with (6.69) and (6.68),

$$\left[i\omega {\mathsf{K}_2}^T(\mathbf{r}') - \mathsf{D}(\mathbf{r}') \right] \bar{\Gamma} \mathrm{e}(\mathbf{r}) = -\bar{\Gamma}\, \mathrm{j}(\mathbf{r}) \tag{6.71}$$

This is just the adjoint Maxwell system in the conjugate medium, in which we may identify

$$\mathsf{K}_2(\mathbf{r}') \equiv \mathsf{K}_2^c(\mathbf{r}') \,, \quad \bar{\Gamma} \mathrm{e}(\mathbf{r}) = \bar{\mathrm{e}}^c(\mathbf{r}') \,, \quad \bar{\Gamma} \mathrm{j}(\mathbf{r}) = \bar{\mathrm{j}}^c(\mathbf{r}') \tag{6.72}$$

and recalling that the adjoint operation is involutary, i.e. $\bar{\bar{\mathrm{e}}}^c(\mathbf{r}') = \mathrm{e}^c(\mathbf{r}')$, we have with (6.67),

$$\bar{\Gamma}^T \mathrm{e}^c(\mathbf{r}') = \bar{\mathrm{e}}(\mathbf{r}) \,, \qquad \bar{\Gamma}^T \mathrm{j}^c(\mathbf{r}') = \bar{\mathrm{j}}(\mathbf{r}) \tag{6.73}$$

by analogy with (4.76).

It will be convenient in what follows to work, as in Sec. 4.4.3, with physical rather than with adjoint fields and currents. If we start with the (physical) Lorentz-adjoint Maxwell equations in V_1 and map them into V_2, by analogy with eqs. (4.71)–(4.74), we obtain with (4.37) and use of the orthonormality of $\boldsymbol{\Gamma}$, $\boldsymbol{\Gamma}^T\boldsymbol{\Gamma}=\mathsf{I}^{(6)}$ (6.47),

$$\boldsymbol{\Gamma}\mathsf{L}^{(L)}\mathbf{e}^{(L)}(\mathbf{r}) = \boldsymbol{\Gamma}\left[i\omega\,\bar{\mathsf{I}}\,\mathsf{K}_1{}^T(\mathbf{r})\,\bar{\mathsf{I}} + \mathsf{D}(\mathbf{r})\right]\boldsymbol{\Gamma}^T\boldsymbol{\Gamma}\mathbf{e}^{(L)}(\mathbf{r}) = -\boldsymbol{\Gamma}\mathbf{j}^{(L)}(\mathbf{r}) \tag{6.74}$$

where $\bar{\mathsf{I}}\,\mathsf{K}_1{}^T(\mathbf{r})\,\bar{\mathsf{I}} \equiv \mathsf{K}_1^{(L)}(\mathbf{r})$. Thus, with (6.70) and (6.72), we get

$$\left[i\omega\mathsf{K}^c(\mathbf{r}') + \mathsf{D}(\mathbf{r}')\right]\,\mathbf{e}^c(\mathbf{r}') = -\mathbf{j}^c(\mathbf{r}') \tag{6.75}$$

in which we have identified

$$\boldsymbol{\Gamma}\mathbf{e}^{(L)}(\mathbf{r}) = \mathbf{e}^c(\mathbf{r}')\,, \quad \boldsymbol{\Gamma}\mathbf{j}^{(L)}(\mathbf{r}) = \mathbf{j}^c(\mathbf{r}')\,, \quad \boldsymbol{\Gamma}\,\mathsf{K}^{(L)}(\mathbf{r})\,\boldsymbol{\Gamma}^T = \mathsf{K}^c(\mathbf{r}') \tag{6.76}$$

The inverse mappings from V_2 to V_1 with $\boldsymbol{\Gamma}^{-1}=\boldsymbol{\Gamma}^T$ (6.47), are

$$\mathbf{e}^{(L)}(\mathbf{r}) = \boldsymbol{\Gamma}^T\mathbf{e}^c(\mathbf{r}')\,, \quad \mathbf{j}^{(L)}(\mathbf{r}) = \boldsymbol{\Gamma}^T\mathbf{j}^c(\mathbf{r}')\,, \quad \mathsf{K}^{(L)}(\mathbf{r}) = \boldsymbol{\Gamma}^T\mathsf{K}^c(\mathbf{r}')\boldsymbol{\Gamma} \tag{6.77}$$

and the mappings of fields and currents are equivalent to those in (6.73), with $\mathbf{e}^{(L)}(\mathbf{r})=\bar{\mathsf{I}}\,\bar{\mathbf{e}}(\mathbf{r})$, $\mathbf{j}^{(L)}(\mathbf{r})=\bar{\mathsf{I}}\,\bar{\mathbf{j}}(\mathbf{r})$ as in (4.34), thereby justifying the definitions in (6.76).

A necessary condition for the identification of the mapped with the conjugate fields and currents, (6.72) or (6.76), is that the mapped and conjugate surface impedances are the same. In Sec. 4.3.2 we found the relation between the Lorentz-adjoint and the given surface impedances, $\mathsf{Z}^{(L)}$ and $\mathsf{Z}\equiv\mathsf{Z}_1$ (4.57) and (4.58),

$$\hat{\mathbf{n}}(\mathbf{r})\times\mathbf{E}^{(L)}(\mathbf{r}) = \mathsf{Z}^{(L)}(\mathbf{r})\,\mathbf{H}^{(L)}(\mathbf{r})\,, \qquad \mathsf{Z}^{(L)}(\mathbf{r}) = \mathsf{Z}^T(\mathbf{r}) \tag{6.78}$$

in which we have dropped the subscript s. With $\mathbf{e}^{(L)}(\mathbf{r})=\bar{\mathsf{I}}\,\bar{\mathbf{e}}(\mathbf{r})$ (4.34), viz.

$$\mathbf{E}^{(L)}(\mathbf{r}) = \overline{\mathbf{E}}(\mathbf{r})\,, \qquad \mathbf{H}^{(L)}(\mathbf{r}) = -\overline{\mathbf{H}}(\mathbf{r}) \tag{6.79}$$

we find from (6.78) that

$$\hat{\mathbf{n}}(\mathbf{r})\times\overline{\mathbf{E}}(\mathbf{r}) = -\mathsf{Z}^{(L)}(\mathbf{r})\,\overline{\mathbf{H}}(\mathbf{r}) = \overline{\mathsf{Z}}(\mathbf{r})\,\overline{\mathbf{H}}(\mathbf{r}) \tag{6.80}$$

so that

$$-\,\overline{\mathsf{Z}}(\mathbf{r}) = \mathsf{Z}^{(L)}(\mathbf{r}) = \mathsf{Z}^T(\mathbf{r}) \tag{6.81}$$

The conjugate surface impedance $\mathsf{Z}^c(\mathbf{r}')$ is defined by analogy with (4.52)

$$\hat{\mathbf{n}}(\mathbf{r}') \times \mathbf{E}^c(\mathbf{r}') = \mathsf{Z}^c(\mathbf{r}')\,\mathbf{H}^c(\mathbf{r}') \tag{6.82}$$

Expressing the axial vectors $\hat{\mathbf{n}}(\mathbf{r}')\times\mathbf{E}^c(\mathbf{r}')$ and $\mathbf{H}^c(\mathbf{r}')$ as orthogonal mappings of $\hat{\mathbf{n}}(\mathbf{r})\times\mathbf{E}^{(L)}(\mathbf{r})$ and $\mathbf{H}^{(L)}(\mathbf{r})$ respectively, we obtain, with $\mathbf{e}^c(\mathbf{r}')=\varGamma\mathbf{e}^{(L)}(\mathbf{r})$, (6.76),

$$(\det\gamma)\,\gamma\left[\hat{\mathbf{n}}(\mathbf{r})\times\mathbf{E}^{(L)}(\mathbf{r})\right] = \mathsf{Z}^c(\mathbf{r}')(\det\gamma)\,\gamma\,\mathbf{H}^{(L)}(\mathbf{r})$$

With (6.78) and (6.81) this yields

$$\mathsf{Z}^c(\mathbf{r}') = \gamma\,\mathsf{Z}^{(L)}(\mathbf{r})\,\gamma^T = \gamma\,\mathsf{Z}^T(\mathbf{r})\,\gamma^T \tag{6.83}$$

We summarize as follows. If for all $\mathbf{r}\in V_1$, $\mathbf{r}'\in V_2$, the constitutive tensors $\mathsf{K}(\mathbf{r})\equiv\mathsf{K}_1(\mathbf{r})$ and $\mathsf{K}^c(\mathbf{r}')\equiv\mathsf{K}_2(\mathbf{r}')$, and the dyadic surface impedances (if any), $\mathsf{Z}(\mathbf{r})$ and $\mathsf{Z}^c(\mathbf{r}')$, are linked by the relations

$$\mathsf{K}^c(\mathbf{r}') = \bar{\varGamma}\mathsf{K}^T(\mathbf{r})\bar{\varGamma}^T\,, \qquad \mathsf{Z}^c(\mathbf{r}') = \gamma\,\mathsf{Z}^T(\mathbf{r})\,\gamma^T\,, \qquad \mathbf{r}' = \gamma\mathbf{r} \tag{6.84}$$

then the orthogonal mapping (6.74) of the Maxwell currents and fields in the Lorentz-adjoint medium, $\bar{\mathsf{I}}\,\mathsf{K}^T\,\bar{\mathsf{I}}$, yields solutions of the Maxwell system (6.75) in the conjugate medium. We shall say that the Maxwell systems in the two orthogonally mapped regions display *conjugation symmetry.*

6.2.3 Lorentz reciprocity in regions possessing conjugation symmetry

Consider the Maxwell fields and currents $\mathbf{e}_1(\mathbf{r})$ and $\mathbf{j}_1(\mathbf{r})$ in a medium $\mathsf{K}(\mathbf{r})$, $\mathbf{r}\in V_1$, and $\mathbf{e}_2^c(\mathbf{r}')$ in a medium $\mathsf{K}^c(\mathbf{r}')$, $\mathbf{r}'\in V_2$. Suppose that the constitutive tensors and surface impedances (if any) possess conjugation symmetry, i.e. that they are related as in (6.84). Maxwell's equations in V_1 and V_2 respectively are

$$\mathsf{L}(\mathbf{r})\,\mathbf{e}_1(\mathbf{r}) := \left[i\omega\mathsf{K}(\mathbf{r}) + \mathsf{D}(\mathbf{r})\right]\,\mathbf{e}_1(\mathbf{r}) = -\mathbf{j}_1(\mathbf{r}) \tag{6.85}$$

$$\mathsf{L}(\mathbf{r}')\,\mathbf{e}_2^c(\mathbf{r}') := \left[i\omega\mathsf{K}^c(\mathbf{r}') + \mathsf{D}(\mathbf{r}')\right]\,\mathbf{e}_2^c(\mathbf{r}') = -\mathbf{j}_2^c(\mathbf{r}') \tag{6.86}$$

An orthonormal mapping of (6.86) from V_2 to V_1:

$$\varGamma^T\mathsf{L}(\mathbf{r}')\,\mathbf{e}_2^c(\mathbf{r}') = \varGamma^T\mathsf{L}(\mathbf{r}')\varGamma\,\varGamma^T\mathbf{e}_2^c(\mathbf{r}') = -\varGamma^T\mathbf{j}_2^c(\mathbf{r}')$$

yields, with the aid of (6.77),

$$\left[i\omega\mathsf{K}^{(L)}(\mathbf{r}) + \mathsf{D}(\mathbf{r})\right]\,\mathbf{e}_2^{c\prime}(\mathbf{r}) = -\mathbf{j}_2^{c\prime}(\mathbf{r}) \tag{6.87}$$

where

$$\mathbf{e}_2^{c\prime}(\mathbf{r}) := \boldsymbol{\Gamma}^T \mathbf{e}_2^c(\mathbf{r}') = \mathbf{e}_2^{(L)}(\mathbf{r}), \qquad \mathbf{j}_2^{c\prime}(\mathbf{r}) := \boldsymbol{\Gamma}^T \mathbf{j}_2^c(\mathbf{r}') = \mathbf{j}_2^{(L)}(\mathbf{r}) \tag{6.88}$$

by analogy with (4.82) and (4.83). Just as in Sec. 4.4.3 we use the notation $\mathbf{e}_2^{c\prime}(\mathbf{r})$ and $\mathbf{j}_2^{c\prime}(\mathbf{r})$, rather than the equivalent Lorentz-adjoint quantities, $\mathbf{e}^{(L)}(\mathbf{r})$ and $\mathbf{j}^{(L)}(\mathbf{r})$, in order to emphasize that they are derived by an orthogonal mapping of physical fields and currents in the conjugate medium. Now the given and Lorentz-adjoint fields and currents, (6.85) and (6.87), have been shown in Sec. 4.2.2, eqs. (4.32)–(4.37), to be related by a reciprocity relation, which in the present case takes the form

$$\begin{aligned}
\int_{V_1} \left(\tilde{\mathbf{e}}_1 \bar{\mathbf{I}} \mathbf{j}_2^{c\prime} - \tilde{\mathbf{e}}_2^{c\prime} \bar{\mathbf{I}} \mathbf{j}_1\right) d^3r &= \int_{S_1} \left(\mathbf{E}_2^{c\prime} \times \mathbf{H}_1 - \mathbf{E}_1 \times \mathbf{H}_2^{c\prime}\right) \cdot \hat{\mathbf{n}}\, dS \\
&= \int \left\{(\hat{\mathbf{n}} \times \mathbf{E}_2^{c\prime}) \cdot \mathbf{H}_1 - (\hat{\mathbf{n}} \times \mathbf{E}_1) \cdot \mathbf{H}_2^{c\prime}\right\} dS \\
&= \int \left(\tilde{\mathbf{H}}_1 \mathsf{Z}^{(L)} \mathbf{H}_2^{c\prime} - \tilde{\mathbf{H}}_2^{c\prime} \mathsf{Z} \mathbf{H}_1\right) dS \\
&= \int \tilde{\mathbf{H}}_1 \left(\mathsf{Z}^{(L)} - \mathsf{Z}^T\right) \mathbf{H}_2^{c\prime}\, dS
\end{aligned} \tag{6.89}$$

In the third line the vector products, $\hat{\mathbf{n}}\times\mathbf{E}$, have been replaced by surface impedance terms, (6.78) and (6.57), and the second term in the integrand has been transposed in the last line. The right-hand side equals zero, since $\mathsf{Z}^{(L)}=\mathsf{Z}^T$ (4.58), and so finally

$$\int_{V_1} \left(\tilde{\mathbf{e}}_1 \bar{\mathbf{I}} \mathbf{j}_2^{c\prime} - \tilde{\mathbf{e}}_2^{c\prime} \bar{\mathbf{I}} \mathbf{j}_1\right) d^3r = 0 \tag{6.90}$$

as in (4.84). Here too it is convenient to map the second term in the integrand, a scalar (invariant), into the conjugate region V_2, with the aid of (6.88),

$$\begin{aligned}
\tilde{\mathbf{e}}_2^{c\prime}(\mathbf{r}) \bar{\mathbf{I}} \mathbf{j}_1(\mathbf{r}) = \tilde{\mathbf{e}}_2^c(\mathbf{r}') \boldsymbol{\Gamma} \bar{\mathbf{I}} \mathbf{j}_1(\mathbf{r}) &= \tilde{\mathbf{e}}_2^c(\mathbf{r}') \bar{\mathbf{I}} \boldsymbol{\Gamma} \mathbf{j}_1(\mathbf{r}) \\
&= \tilde{\mathbf{e}}_2^c(\mathbf{r}') \bar{\mathbf{I}} \mathbf{j}_1'(\mathbf{r}')
\end{aligned} \tag{6.91}$$

with $\mathbf{j}_1'(\mathbf{r}'):=\boldsymbol{\Gamma}\mathbf{j}_1(\mathbf{r})$, so that (6.90) becomes, in mixed vector-matrix notation,

$$\int_{V_1} \mathbf{e}_1 \cdot \bar{\mathbf{I}} \mathbf{j}_2^{c\prime}\, d^3r = \int_{V_2} \mathbf{e}_2^c \cdot \bar{\mathbf{I}} \mathbf{j}_1'\, d^3r' \tag{6.92}$$

as in (4.86). The subsequent conclusions and applications discussed in Chap. 4 are all applicable here, but the mapping from region V_1 to V_2 is no longer restricted to reflection or other elements of the $\mathrm{D_{2h}}$ symmetry group, but includes all orthogonal mappings in the full rotation group. It should be noted that the *identity mapping*, generated by $\gamma \to \mathbf{q}_0 \equiv \mathbf{I}^{(3)}$ (6.8), belongs to this general category. With this mapping the conjugate constitutive tensor (6.84) is simply

$$\mathbf{K}^c(\mathbf{r}') = \bar{\mathbf{I}}\mathbf{K}^T(\mathbf{r})\bar{\mathbf{I}} \equiv \mathbf{K}^{(L)}(\mathbf{r}), \qquad \mathbf{r}' = \mathbf{r}, \quad V_1 = V_2,$$

the conjugate fields and currents reduce to the Lorentz-adjoint quantities, and (6.90) reduces to the usual Lorentz reciprocity theorem

$$\int \left(\mathbf{e}_1 \cdot \bar{\mathbf{I}}\mathbf{j}_2^{(L)} - \mathbf{e}_2^{(L)} \cdot \bar{\mathbf{I}}\mathbf{j}_1\right) d^3r = 0 \tag{6.93}$$

6.3 Mapping the Green's functions and scattering matrices

6.3.1 Mapping the Green's functions

In Chap. 5 the structure of the Green's function in transverse-**k** space was found in terms of the eigenmodes and transfer matrices for any two levels in the medium (5.90). The relation between the Green's functions in the conjugate and given media was then determined, first in transverse-**k** space (5.99), and then by Fourier synthesis in physical space (5.100) The relationship between them was then used to prove the Lorentz reciprocity theorem for currents and fields in real (physical) space. In the present section we proceed in the opposite direction. It will be shown that if Lorentz reciprocity is assumed, then the relation between the given and conjugate Green's functions is very simply derived. The derivation relies on the form of the adjoint Green's function, which was found in transverse-**k** space in Chap. 5, eq. (5.55), although details of the derivation were not given. In this section the form of the adjoint Green's function in terms of the given function is inferred as an immediate consequence of the Lorentz reciprocity relation.

Orthogonal mapping of the Green's function

The dyadic (tensor) Green's function $\mathbf{G}(\mathbf{r},\mathbf{r}_0)$ relates the field $\mathbf{e}(\mathbf{r})$ at an observation point $\mathbf{r}$, to the integrated contributions of currents $\mathbf{j}(\mathbf{r}_0)$ at all points in the region V_1 (5.35), which may be bounded or unbounded,

$$\mathbf{e}(\mathbf{r}) = \int_{V_1} \mathbf{G}(\mathbf{r},\mathbf{r}_0)\mathbf{j}(\mathbf{r}_0)\, d^3r_0 \tag{6.94}$$

(Source points are denoted here by $\mathbf{r}_0$ rather than $\mathbf{r}'$, as in Chap. 5, the primes being reserved in this chapter to denote mapped vectors and tensors.) If the orthogonally mapped fields and currents

$$\mathbf{e}'(\mathbf{r}') = \boldsymbol{\Gamma}\mathbf{e}(\mathbf{r}), \quad \mathbf{j}'(\mathbf{r}') = \boldsymbol{\Gamma}\mathbf{j}(\mathbf{r}), \quad \mathbf{r}' = \gamma\mathbf{r}, \quad \mathbf{r} \in V_1, \quad \mathbf{r}' \in V_2$$

are solutions of Maxwell's equations in the region V_2, implying that the constitutive tensor $\mathbf{K}'(\mathbf{r}')$ and the surface impedance tensor $\mathbf{Z}'(\mathbf{r}')$, if there is one, are related to those in the given region V_1 by an equivalence mapping (6.62), then we may map eq. (6.94) from V_1 to V_2, recalling that $\boldsymbol{\Gamma}^T\boldsymbol{\Gamma}=\mathbf{I}^{(6)}$ (6.47),

$$\begin{aligned} \mathbf{e}'(\mathbf{r}') &= \boldsymbol{\Gamma}\mathbf{e}(\mathbf{r}) = \int \boldsymbol{\Gamma}\mathbf{G}(\mathbf{r},\mathbf{r}_0)\boldsymbol{\Gamma}^T\,\boldsymbol{\Gamma}\mathbf{j}(\mathbf{r}_0)\,d^3r_0 \\ &= \int \boldsymbol{\Gamma}\mathbf{G}(\mathbf{r},\mathbf{r}_0)\boldsymbol{\Gamma}^T\,\mathbf{j}'(\mathbf{r}_0')\,d^3r_0', \qquad \mathbf{r}' = \gamma\mathbf{r}, \quad \mathbf{r}_0' = \gamma\mathbf{r}_0 \\ &:= \int_{V_2} \mathbf{G}'(\mathbf{r}',\mathbf{r}_0')\,\mathbf{j}'(\mathbf{r}_0')\,d^3r_0' \end{aligned} \tag{6.95}$$

Comparison of the last two lines yields finally

$$\mathbf{G}'(\mathbf{r}',\mathbf{r}_0') = \boldsymbol{\Gamma}\mathbf{G}(\mathbf{r},\mathbf{r}_0)\boldsymbol{\Gamma}^T, \qquad \mathbf{r}' = \gamma\mathbf{r} \quad \mathbf{r}_0' = \gamma\mathbf{r}_0 \tag{6.96}$$

The adjoint Green's functions

Suppose that the fields and currents, $\mathbf{e}(\mathbf{r})$ and $\mathbf{j}(\mathbf{r})$, satisfy Maxwell's equations (4.28) in a region V, and that $\bar{\mathbf{e}}(\mathbf{r})$ and $\bar{\mathbf{j}}(\mathbf{r})$ satisfy the adjoint equations (4.29) in the same region. Then the Lorentz reciprocity theorem (4.51) takes the form

$$\int \{\mathbf{e}(\mathbf{r})\cdot\bar{\mathbf{j}}(\mathbf{r}) - \bar{\mathbf{e}}(\mathbf{r})\cdot\mathbf{j}(\mathbf{r})\}\,d^3r = 0 \tag{6.97}$$

If the fields $\mathbf{e}(\mathbf{r})$ and $\bar{\mathbf{e}}(\mathbf{r})$ are related to their sources through the corresponding Green's functions, as in (6.94), this becomes

$$\iint \left\{\bar{\mathbf{j}}^T(\mathbf{r})\,\mathbf{G}(\mathbf{r},\mathbf{r}_0)\,\mathbf{j}(\mathbf{r}_0) - \mathbf{j}^T(\mathbf{r})\,\bar{\mathbf{G}}(\mathbf{r},\mathbf{r}_0)\,\bar{\mathbf{j}}(\mathbf{r}_0)\right\}\,d^3r\,d^3r_0 \tag{6.98}$$

If we transpose the first term, and then interchange the variables $\mathbf{r} \leftrightarrow \mathbf{r}_0$ in the second term (which is legitimate since we integrate over all $\mathbf{r}$ and all $\mathbf{r}_0$ in the same space), (6.98) becomes

$$\iint \mathbf{j}^T(\mathbf{r}_0)\left\{\mathbf{G}^T(\mathbf{r},\mathbf{r}_0) - \bar{\mathbf{G}}(\mathbf{r}_0,\mathbf{r})\right\}\bar{\mathbf{j}}(\mathbf{r})\,d^3r\,d^3r_0 = 0 \tag{6.99}$$

Since this result is true for arbitrary currents, $\mathbf{j}$ and $\bar{\mathbf{j}}$, we conclude, as in [53, eq. 1.1.28], that

$$\bar{\mathbf{G}}(\mathbf{r}_0,\mathbf{r}) = \mathbf{G}^T(\mathbf{r},\mathbf{r}_0) \tag{6.100}$$

which is the same result given in (5.55) for transverse-$\mathbf{k}$ space.

The conjugate Green's function

Let us premultiply the defining equation for the adjoint Green's function by the adjoint operator $\bar{\boldsymbol{\Gamma}}$, remembering that $\bar{\boldsymbol{\Gamma}}^T\bar{\boldsymbol{\Gamma}}=\mathsf{I}^{(6)}$ (6.67),

$$\bar{\boldsymbol{\Gamma}}\bar{\mathsf{e}}(\mathbf{r}_0) = \int_{V_1} \bar{\boldsymbol{\Gamma}}\bar{\mathsf{G}}(\mathbf{r}_0,\mathbf{r})\bar{\boldsymbol{\Gamma}}^T\,\bar{\boldsymbol{\Gamma}}\,\bar{\mathsf{j}}(\mathbf{r})\,d^3r\,, \qquad \mathbf{r}\,,\mathbf{r}_0 \in V_1 \tag{6.101}$$

Now since, by (6.73),

$$\mathsf{e}^c(\mathbf{r}') = \bar{\boldsymbol{\Gamma}}\bar{\mathsf{e}}(\mathbf{r})\,, \qquad \mathsf{j}^c(\mathbf{r}') = \bar{\boldsymbol{\Gamma}}\bar{\mathsf{j}}(\mathbf{r})\,, \qquad \mathbf{r}' = \gamma\mathbf{r} \tag{6.102}$$

we obtain from (6.101), with the aid of (6.100),

$$\begin{aligned}
\mathsf{e}^c(\mathbf{r}_0') &= \int \bar{\boldsymbol{\Gamma}}\bar{\mathsf{G}}(\mathbf{r}_0,\mathbf{r})\bar{\boldsymbol{\Gamma}}^T\,\mathsf{j}^c(\mathbf{r}')\,d^3r'\,, \qquad \mathbf{r}_0' = \gamma\mathbf{r}_0 \\
&= \int \bar{\boldsymbol{\Gamma}}\mathsf{G}^T(\mathbf{r},\mathbf{r}_0)\bar{\boldsymbol{\Gamma}}^T\,\mathsf{j}^c(\mathbf{r}')\,d^3r' \\
&=: \int \mathsf{G}^c(\mathbf{r}_0',\mathbf{r}')\,\mathsf{j}^c(\mathbf{r}')\,d^3r'
\end{aligned} \tag{6.103}$$

the last line being the defining equation for G^c. The equality of the integrands in all three equations, which holds for arbitrary j^c, gives

$$\mathsf{G}^c(\mathbf{r}_0',\mathbf{r}') = \bar{\boldsymbol{\Gamma}}\,\mathsf{G}^T(\mathbf{r},\mathbf{r}_0)\bar{\boldsymbol{\Gamma}}^T = \bar{\boldsymbol{\Gamma}}\,\bar{\mathsf{G}}(\mathbf{r}_0,\mathbf{r})\bar{\boldsymbol{\Gamma}}^T \tag{6.104}$$

with $\mathbf{r}'=\gamma\mathbf{r}$, $\mathbf{r}_0'=\gamma\mathbf{r}_0$, which is the generalization of the result (5.100) found for plane-stratified media under adjoint reflection mappings.

If the defining equation for $\bar{\mathsf{G}}$ is premultiplied by $\bar{\mathsf{I}}$, rather than by $\bar{\boldsymbol{\Gamma}}$ as in (6.101), we get with $\mathsf{e}^{(L)}=\bar{\mathsf{I}}\,\bar{\mathsf{e}}$, $\mathsf{j}^{(L)}=\bar{\mathsf{I}}\,\bar{\mathsf{j}}$ (4.34) and $\bar{\mathsf{I}}=\bar{\mathsf{I}}^{-1}$ (2.81),

$$\bar{\mathsf{I}}\,\bar{\mathsf{e}}(\mathbf{r}_0) = \mathsf{e}^{(L)}(\mathbf{r}_0) = \int \mathsf{G}^{(L)}(\mathbf{r}_0,\mathbf{r})\,\mathsf{j}^{(L)}(\mathbf{r})\,d^3r \tag{6.105}$$

so that the Lorentz-adjoint Green's function is given by

$$\mathsf{G}^{(L)}(\mathbf{r}_0,\mathbf{r}) = \bar{\mathsf{I}}\,\bar{\mathsf{G}}(\mathbf{r}_0,\mathbf{r})\,\bar{\mathsf{I}} \tag{6.106}$$

Comparison with (6.104) yields the conjugate Green's function as an orthogonal mapping of the Lorentz-adjoint Green's function

$$\mathsf{G}^c(\mathbf{r}_0',\mathbf{r}') = \boldsymbol{\Gamma}\mathsf{G}^{(L)}(\mathbf{r}_0,\mathbf{r})\boldsymbol{\Gamma}^T \tag{6.107}$$

We note finally from (6.100), (6.104) and (6.106) that the adjoint, the conjugate and the Lorentz-adjoint Green's functions are mutually related, not unexpectedly, in the same way as the corresponding constitutive tensors.

6.3.2 Mapping the scattering matrices

In Chaps. 2, 3, and 4 the scattering matrices **S** and $\mathbf{S}^c$ were determined for stratified media possessing conjugation symmetry, and were found to be related by a scattering theorem (2.112). A conjugate medium was understood to be one whose constitutive tensor was formed by a reflection mapping with transposition (or time-reversal) of the constitutive tensor of the given medium. In Sec. 6.2.2 the concept of a pair of media conjugate to each other, cf. Secs. 2.4.1 and 2.4.2, was extended to include any pair of media whose constitutive tensors and surface impedances (if any) were mutually transposed when subject to an adjoint mapping, as in (6.84).

We would expect that the scattering theorem (2.112) could also be generalized to accommodate such adjoint transformations. Now the scattering matrices link the amplitudes of incoming and outgoing eigenmodes (2.103), and in order to treat the transformation of scattering matrices, we must first investigate the transformation of eigenmodes and their amplitudes under general conjugation transformations.

Orthogonal transformations in equivalent media

In the 'active' transformations we have been using, the coordinate axes remained fixed while the constitutive tensors, the fields and current distributions were mapped. On the other hand, in the formalism used in Chaps. 2–4 to treat eigenmodes in plane-stratified media, the z-axis was always taken normal to the stratification. In order to combine both formalisms, we shall restrict the mappings to rotations and reflections within the transverse (stratification) plane. The most general transformation or mapping of the plane-stratified system will then be an arbitrary rotation $\boldsymbol{\phi}=\varphi\hat{\mathbf{z}}$ about the z-axis, with or without reflection with respect to any plane containing the z-axis. This reflection plane can be taken as the y=0 plane, without loss of generality, since the reflection of the vector $\mathbf{r}$, and hence of any polar-vector field $\boldsymbol{v}(\mathbf{r})$ with respect to the $\varphi=\varphi_0$ plane (φ_0=const.), is equivalent to a reflection with respect to the φ=0 plane (i.e. the y=0 plane) followed by a rotation of $2\varphi_0$ about the z-axis. Thus, with $\boldsymbol{\phi}=\varphi\hat{\mathbf{z}}$, the transformation matrix has the form, cf. (6.34),

$$\gamma \equiv \begin{bmatrix} 1 & 0 & 0 \\ 0 & \pm 1 & 0 \\ 0 & 0 & 1 \end{bmatrix} \begin{bmatrix} \cos\varphi & -\sin\varphi & 0 \\ \sin\varphi & \cos\varphi & 0 \\ 0 & 0 & 1 \end{bmatrix} = \begin{bmatrix} \gamma^{(2)}(\boldsymbol{\phi}) & 0 \\ 0 & 1 \end{bmatrix}, \quad \det\gamma = \pm 1 \tag{6.108}$$

the plus/minus sign depending on whether or not a reflection is involved. The 2×2 matrix $\gamma^{(2)}$ operates only on the transverse components of a vector.

Consider the homogeneous Maxwell system which, with the plane-wave ansatz $\mathbf{e}(\mathbf{r}) \sim \exp(-i\mathbf{k}\cdot\mathbf{r})$, takes the form, cf. (4.42) and (4.43),

$$\left[c\mathsf{K}(z) - \hat{\mathcal{K}}\right] \mathbf{e}(\mathbf{r}) = 0\,, \qquad \hat{\mathcal{K}} := \frac{1}{k}\mathcal{K} := \begin{bmatrix} 0 & -\hat{\mathbf{k}} \times \mathsf{I}^{(3)} \\ \hat{\mathbf{k}} \times \mathsf{I}^{(3)} & 0 \end{bmatrix} \tag{6.109}$$

where $\mathbf{k}{=}(\mathbf{k}_t\,, k_z)$; in plane-stratified media, K is a function of z only, and $\hat{\mathbf{k}}_t{:=}(s_x\,, s_y)$ is a constant (Snell's law), so that (6.109) becomes an eigenmode equation. If the operator $\hat{\mathcal{K}}$ is split for convenience into three cartesian components, cf. (2.24) and (2.25), and $\mathbf{k}\cdot\mathbf{r}$ in the exponent of the plane-wave ansatz for $\mathbf{e}_\alpha(\mathbf{r})$ is also decomposed, $\mathbf{e}_\alpha(\mathbf{r}) = \mathbf{e}_\alpha(\mathbf{k}_t, z)\exp[-ik_0(s_x x + s_y y)]$, eq. (6.109) becomes

$$\left[c\mathsf{K}(z) - s_x\mathsf{U}_x - s_y\mathsf{U}_y - q_\alpha\mathsf{U}_z\right] \mathbf{e}_\alpha(\mathbf{k}_t, z) = 0 \tag{6.110}$$

in which, with $\hat{\mathbf{k}}_\alpha{:=}(s_x, s_y, q_\alpha)$, the operator $\hat{\mathcal{K}} \to \hat{\mathcal{K}}_\alpha$ in (6.109) is given by

$$\hat{\mathcal{K}}_\alpha := \begin{bmatrix} 0 & -\hat{\mathbf{k}}_\alpha \times \mathsf{I}^{(3)} \\ \hat{\mathbf{k}}_\alpha \times \mathsf{I}^{(3)} & 0 \end{bmatrix} = s_x\mathsf{U}_x + s_y\mathsf{U}_y + q_\alpha\mathsf{U}_z \tag{6.111}$$

When eq. (6.109) with $\mathbf{k}{=}\mathbf{k}_\alpha$, or equivalently eq. (6.110), is mapped from V_1 to V_2 in the transverse plane by means of $\mathit{\Gamma}$ (6.46), in which the constituent 3×3 operator $\gamma{:=}\gamma(\phi)$, as in (6.108), acts only on the transverse vector components, we obtain

$$\begin{aligned} &\mathit{\Gamma}\left[c\mathsf{K} - s_x\mathsf{U}_x - s_y\mathsf{U}_y - q_\alpha\mathsf{U}_z\right]\mathit{\Gamma}^T\mathit{\Gamma}\,\mathbf{e}_\alpha(\mathbf{k}_t, z) \\ &\qquad = \left[c\mathsf{K}' - s'_x\mathsf{U}_x - s'_y\mathsf{U}_y - q'_\alpha\mathsf{U}_z\right]\mathbf{e}'_\alpha(\mathbf{k}'_t, z) \;=\; 0 \end{aligned} \tag{6.112}$$

in which

$$\mathsf{K}'(z) = \mathit{\Gamma}\mathsf{K}(z)\mathit{\Gamma}^T\,, \qquad \hat{\mathcal{K}}'_\alpha = \mathit{\Gamma}\hat{\mathcal{K}}_\alpha\mathit{\Gamma}^T = \begin{bmatrix} 0 & -\hat{\mathbf{k}}'_\alpha \times \mathsf{I}^{(3)} \\ \hat{\mathbf{k}}'_\alpha \times \mathsf{I}^{(3)} & 0 \end{bmatrix}\,;$$

$$\hat{\mathbf{k}}'_\alpha = \gamma\hat{\mathbf{k}}_\alpha \quad \text{with} \quad \begin{bmatrix} s'_x \\ s'_y \end{bmatrix} \equiv \hat{\mathbf{k}}'_t = \gamma^{(2)}\hat{\mathbf{k}}_t \equiv \gamma^{(2)}\begin{bmatrix} s_x \\ s_y \end{bmatrix} \quad \text{and}$$

$$\hat{k}'_z \equiv q'_\alpha = q_\alpha \equiv \hat{k}_z\,; \qquad \mathbf{e}'_\alpha(\mathbf{r}') = \mathit{\Gamma}\mathbf{e}_\alpha(\mathbf{r})\,, \quad \mathbf{r}' = \gamma\mathbf{r}$$

The adjoint eigenmodes are similarly mapped.

The z-component of the concomitant vector, P_z (2.39), is invariant under this mapping, since

$$P_z := \tilde{\bar{\mathbf{e}}}\,\mathsf{U}_z\,\mathbf{e} = \tilde{\bar{\mathbf{e}}}'\mathit{\Gamma}\mathsf{U}_z\mathit{\Gamma}^T\mathbf{e}' = \tilde{\bar{\mathbf{e}}}'\,\mathsf{U}_z\,\mathbf{e}' = P'_z \tag{6.113}$$

as can be seen when U_z is written in terms of the antisymmetric matrices $\mp\hat{\mathbf{z}} \times \mathsf{I}^{(3)}$ (2.25). Thus, with the aid of (6.42), we have

$$\begin{aligned} \boldsymbol{\Gamma}\mathsf{U}_z\boldsymbol{\Gamma}^T &:= \boldsymbol{\Gamma}\begin{bmatrix} 0 & -\hat{\mathbf{z}} \times \mathsf{I}^{(3)} \\ \hat{\mathbf{z}} \times \mathsf{I}^{(3)} & 0 \end{bmatrix}\boldsymbol{\Gamma}^T \\ &= \begin{bmatrix} 0 & -(\det\gamma)\gamma(\hat{\mathbf{z}} \times \mathsf{I}^{(3)})\gamma^T \\ (\det\gamma)\gamma(\hat{\mathbf{z}} \times \mathsf{I}^{(3)})\gamma^T & 0 \end{bmatrix} \\ &= \begin{bmatrix} 0 & -\hat{\mathbf{z}}' \times \mathsf{I}^{(3)} \\ \hat{\mathbf{z}}' \times \mathsf{I}^{(3)} & 0 \end{bmatrix} = \mathsf{U}_z \end{aligned} \tag{6.114}$$

since $\hat{\mathbf{z}}' = \hat{\mathbf{z}}$ under this mapping, as assumed. We could have reached this conclusion directly by noting that the bilinear concomitant vector $\mathbf{P}$ (2.42) is a polar vector, and therefore its z-component is invariant under the assumed mapping.

The modal amplitudes, a_α and a'_α, are thus equal, cf. (2.78), as are also the adjoint amplitudes, $\bar{a}_\alpha$ and $\bar{a}'_\alpha$. The given and the mapped problems are *equivalent*, and the scattering matrices are equal

$$\mathsf{S}'(\mathbf{k}'_t) = \mathsf{S}(\mathbf{k}_t) , \qquad \mathbf{k}'_t = \gamma^{(2)}\mathbf{k}_t \tag{6.115}$$

where $\gamma^{(2)}$ (6.108) is an arbitrary 2×2 orthogonal transformation matrix that maps transverse vector components in the stratification plane.

Adjoint transformations in conjugate media

Consider next the mapping of eigenmodes from a given plane-stratified medium into a conjugate medium, related by [cf. (6.70) and (6.72)]

$$\mathsf{K}^c(z') = \bar{\boldsymbol{\Gamma}}\mathsf{K}^T(z)\bar{\boldsymbol{\Gamma}}^T , \qquad z' = z \tag{6.116}$$

in which $\bar{\boldsymbol{\Gamma}}=\boldsymbol{\Gamma}\bar{\mathsf{I}}=\bar{\mathsf{I}}\,\boldsymbol{\Gamma}$ (6.67) and γ, given by (6.108), generate rotations about the z-axis, with or without reflections with respect to planes containing the z-axis. The z-components (normal to the stratification) of polar-vector fields are thus, as before, unaffected by the mapping. (The z-component of axial-vector fields would be reversed under reflection.)

The adjoint eigenmode equation in the given medium, (2.44), with the plane-wave ansatz $\bar{\mathbf{e}}_\alpha \sim \exp(ik_0\,\bar{q}_\alpha z)$ (2.43), is

$$\left[c\mathsf{K}^T(z) - s_x\mathsf{U}_x - s_y\mathsf{U}_y - \bar{q}_\alpha\mathsf{U}_z\right]\bar{\mathbf{e}}_\alpha(\mathbf{k}_t, z) = 0 \tag{6.117}$$

with $\bar{q}_\alpha = q_\alpha$(2.46). When transforming (mapping) this equation with the aid of $\bar{\Gamma}$ we note, with (6.67) and (2.82), that

$$\bar{\Gamma}\mathsf{U}_i\bar{\Gamma}^T = \Gamma\bar{\mathsf{I}}\,\mathsf{U}_i\bar{\mathsf{I}}\,\Gamma^T = -\Gamma\mathsf{U}_i\Gamma^T\,, \qquad i = x\,,\, y \text{ or } z \tag{6.118}$$

and so, using (6.114) and (6.112), we obtain

$$\begin{aligned} &\bar{\Gamma}\Big[c\mathsf{K}^T - s_x\mathsf{U}_x - s_y\mathsf{U}_y - \bar{q}_\alpha\mathsf{U}_z\Big]\,\bar{\Gamma}^T\,\bar{\Gamma}\bar{\mathbf{e}}_\alpha(\mathbf{k}_t, z) \\ &\qquad = \Big[c\mathsf{K}^c + s'_x\mathsf{U}_x + s'_y\mathsf{U}_y + \bar{q}_\alpha\mathsf{U}_z\Big]\,\bar{\Gamma}\bar{\mathbf{e}}_\alpha(\mathbf{k}_t, z) = 0 \end{aligned} \tag{6.119}$$

Note that when the field $\bar{\mathbf{e}}_\alpha(\mathbf{k}_t,z)$ is mapped with $\bar{\Gamma}$, the transverse propagation vector $\mathbf{k}_t$ is mapped too into $\mathbf{k}'_t = \gamma\mathbf{k}_t$, with $\mathbf{k}'_t = k_0(s'_x\,, s'_y)$ and $\mathbf{k}_t = k_0(s_x\,, s_y)$. Comparison of (6.119) with the conjugate eigenmode equation

$$\Big[c\mathsf{K}^c - s^c_x\mathsf{U}_x - s^c_y\mathsf{U}_y - q^c_{-\alpha}\mathsf{U}_z\Big]\,\mathbf{e}^c_{-\alpha}(\mathbf{k}^c_t, z) = 0 \tag{6.120}$$

cf. (2.90), yields finally

$$-\,q^c_{-\alpha} = \bar{q}_\alpha = q_\alpha\,, \qquad \mathbf{e}^c_{-\alpha}(-s'_x, -s'_y) = \bar{\Gamma}\bar{\mathbf{e}}_\alpha(s_x, s_y) \tag{6.121}$$

by analogy with (2.91), where under reflection with respect to the $y=0$ plane we had $(-s'_x\,, -s'_y) \rightarrow (-s_x\,, s_y)$. Since the adjoint operation is involutary, we have also

$$\bar{\mathbf{e}}^c_{-\alpha}(-s'_x, -s'_y) = \bar{\Gamma}\mathbf{e}_\alpha(s_x, s_y) \tag{6.122}$$

by analogy with (2.94). The conjugate modal amplitudes are directly related to the adjoint modal amplitudes, as in (2.96) with $\bar{\mathbf{e}} = \bar{\mathbf{e}}_\alpha$,

$$\begin{aligned} \bar{a}_\alpha &= \hat{\mathbf{e}}^T_\alpha\mathsf{U}_z\bar{\mathbf{e}}_\alpha\,\mathrm{sgn}(\alpha) = (\hat{\bar{\mathbf{e}}}^c_{-\alpha})^{\mathrm{T}}\,\bar{\Gamma}\mathsf{U}_z\bar{\Gamma}^T\,\mathbf{e}^c_{-\alpha}\,\mathrm{sgn}(\alpha) \\ &= -(\hat{\bar{\mathbf{e}}}^c_{-\alpha})^T\mathsf{U}_z\,\mathbf{e}^c_{-\alpha}\,\mathrm{sgn}(\alpha) = a^c_{-\alpha} \end{aligned} \tag{6.123}$$

with the aid of (6.114) and (6.118), and with $\mathrm{sgn}(\alpha) = -\mathrm{sgn}(-\alpha)$. We conclude then that

$$a^c_\alpha = \bar{a}_{-\alpha}\,, \qquad \bar{a}^c_\alpha = a_{-\alpha} \tag{6.124}$$

From here on the proof of the scattering theorem follows the same lines as in Sec. 2.5. The constancy of the z-component of the concomitant vector, (2.39) and (2.78), leads to a scattering theorem relating the given and adjoint scattering matrices, as in (2.108),

$$\overline{\mathsf{S}}^T\mathsf{S} = \mathsf{I}^{(4)} = \mathsf{S}\overline{\mathsf{S}}^T\,, \qquad \mathsf{S} \equiv \mathsf{S}(\mathbf{k}_t)\,, \quad \overline{\mathsf{S}} \equiv \overline{\mathsf{S}}(\mathbf{k}_t) \tag{6.125}$$

This theorem, together with the relation between conjugate and adjoint modal amplitudes (2.109), then leads to the scattering theorem relating given and conjugate scattering matrices, as in (2.112),

$$\mathsf{S}^c(\mathbf{k}_t^c) = \mathsf{S}^T(\mathbf{k}_t)\,, \qquad \mathbf{k}_t^c = \gamma^{(2)}\mathbf{k}_t \tag{6.126}$$

where $\gamma^{(2)}$ effects an arbitrary rotation or reflection in the tranverse, stratification plane.

We remark finally that the scattering theorems, (2.108) and (2.112), have been generalized in two important respects. In Sec. 2.6 the constancy of the normal (to the stratification) components of the bilinear concomitant vector $\mathbf{P}$ was shown to apply also to curved stratifications, and hence the scattering theorems may be generalized to include curved stratified media. In the present section the concept of a conjugate medium has been generalized to include any plane-stratified medium whose transposed constitutive tensor can be derived, as in (6.116), by any orthogonal adjoint mapping in the transverse, stratification plane.

6.4 Mapping the constitutive tensors

6.4.1 Uniaxial crystalline media

These media are self-adjoint, as we have seen in Sec. 4.1.1, with $\mathsf{K}=\mathsf{K}^T$ (4.1), and are also Lorentz self-adjoint, viz.

$$\mathsf{K}(\mathbf{r}) = \mathsf{K}^{(L)}(\mathbf{r}) := \bar{\mathsf{I}}\,\mathsf{K}^T(\mathbf{r})\,\bar{\mathsf{I}} \tag{6.127}$$

cf. (4.37). Lorentz reciprocity (4.94) thus applies to currents and fields in any spatial region V_1 containing the medium $\mathsf{K}(\mathbf{r})$:

$$\langle \mathbf{e}_a\,,\,\bar{\mathsf{I}}\,\mathbf{j}_b\rangle = \langle \mathbf{e}_b^{(L)}\,,\,\bar{\mathsf{I}}\,\mathbf{j}_a^{(L)}\rangle = \langle \mathbf{e}_b\,,\,\bar{\mathsf{I}}\,\mathbf{j}_a\rangle \tag{6.128}$$

where

$$\langle \mathbf{e}_a\,,\,\bar{\mathsf{I}}\,\mathbf{j}_b\rangle := \int_{V_1} \mathbf{e}_a(\mathbf{r})\cdot\bar{\mathsf{I}}\,\mathbf{j}_b(\mathbf{r})\,d^3r \tag{6.129}$$

Any orthogonal mapping (rotation, reflection or inversion) of the right-hand side of (6.129) from V_1 into V_2, ($\mathbf{r}\in V_1$, $\mathbf{r}'\in V_2$), by means of the matrix $\boldsymbol{\Gamma}$ (6.46) gives, with $\boldsymbol{\Gamma}^T\boldsymbol{\Gamma}=\mathsf{I}^{(6)}$ (6.47) and $\boldsymbol{\Gamma}\bar{\mathsf{I}}=\bar{\mathsf{I}}\boldsymbol{\Gamma}$ (6.67),

$$\begin{aligned}\langle \mathbf{e}_b\,,\,\bar{\mathsf{I}}\,\mathbf{j}_a\rangle &:= \int_{V_1} \tilde{\mathbf{e}}_b(\mathbf{r})\boldsymbol{\Gamma}^T\,\boldsymbol{\Gamma}\,\bar{\mathsf{I}}\,\mathbf{j}_a(\mathbf{r})\,d^3r = \int_{V_2} \tilde{\mathbf{e}}'_b(\mathbf{r}')\,\bar{\mathsf{I}}\,\mathbf{j}'_a(\mathbf{r}')\,d^3r' \\ &= \int_{V_2} \tilde{\mathbf{e}}_{b'}(\mathbf{r}')\,\bar{\mathsf{I}}\,\mathbf{j}_{a'}(\mathbf{r}')\,d^3r' = \langle \mathbf{e}_{b'}\,,\,\bar{\mathsf{I}}\,\mathbf{j}_{a'}\rangle\end{aligned} \tag{6.130}$$

where a, b and a', b' denote symbolically the given and mapped sources respectively: $\mathbf{j}_a \to a$, $\mathbf{j}'_a \equiv \mathbf{j}_{a'} \to a'$. Note that the passage from $\mathbf{e}'_b$ in the first line of the above equation to $\mathbf{e}_{b'}$ in the second line, is not just an alternative notation for the same thing, but expresses the equality between the mapped field $\mathbf{e}'_b(\mathbf{r}')=\boldsymbol{\Gamma}\mathbf{e}(\mathbf{r})$ and the field $\mathbf{e}_{b'}(\mathbf{r}')$ radiated by the mapped current $\mathbf{j}'_b(\mathbf{r}')$ in V_2, as in (4.90). This is equivalent to saying that the mapped field and currents solve Maxwell's equations in the mapped region. Eq. (6.130) states simply that the inner product of $\mathbf{e}_b$ and $\mathbf{j}_a$, a scalar, is invariant under an orthogonal mapping, which is not surprising. In view of (6.128), we deduce from (6.130) that

$$\langle \mathbf{e}_a \, , \bar{\mathbf{I}}\,\mathbf{j}_b \rangle = \langle \mathbf{e}_{b'} \, , \bar{\mathbf{I}}\,\mathbf{j}_{a'} \rangle \tag{6.131}$$

Now we could just as well have have mapped the (scalar) inner product on the left-hand side of (6.128), and obtained directly

$$\langle \mathbf{e}_a \, , \bar{\mathbf{I}}\,\mathbf{j}_b \rangle = \langle \mathbf{e}_{a'} \, , \bar{\mathbf{I}}\,\mathbf{j}_{b'} \rangle \tag{6.132}$$

and we see from (6.131) and (6.132) that the given and mapped sources and fields are both Lorentz-reciprocal and equivalent. Using Rumsey's *reaction* notation, cf. (4.89), we may combine (6.131) and (6.132) in the compact form

$$\langle a \, , b \rangle = \langle b' \, , a' \rangle = \langle a' \, , b' \rangle \tag{6.133}$$

6.4.2 Gyrotropic media

Gyrotropic media have been discussed in some detail in Sec. 4.4. The Lorentz-adjoint medium is characterized by the field-reversed (time-reversed) constitutive tensor, cf. (4.37), (2.36), (2.81) and (2.35),

$$\mathsf{K}^{(L)}(\mathbf{r}) := \bar{\mathbf{I}}\,\mathsf{K}^T(\mathbf{b},\mathbf{r})\,\bar{\mathbf{I}} = \bar{\mathbf{I}}\,\mathsf{K}(-\mathbf{b},\mathbf{r})\,\bar{\mathbf{I}} = \mathsf{K}(-\mathbf{b},\mathbf{r}) \tag{6.134}$$

If $\mathbf{e}_a(\mathbf{r})$ and $\mathbf{j}_a(\mathbf{r})$ satisfy Maxwell's equations in V_1, and $\mathbf{e}_b^{(L)}(\mathbf{r})$ and $\mathbf{j}_b^{(L)}(\mathbf{r})$ satisfy them in the Lorentz-adjoint medium, also in V_1, then as we have seen in (4.94) or (6.128), the two sets of currents and fields are reciprocal. The medium $\mathsf{K}^c(\mathbf{r}')$ formed by any orthogonal mapping of the Lorentz-adjoint medium from V_1 into another region V_2 ($\mathbf{r}' \in V_2$) is the *conjugate* medium:

$$\mathsf{K}^c(\mathbf{r}') = \boldsymbol{\Gamma}\mathsf{K}^{(L)}(\mathbf{r})\boldsymbol{\Gamma}^T \, , \qquad \mathbf{r}' = \gamma\mathbf{r} \tag{6.135}$$

The Lorentz-adjoint fields and currents which are mapped into the conjugate medium $\mathsf{K}^c(\mathbf{r}')$,

$$\mathbf{e}_b^c(\mathbf{r}') = \boldsymbol{\Gamma}\,\mathbf{e}_b^{(L)}(\mathbf{r}) \, , \qquad \mathbf{j}_b^c(\mathbf{r}') = \boldsymbol{\Gamma}\,\mathbf{j}_b^{(L)}(\mathbf{r}) \, , \qquad \mathbf{r}' = \gamma\mathbf{r} \tag{6.136}$$

satisfy Maxwell's equations in that medium, as we have seen in Sec 6.2.1, and are reciprocal to the fields and currents in the given medium,

$$\langle \mathbf{e}_a \, , \bar{\mathbf{I}}\, \mathbf{j}_b \rangle = \langle \mathbf{e}^c_{b'} \, , \bar{\mathbf{I}}\, \mathbf{j}^c_{a'} \rangle \tag{6.137}$$

This result follows immediately from the mapping of the (scalar) inner product $\langle \mathbf{e}^{(L)}_b \, , \bar{\mathbf{I}}\, \mathbf{j}^{(L)}_a \rangle$ (6.128) into the conjugate region.

Orthogonal mappings of the given, rather than the Lorentz-adjoint constitutive tensor, currents and fields, yields an equivalence relation

$$\langle \mathbf{e}_a \, , \bar{\mathbf{I}}\, \mathbf{j}_b \rangle = \langle \mathbf{e}_{a'} \, , \bar{\mathbf{I}}\, \mathbf{j}_{b'} \rangle \, , \qquad \mathsf{K}'(\mathbf{r}') = \boldsymbol{\Gamma} \mathsf{K}(\mathbf{r}) \boldsymbol{\Gamma}^T \tag{6.138}$$

6.4.3 Bianisotropic magnetoelectric media

We consider magnetoelectric crystalline media whose constitutive tensors have the form, as in (4.2) and (4.3),

$$\mathsf{K} \equiv \begin{bmatrix} \boldsymbol{\varepsilon} & \boldsymbol{\xi} \\ \boldsymbol{\xi} & \boldsymbol{\mu} \end{bmatrix} = \mathsf{K}^T , \qquad \boldsymbol{\varepsilon} = \boldsymbol{\varepsilon}^T , \quad \boldsymbol{\mu} = \boldsymbol{\mu}^T , \quad \boldsymbol{\xi} = \boldsymbol{\xi}^T \tag{6.139}$$

in which the 3×3 uniaxial tensors $\boldsymbol{\varepsilon}$, $\boldsymbol{\mu}$ and $\boldsymbol{\xi}$ are all symmetric, and have a common symmetry axis.

The Lorentz-adjoint constitutive tensor

$$\mathsf{K}^{(L)} := \bar{\mathbf{I}}\, \mathsf{K}^T \, \bar{\mathbf{I}} = \bar{\mathbf{I}}\, \mathsf{K} \, \bar{\mathbf{I}} = \begin{bmatrix} \boldsymbol{\varepsilon} & -\boldsymbol{\xi} \\ -\boldsymbol{\xi} & \boldsymbol{\mu} \end{bmatrix} \tag{6.140}$$

is not 'physical' in the following sense. It was pointed out in Sec. 4.1.3 that if K in (6.139) corresponds to a medium in which, under the influence of an external electric or magnetic field, both the electric and magnetic dipole elements become aligned in a direction parallel to the external field, then $\mathsf{K}^{(L)}$ (6.140) corresponds to a medium in which the respective electric and magnetic dipoles become aligned in opposite directions.

Two types of mapping will lead to the same physical constitutive tensor (in the sense just used) in the mapped region. First, a rotation mapping of the given medium,

$$\mathsf{K}'(\mathbf{r}') \equiv \begin{bmatrix} \boldsymbol{\varepsilon}' & \boldsymbol{\xi}' \\ \boldsymbol{\xi}' & \boldsymbol{\mu}' \end{bmatrix} = \boldsymbol{\Gamma} \mathsf{K}(\mathbf{r}) \boldsymbol{\Gamma}^T \tag{6.141}$$

with, cf. (6.29),

$$\mathbf{r}' = \boldsymbol{\gamma} \mathbf{r} \, , \qquad \boldsymbol{\gamma} \equiv \boldsymbol{\gamma}_+(\boldsymbol{\phi}) := \exp\left[\boldsymbol{\phi} \times \mathbf{I}\right] , \qquad \det \boldsymbol{\gamma} = 1 \tag{6.142}$$

leading to

$$\boldsymbol{\Gamma} := \begin{bmatrix} \gamma & 0 \\ 0 & (\det\gamma)\gamma \end{bmatrix} = \begin{bmatrix} \gamma_+(\phi) & 0 \\ 0 & \gamma_+(\phi) \end{bmatrix} \tag{6.143}$$

The 3×3 matrices composing K' in (6.141) are given by

$$\boldsymbol{\varepsilon}'(\mathbf{r}') = \gamma_+ \boldsymbol{\varepsilon}(\mathbf{r}) {\gamma_+}^T, \qquad \boldsymbol{\mu}'(\mathbf{r}') = \gamma_+ \boldsymbol{\mu}(\mathbf{r}) {\gamma_+}^T, \qquad \boldsymbol{\xi}'(\mathbf{r}') = \gamma_+ \boldsymbol{\xi}(\mathbf{r}) {\gamma_+}^T$$

with $\mathbf{r}'{=}\gamma_+\mathbf{r}$, and the mapped matrix K' thus preserves its original form.

Second, a mapping of the Lorentz-adjoint medium consisting of a rotation through an angle $\boldsymbol{\phi}$, followed by inversion:

$$\mathsf{K}'(\mathbf{r}') = \boldsymbol{\Gamma}\mathsf{K}^{(L)}(\mathbf{r})\boldsymbol{\Gamma}^T = \bar{\boldsymbol{\Gamma}}\mathsf{K}^T(\mathbf{r})\bar{\boldsymbol{\Gamma}}^T = \bar{\boldsymbol{\Gamma}}\mathsf{K}(\mathbf{r})\bar{\boldsymbol{\Gamma}}^T, \qquad \mathbf{r}' = \gamma\mathbf{r} \tag{6.144}$$

with $\mathsf{K}^{(L)}$ given by (6.140), $\bar{\boldsymbol{\Gamma}}{:=}\boldsymbol{\Gamma}\,\bar{\mathsf{I}}$ (6.67) and γ, with the aid of (6.38), given by

$$\gamma = \gamma_-(\phi) := -\exp\left[\boldsymbol{\phi} \times \mathsf{I}\right] = -\gamma_+(\phi)\,, \qquad \det\gamma = -1 \tag{6.145}$$

This leads to

$$\bar{\boldsymbol{\Gamma}} := \begin{bmatrix} \gamma & 0 \\ 0 & -(\det\gamma)\gamma \end{bmatrix} = \begin{bmatrix} \gamma_-(\phi) & 0 \\ 0 & \gamma_-(\phi) \end{bmatrix} \tag{6.146}$$

and, with $\gamma_-(\phi) = -\gamma_+(\phi)$ as in (6.145),

$$\begin{gathered} \boldsymbol{\varepsilon}'(\mathbf{r}') = \gamma_- \boldsymbol{\varepsilon}(\mathbf{r}) {\gamma_-}^T = \gamma_+ \boldsymbol{\varepsilon}(\mathbf{r}) {\gamma_+}^T, \qquad \boldsymbol{\mu}'(\mathbf{r}') = \gamma_- \boldsymbol{\mu}(\mathbf{r}) {\gamma_-}^T = \gamma_+ \boldsymbol{\mu}(\mathbf{r}) {\gamma_+}^T \\ \boldsymbol{\xi}'(\mathbf{r}') = \gamma_- \boldsymbol{\xi}(\mathbf{r}) {\gamma_-}^T = \gamma_+ \boldsymbol{\xi}(\mathbf{r}) {\gamma_+}^T, \qquad \mathbf{r}' = \gamma_-\mathbf{r} \end{gathered} \tag{6.147}$$

Thus the constitutive tensor $\mathsf{K}'(\mathbf{r}'{=}\gamma_+\mathbf{r})$ in (6.141), with $\boldsymbol{\Gamma}$ given by (6.143), is identical to $\mathsf{K}'(\mathbf{r}'{=}\gamma_-\mathbf{r})$ in (6.144), where the constituent tensors of $\boldsymbol{\Gamma}$ are given by (6.147), although the spatial structure of the two media is different. But the different spatial structure of $\mathsf{K}'(\mathbf{r}')$ in the two cases, in relation to that of $\mathsf{K}(\mathbf{r})$ in the given medium, leads to the striking difference that in the case of a rotation mapping only an equivalence relation with the given medium is obtained,

$$\langle \mathbf{e}_a\,, \bar{\mathsf{I}}\,\mathbf{j}_b \rangle = \langle \mathbf{e}_{a'}\,, \bar{\mathsf{I}}\,\mathbf{j}_{b'} \rangle \tag{6.148}$$

whereas the rotation plus inversion (or reflection) mapping gives a reciprocity relation only:

$$\langle \mathbf{e}_a\,, \bar{\mathsf{I}}\,\mathbf{j}_b \rangle = \langle \mathbf{e}_{b'}\,, \bar{\mathsf{I}}\,\mathbf{j}_{a'} \rangle \tag{6.149}$$

6.4.4 Bianisotropic moving media

The essential features specific to this type of bianisotropy is exhibited by a medium which is isotropic in its rest frame,

$$\boldsymbol{\varepsilon}(0) = \varepsilon \mathsf{I}^{(3)}, \qquad \boldsymbol{\mu}(0) = \mu \mathsf{I}^{(3)}, \qquad \boldsymbol{\xi}(0) = \boldsymbol{\eta}(0) = 0$$

In the laboratory frame, in which the medium is moving with a velocity $v = \beta c$ in the x-direction, the constitutive tensor, given in (4.4), becomes

$$\mathsf{K} = \begin{bmatrix} \boldsymbol{\varepsilon} & \boldsymbol{\xi} \\ \boldsymbol{\xi}^T & \boldsymbol{\mu} \end{bmatrix} = \mathsf{K}^T, \qquad \boldsymbol{\xi} = \begin{bmatrix} 0 & 0 & 0 \\ 0 & 0 & -\xi \\ 0 & \xi & 0 \end{bmatrix} = -\boldsymbol{\xi}^T \tag{6.150}$$

with $\boldsymbol{\varepsilon}$ and $\boldsymbol{\mu}$ diagonal and $\xi = v(c^2\varepsilon\mu - 1)/c^2(1 - v^2\varepsilon\mu)$.

As in the case of the magnetoelectric medium, the Lorentz-adjoint tensor is given by

$$\mathsf{K}^{(L)} := \bar{\mathsf{I}}\, \mathsf{K}^T\, \bar{\mathsf{I}} = \begin{bmatrix} \boldsymbol{\varepsilon} & -\boldsymbol{\xi} \\ -\boldsymbol{\xi}^T & \boldsymbol{\mu} \end{bmatrix} \tag{6.151}$$

but in this case the medium is 'physical', corresponding to a reversal of its direction of motion with respect to the laboratory frame, $v \to -v$. If everywhere in the medium $\boldsymbol{v} = v\hat{\mathbf{x}}$, then the salient features of the different mappings are demonstrated by simple transformations.

The constitutive tensor K in the given medium (6.150) can be retrieved by a number of mappings. Equivalence relations will result either from a rotation mapping about the x-axis, which is the direction of the relative velocity vector $\boldsymbol{v}$, generated by q_1:

$$\mathbf{r}' = \mathsf{q}_1 \mathbf{r}, \qquad \mathsf{q} = \mathsf{q}_1 := \begin{bmatrix} 1 & 0 & 0 \\ 0 & -1 & 0 \\ 0 & 0 & -1 \end{bmatrix}, \qquad \det \mathsf{q} = 1 \tag{6.152}$$

or by a reflection mapping with respect to the y=0 or z=0 planes, generated by q_{-2} or q_{-3},

$$\mathsf{q} = \mathsf{q}_{-2} := \begin{bmatrix} 1 & 0 & 0 \\ 0 & -1 & 0 \\ 0 & 0 & 1 \end{bmatrix}, \qquad \mathsf{q} = \mathsf{q}_{-3} := \begin{bmatrix} 1 & 0 & 0 \\ 0 & 1 & 0 \\ 0 & 0 & -1 \end{bmatrix}, \qquad \det \mathsf{q} = -1 \tag{6.153}$$

In both cases

$$\mathsf{K}'(\mathbf{r}') = \mathsf{Q}\mathsf{K}(\mathbf{r})\mathsf{Q} = \mathsf{K}(\mathbf{r}), \quad \text{with} \quad \mathsf{Q} := \begin{bmatrix} \mathsf{q} & 0 \\ 0 & (\det \mathsf{q})\mathsf{q} \end{bmatrix} \tag{6.154}$$

and fields and currents in the given medium will be equivalent to the mapped fields and currents in the mapped medium.

The constitutive tensor $\mathsf{K}(\mathbf{r})$ in the given system can also be retrieved by a number of mappings of the Lorentz-adjoint (velocity reversed) system, and the mapped fields and currents will be reciprocal to those in the given system. Thus

$$\mathsf{K}'(\mathbf{r}') = \mathsf{Q}\mathsf{K}^{(L)}(\mathbf{r})\mathsf{Q} = \overline{\mathsf{Q}}\mathsf{K}^{T}(\mathbf{r})\overline{\mathsf{Q}} = \overline{\mathsf{Q}}\mathsf{K}(\mathbf{r})\overline{\mathsf{Q}} = \mathsf{K}(\mathbf{r}) \tag{6.155}$$

if $\mathsf{q}=\mathsf{q}_2$, q_3 or q_{-1}, corresponding to rotations about the y- or z-axes or reflections with respect to the x=0 plane. The three transformations generated by q in these cases reverse the sign of $\boldsymbol{v}$, but this is restored to its original value by the adjoint operator $\overline{\mathsf{Q}}:=\mathsf{Q}\bar{\mathsf{I}}$ (6.63). The mapped fields and currents

$$\mathbf{e}^{c}(\mathbf{r}') = \mathsf{Q}\mathbf{e}^{(L)}(\mathbf{r})\,, \qquad \mathbf{j}^{c}(\mathbf{r}') = \mathsf{Q}\,\mathbf{j}^{(L)}(\mathbf{r})\,, \qquad \mathbf{r}' = \mathsf{q}\,\mathbf{r}$$

are reciprocal to those in the given system,

$$\langle \mathbf{e}_a\,, \bar{\mathsf{I}}\,\mathbf{j}_b\rangle = \langle \mathbf{e}_b^c\,, \bar{\mathsf{I}}\,\mathbf{j}_a^c\rangle \tag{6.156}$$

Chapter 7

Time reversal and reciprocity

A number of phenomena discussed in earlier chapters indicated a close relationship between Lorentz adjointness and time reversal. In Secs. 2.4.1–3, Secs. 6.2.2–3 and also in Secs. 4.4.1, 4.4.3 and 4.5.3 (Figs. 4.2, 4.4 and 4.6) we noted that spatially transformed time-reversed gyrotropic media were 'conjugate' or 'reciprocal' to the given medium. The spatial transformation needed to be no more than an identity transformation, so that it was the time reversal (specifically, the reversal in direction of the external magnetic field) that rendered the medium 'reciprocal'. Furthermore, the Lorentz-adjoint fields and currents that obeyed Maxwell's equations in this reciprocal medium were found to be related to the fields and currents in the given medium by a Lorentz reciprocity relationship, one of the consequences of which was the interchange of roles of receiving and transmitting antennas. This too could be seen as an expression of time reversal of the transmitting-receiving process.

In order to analyze this relationship more closely, we shall develop the time-reversal transformation of time-harmonic vector quantities such as fields and currents, and of tensors such as the constitutive tensor $\mathbf{K}$. We then demonstrate that the time-reversed fields and currents obey Maxwell's equations in the time-reversed medium; in other words, Maxwell's equations are invariant under time reversal. It is shown that the Maxwell time-reversed and Lorentz-adjoint equations are identical, and in source free media the solutions are identical too. In this case time reversal may be employed to give useful physical results.

The concept of time-reversed wave fields is applied to the mapping of ray paths (the trajectory of wave packets) from a given to a reciprocal medium; to the rederivation of the eigenmode scattering theorem for plane-stratified media; and finally, to the generalization of Kerns' scattering theorem to scattering objects imbedded in plane-stratified anisotropic (and possibly absorbing) media which, as a special case, reduces to the problem of a scattering object imbedded in a homogeneous anisotropic medium.

To provide further insight into the relationship between the time-reversed and the Lorentz-adjoint equations, a compressible magnetoplasma is considered, which can support both electromagnetic and acoustic field variables, $\mathbf{E}$, $\mathbf{H}$, $\boldsymbol{v}$ and p, where $\boldsymbol{v}$ and p are the macroscopic velocity and pressure variables of the 'electron gas'. The 10-component electromagnetic-acoustic field, with the corresponding 10-component source terms, have their Lorentz-adjoint and time-reversed counterparts, and a comparison between them provides additional insights. The application of time reversal to a compressible magnetoplasma to obtain a reciprocal medium, and thence a Lorentz reciprocity theorem, has already been employed by Deschamps and Kesler [48, Sec. 6].

All media considered hitherto have been frequency dispersive (implying nonlocality in time in the constitutive relations). For the sake of completeness we consider Lorentz adjointness, reciprocity and time reversal in optically active, chiral media which exhibit spatial dispersion (implying nonlocality in space in the constitutive relations).

Finally, we compare the Lorentz-adjoint and time-reversed Maxwell equations in media *containing sources.* Although the equations are identical, the solutions are different! One is 'causal' and the other is 'non-causal'. Nevertheless, if we consider the 'reaction' of one current system (antenna) on another, via the field it radiates, then the time-reversed wave fields have only one thing in common with the reciprocal, Lorentz-adjoint wave fields—the rays that link the two current systems or antennas. But bearing in mind that any reaction between the two current systems is mediated precisely by these (time-reversible) rays that link the two systems, we may expect the time-reversed solution to produce precisely the response predicted by Lorentz reciprocity.

7.1 Time reversal of time-harmonic quantities

Until now we have confined our attention to fields, currents and related quantities having harmonic $\exp(i\omega t)$ time dependence. These could be regarded as the Fourier spectral components of quantities such as $\mathbf{E}(t)$ or $\mathbf{J}_{\mathrm{m}}(t)$, having arbitrary time dependence, so that typically

$$\mathbf{E}(t) = \frac{1}{2\pi}\int_{-\infty}^{\infty} \mathbf{E}(\omega)\, \exp(i\omega t)\, d\omega \tag{7.1}$$

$$\mathbf{E}(\omega) = \int_{-\infty}^{\infty} \mathbf{E}(t)\, \exp(-i\omega t)\, dt \tag{7.2}$$

Insofar as we have hitherto been discussing time-harmonic fields and currents, we should strictly speaking have denoted them $\mathbf{e}(\omega)$ and $\mathbf{j}(\omega)$ to emphasize that they were quantities specified in the frequency, rather than time, domain. The same is true of the constitutive tensors $\mathbf{K}(\omega)$ that relate the time-harmonic fields, as in (2.20):

$$\mathbf{D}(\omega) = \boldsymbol{\varepsilon}(\omega)\mathbf{E}(\omega) + \boldsymbol{\xi}(\omega)\mathbf{H}(\omega), \qquad \mathbf{B}(\omega) = \boldsymbol{\eta}(\omega)\mathbf{E}(\omega) + \boldsymbol{\mu}(\omega)\mathbf{H}(\omega) \tag{7.3}$$

We suppose, for the present, that the media are not spatially dispersive, so that (7.3) represents a local, point relation in space, and the dependence on the spatial coordinate $\mathbf{r}$ of all quantities in (7.3) has been omitted.

Let us consider, for simplicity, an isotropic medium in which $\mathbf{D}(\omega)$ in (7.3) is given by

$$\mathbf{D}(\omega) = \varepsilon(\omega)\mathbf{E}(\omega) = \varepsilon_0\{1 + \chi(\omega)\}\mathbf{E}(\omega) \tag{7.4}$$

where $\varepsilon(\omega)$ is the scalar permittivity, and $\chi(\omega)$ the scalar susceptibility, of the medium. In this and in the subsequent treatment ω is taken to be real. Fourier transforming back to the time domain yields $\mathbf{D}(t)$ in terms of a convolution integral,

$$\mathbf{D}(t) = \varepsilon_0\mathbf{E}(t) + \varepsilon_0 \int_{-\infty}^{\infty} G(\tau)\mathbf{E}(t-\tau)\, d\tau \tag{7.5}$$

where the susceptibility kernel $G(\tau)$, given by

$$G(\tau) = \frac{1}{2\pi}\int_{-\infty}^{\infty} \chi(\omega)\exp(i\omega\tau)\, d\omega \tag{7.6}$$

may be expected to vanish for $\tau < 0$, cf. [72, Sec. 7.10], because of causality requirements. Eqs. (7.4)–(7.6) imply that if a medium is frequency dispersive, i.e. $\varepsilon(\omega) \neq \text{const}$, then the polarization of the medium given by the nonlocal part of $\mathbf{D}(t)$ in (7.5) depends on the value of the field $\mathbf{E}(t-\tau)$ at earlier times, $t - \tau < t$.

Consider next the time reversal of a time-harmonic field or, what amounts to the same thing, the Fourier transform of a time-reversed field $\mathbf{E}'(t')$, where

$$\mathbf{E}'(t') := \mathcal{T}\mathbf{E}(t) := \mathbf{E}(-t), \qquad t' := -t \tag{7.7}$$

$$\begin{aligned} \mathbf{E}'(\omega) := \mathcal{T}\mathbf{E}(\omega) &= \int_{-\infty}^{\infty} \mathbf{E}(-t)\exp(-i\omega t)\, dt \\ &= \int_{-\infty}^{\infty} \mathbf{E}(t)\exp(i\omega t)\, dt\,, \qquad t \to -t \\ &= \mathbf{E}(-\omega) = \mathbf{E}^*(\omega) \end{aligned} \tag{7.8}$$

Thus the time-reversed field is just the complex conjugate of the given field, $\mathbf{E}'(\omega)=\mathbf{E}^\star(\omega)$. We could just as well use $\mathbf{E}'(\omega) = \mathbf{E}(-\omega)$, but because ω is not always displayed explicitly, it is better practice to use complex conjugation as a general prescription for time reversal. A plane wave propagating in the $\hat{\mathbf{x}}$-direction

$$\mathbf{E}(x,\omega) = \mathbf{E}(\omega)\exp(-ikx)\ , \qquad k = \omega(\varepsilon\mu)^{1/2} \tag{7.9}$$

with ω real, is time reversed to

$$\mathbf{E}'(x,\omega) = \mathbf{E}^\star(\omega)\exp(ik^\star x)\ , \qquad k^\star = \omega(\varepsilon^\star\mu^\star)^{1/2} \tag{7.10}$$

the real part of k and $k^\star$ taken as positive in (7.9) and (7.10). Thus an attenuated plane wave propagating in a given direction is transformed under time reversal into a growing plane wave propagating in the opposite direction. It should be noted that in the above formalism, the harmonic time dependence, $\exp(i\omega t)$, is the same for both given and time-reversed quantities, as can be seen in the Fourier transformation of $\mathbf{E}(t)$ and $\mathbf{E}(-t)$ respectively in (7.2) and (7.8).

Now in the above example we advisedly used the electric field $\mathbf{E}(t)$ which is even under time reversal. Let us explain more precisely what we mean by 'time reversal'. We perform the following thought experiment, already mentioned in Sec. 2.4.1. We imagine the given process, say the emission of an electromagnetic pulse by an antenna into a magnetoplasma, to have been recorded on a movie film. If the film were run *back to front*, we would observe a *reflected* version of the process which would nevertheless obey the laws of physics, since Maxwell's equations are invariant under reflection. But suppose next that the film is run *backwards*, i.e. from the end to the beginning. We would then observe the process with *time reversed.* The original pulse which propagated away from the antenna, would be seen to converge on it. Electrons that moved with a velocity $\boldsymbol{v}$ would appear to move with a velocity $-\boldsymbol{v}$, and by the same token an electric current $\mathbf{J}_e$ (composed of moving charges) would also be reversed in direction. Magnetic fields, that are equivalent to rotating electric charges (see for example Fig. 6.1), would also be reversed in direction, since the rotating charges would be seen to rotate in the opposite sense. Thus some physical quantities like ρ (the electric charge density), $\mathbf{E}$, $\mathbf{J}_m$, $\mathbf{f}$ (force) and p (pressure) are *even* (i.e. retain their sign) *under time reversal*, cf. [72, Sec. 6.11], whereas $\boldsymbol{v}$ (velocity), $\mathbf{H}$, $\mathbf{J}_e$ and $\mathbf{S}$ (the Poynting vector) are *odd* (i.e. change sign) under time reversal. Thus, with primes indicating time-reversed quantities, we may write, with $t' := -t$,

$$\begin{aligned} \rho'(t') &= \rho(-t), \quad \mathbf{E}'(t') = \mathbf{E}(-t), \quad \mathbf{J}'_{\mathrm{m}}(t') = \mathbf{J}_{\mathrm{m}}(-t) \\ \boldsymbol{v}'(t') &= -\boldsymbol{v}(-t), \quad \mathbf{H}'(t') = -\mathbf{H}(-t), \quad \mathbf{J}'_{\mathrm{e}}(t') = -\mathbf{J}_{\mathrm{e}}(-t) \end{aligned} \tag{7.11}$$

Under time reversal the 6-field and 6-current vectors, $\mathbf{e}(t)$ and $\mathbf{j}(t)$, become

$$\mathbf{e}'(t') = \bar{\mathbf{I}}^{(6)}\mathbf{e}(-t) = \begin{bmatrix} \mathbf{E}(-t) \\ -\mathbf{H}(-t) \end{bmatrix}, \quad \mathbf{j}'(t') = -\bar{\mathbf{I}}^{(6)}\mathbf{j}(-t) = -\begin{bmatrix} \mathbf{J}_{\mathrm{e}}(-t) \\ -\mathbf{J}_{\mathrm{m}}(-t) \end{bmatrix} \tag{7.12}$$

with $\bar{\mathbf{I}}^{(6)}$ defined in (2.81). We revert now to time-harmonic quantities, bearing in mind that $\mathbf{H}(t)$ and $\mathbf{J}_{\mathrm{e}}(t)$ are odd under time reversal, so that the prescription (7.8), when applied to the 6-component vectors $\mathbf{e}(\omega)$ and $\mathbf{j}(\omega)$, yields

$$\mathbf{e}'(\omega) = \bar{\mathbf{I}}^{(6)}\mathbf{e}^{\star}(\omega) = \begin{bmatrix} \mathbf{E}^{\star}(\omega) \\ -\mathbf{H}^{\star}(\omega) \end{bmatrix}, \quad \mathbf{j}'(\omega) = -\bar{\mathbf{I}}^{(6)}\mathbf{j}^{\star}(\omega) = -\begin{bmatrix} \mathbf{J}^{\star}_{\mathrm{e}}(\omega) \\ -\mathbf{J}^{\star}_{\mathrm{m}}(\omega) \end{bmatrix} \tag{7.13}$$

Consider finally the constitutive tensor $\mathsf{K}(\omega)$ that relates electric and magnetic fields, as in (2.20),

$$\mathbf{d}(\omega) := \begin{bmatrix} \mathbf{D}(\omega) \\ \mathbf{B}(\omega) \end{bmatrix} = \begin{bmatrix} \boldsymbol{\varepsilon}(\omega) & \boldsymbol{\xi}(\omega) \\ \boldsymbol{\eta}(\omega) & \boldsymbol{\mu}(\omega) \end{bmatrix} \begin{bmatrix} \mathbf{E}(\omega) \\ \mathbf{H}(\omega) \end{bmatrix} \equiv \mathsf{K}(\omega)\mathbf{e}(\omega) \tag{7.14}$$

Taking the complex conjugate of this equation and premultiplying by $\bar{\mathbf{I}}^{(6)}$, with $\bar{\mathbf{I}}^{(6)} = (\bar{\mathbf{I}}^{(6)})^{-1}$ (2.81), we find with the aid of (7.13)

$$\begin{aligned} \bar{\mathbf{I}}^{(6)}\mathbf{d}^{\star}(\omega) := \mathbf{d}'(\omega) &= \{\bar{\mathbf{I}}^{(6)}\mathsf{K}^{\star}(\omega)\,\bar{\mathbf{I}}^{(6)}\}\,\bar{\mathbf{I}}^{(6)}\mathbf{e}^{\star}(\omega) \\ &= \mathsf{K}'(\omega)\mathbf{e}'(\omega) \end{aligned} \tag{7.15}$$

in which we have identified

$$\mathsf{K}'(\omega) := \mathcal{T}\,\mathsf{K}(\omega) = \bar{\mathbf{I}}^{(6)}\mathsf{K}^{\star}(\omega)\,\bar{\mathbf{I}}^{(6)} = \begin{bmatrix} \boldsymbol{\varepsilon}^{\star}(\omega) & -\boldsymbol{\xi}^{\star}(\omega) \\ -\boldsymbol{\eta}^{\star}(\omega) & \boldsymbol{\mu}^{\star}(\omega) \end{bmatrix} \tag{7.16}$$

The change in sign of the off-diagonal matrices in K, viz. $\boldsymbol{\xi}$ and $\boldsymbol{\eta}$, on time reversal was to be expected [101, Sec. 8.1], since each of them links a vector which is odd to one which is even under time reversal.

In the next section we consider examples of time-reversed constitutive tensors.

7.2 Time reversal and Lorentz adjointness

7.2.1 The constitutive tensors

In comparing and attempting to identify the time-reversed with the Lorentz-adjoint constitutive tensors, we use a ***restricted time-reversal*** procedure (discussed in Sec. 2.4.1) which provides a recipe for treating collision losses, expressed in the constitutive tensor K through an imaginary term $i\nu$, where ν is an effective collision frequency. The time-reversal procedure, as pointed out above, changes the sign of all imaginary terms, and in particular of $i\nu$, so that damped plane waves propagating in a given direction are transposed into growing plane waves propagating in the opposite direction. To ensure 'physicality' we therefore adopt the ***restricted time-reversal*** procedure whereby complex conjugation is applied to all terms except those expressing absorption losses. It will subsequently be shown that Maxwell's equations are invariant under time reversal. The restricted time-reversal procedure provides another valid solution of Maxwell's equations, but it will be convenient initially to limit the discussion to loss-free media and to adopt the time-reversal procedure of the preceding section.

We compare the time-reversed constitutive tensor $\mathsf{K}'(\omega){:=}\bar{\mathsf{I}}^{(6)}\mathsf{K}^{\star}(\omega)\bar{\mathsf{I}}^{(6)}$ with the Lorentz-adjoint tensor $\mathsf{K}^{(L)}{=}\bar{\mathsf{I}}^{(6)}\mathsf{K}^{T}\bar{\mathsf{I}}^{(6)}$ (4.37). These are equal provided that the constitutive tensor $\mathsf{K}(\omega)$ is hermitian. Indeed it is if the medium is loss-free, as can be proved from requirements of energy conservation [83, Sec. 1.2b], so that for lossless media we have

$$\mathsf{K}^{\star} = \mathsf{K}^{T}, \quad \boldsymbol{\varepsilon}^{\star} = \boldsymbol{\varepsilon}^{T}, \quad \boldsymbol{\mu}^{\star} = \boldsymbol{\mu}^{T}, \quad \boldsymbol{\xi}^{\star} = \boldsymbol{\eta}^{T}$$

and hence

$$\mathsf{K}'(\omega) := \mathcal{T}\,\mathsf{K}(\omega) = \mathsf{K}^{(L)}(\omega) \tag{7.17}$$

Suppose now that an effective collision frequency ν is included in the constitutive relations to account for dissipation. Let us assume, for simplicity, that in the above only the electric permittivity tensor $\boldsymbol{\varepsilon}$, and hence the susceptibility tensor χ, is anisotropic, where $\boldsymbol{\varepsilon}{=}\varepsilon_0(\mathsf{I}^{(3)}{+}\chi)$. ν will appear in the equation of motion (the equation of momentum transfer) of charged particles, cf. (1.1), in the form

$$m\frac{dv}{dt} + \nu m v = -m\omega^2 U\mathbf{r} = q\mathbf{E} + \{\textit{other forces linear in } \mathbf{r} \textit{ or } \dot{\mathbf{r}}\}$$

with $\dot{\mathbf{r}} = i\omega\mathbf{r}$, $dv/dt = -\omega^2\mathbf{r}$ and $U{:=}1 - i\nu/\omega$. If this is solved for $\mathbf{E}$ in terms of $\mathbf{r}$ (or equivalently, in terms of the polarization vector $\mathbf{P}$ of the medium), we obtain an equation of the form

$$\mathbf{E} = \frac{1}{\varepsilon_0}\chi^{-1}\mathbf{P}$$

with ν appearing as an imaginary term along the diagonal of χ^{-1} in the form $(1 - i\nu/\omega)\mathsf{I}^{(3)} =: U\,\mathsf{I}^{(3)}$. In the absence of absorption χ, and hence also χ^{-1}, are hermitian, so that

$$\mathcal{T}\chi = \chi^* = \chi^T, \qquad \mathcal{T}\chi^{-1} = (\chi^{-1})^* = (\chi^*)^{-1} = (\chi^T)^{-1} \tag{7.18}$$

If absorption is present, we let $\mathcal{T}$ represent the restricted time-reversal operator which imposes complex conjugation on all elements in χ^{-1} except the collision factor $i\nu$ which appears in the term $U\mathsf{I}^{(3)}$ along the main diagonal. This is equivalent to regarding U as a real quantity as far as the complex conjugation is concerned, $U^* \to U$, $(iU)^* \to -iU$, and in this sense both χ^{-1} and χ are 'hermitian'. Thus

$$\chi'(\omega) := \mathcal{T}\chi(\omega) = \chi^T(\omega)$$

as in (7.18), and by generalization we again arrive at (7.17), but with $\mathcal{T}$ now representing the restricted time-reversal operator.

One can easily confirm that all the constitutive tensors discussed until now have been hermitian in the absence of dissipation. We note, inter alia, that in the magnetoplasma (2.35) the imaginary off-diagonal terms are proportional to the external magnetic field **b**, and these change sign in the complex conjugation (time reversal) or when the matrix is transposed (in the Lorentz-adjoint medium).

In the case of moving media that are isotropic in the rest frame, the elements of K (4.4) are all real, but the sign of the velocity vector $\boldsymbol{v}$ contained in the off-diagonal elements, $\pm\xi$, is changed by the $\bar{\mathsf{I}}^{(6)}$ matrices (2.81) both in the time-reversal and the Lorentz-adjoint transformations.

In Sec. 7.5, in our discussion of isotropic chiral media, we shall encounter a constitutive tensor of the form:

$$\mathsf{K} = \frac{1}{1 - \beta^2\varepsilon\mu\omega^2}\begin{bmatrix} \varepsilon\mathsf{I}^{(3)} & -i\beta\varepsilon\mu\omega\mathsf{I}^{(3)} \\ i\beta\varepsilon\mu\omega\mathsf{I}^{(3)} & \mu\mathsf{I}^{(3)} \end{bmatrix} \tag{7.19}$$

where β, a small real constant ($\beta k << 1$), is a measure of the chirality or 'handedness' of the medium. (Kong [83, end of Sec. 1.2c] has termed such a medium 'biisotropic'.) The hermiticity is again evident.

7.2.2 Time reversal of Maxwell's equations

Let us take the complex conjugate of Maxwell's equations (2.21)

$$\left[i\omega\mathsf{K}(\omega,\mathbf{r})+\mathsf{D}\right]\mathbf{e}(\omega,\mathbf{r})=-\mathbf{j}(\omega,\mathbf{r}) \tag{7.20}$$

and premultiply by $-\bar{\mathsf{I}}^{(6)}$, with $\bar{\mathsf{I}}^{(6)}=(\bar{\mathsf{I}}^{(6)})^{-1}$ (2.81), to obtain

$$-\bar{\mathsf{I}}^{(6)}\left[-i\omega\mathsf{K}^{\star}(\omega,\mathbf{r})+\mathsf{D}\right]\bar{\mathsf{I}}^{(6)}\bar{\mathsf{I}}^{(6)}\mathbf{e}^{\star}(\omega,\mathbf{r})=\bar{\mathsf{I}}^{(6)}\mathbf{j}^{\star}(\omega,\mathbf{r})$$

or with D (2.22) transformed as in (2.81), $\bar{\mathsf{I}}^{(6)}\mathsf{D}\,\bar{\mathsf{I}}^{(6)}=-\mathsf{D}$,

$$\left[i\omega\bar{\mathsf{I}}^{(6)}\mathsf{K}^{\star}(\omega,\mathbf{r})\bar{\mathsf{I}}^{(6)}+\mathsf{D}\right]\bar{\mathsf{I}}^{(6)}\mathbf{e}^{\star}(\omega,\mathbf{r})=\bar{\mathsf{I}}^{(6)}\mathbf{j}^{\star}(\omega,\mathbf{r}) \tag{7.21}$$

Application of (7.13) and (7.16) then gives

$$\left[i\omega\mathsf{K}'(\omega,\mathbf{r})+\mathsf{D}\right]\mathbf{e}'(\omega,\mathbf{r})=-\mathbf{j}'(\omega,\mathbf{r}) \tag{7.22}$$

and we have retrieved Maxwell's equations, which are obeyed by the time-reversed fields and currents, $\mathbf{e}'(\omega,\mathbf{r})$ and $\mathbf{j}'(\omega,\mathbf{r})$, in the time-reversed medium $\mathsf{K}'(\omega,\mathbf{r})$.

The equation formally adjoint to (7.20), cf.(4.29), is

$$\left[i\omega\mathsf{K}^{T}(\omega,\mathbf{r})-\mathsf{D}\right]\bar{\mathbf{e}}(\omega,\mathbf{r})=-\bar{\mathbf{j}}(\omega,\mathbf{r}) \tag{7.23}$$

and premultiplication by $\bar{\mathsf{I}}^{(6)}$, with $\bar{\mathsf{I}}^{(6)}=(\bar{\mathsf{I}}^{(6)})^{-1}$, gives

$$\left[i\omega\bar{\mathsf{I}}^{(6)}\mathsf{K}^{T}\bar{\mathsf{I}}^{(6)}+\mathsf{D}\right]\bar{\mathsf{I}}^{(6)}\bar{\mathbf{e}}(\omega,\mathbf{r})=-\bar{\mathsf{I}}^{(6)}\bar{\mathbf{j}}(\omega,\mathbf{r})$$

or

$$\left[i\omega\mathsf{K}^{(L)}(\omega,\mathbf{r})+\mathsf{D}\right]\mathbf{e}^{(L)}(\omega,\mathbf{r})=-\mathbf{j}^{(L)}(\omega,\mathbf{r}) \tag{7.24}$$

with $\mathbf{e}^{(L)}:=\bar{\mathsf{I}}^{(6)}\bar{\mathbf{e}}$, $\mathbf{j}^{(L)}:=\bar{\mathsf{I}}^{(6)}\bar{\mathbf{j}}$, $\mathsf{K}^{(L)}=\bar{\mathsf{I}}^{(6)}\mathsf{K}^{T}\bar{\mathsf{I}}^{(6)}$, as in (4.34) and (4.37).

Comparison of (7.22) with (7.24), with $\mathsf{K}'(\omega,\mathbf{r})=\mathsf{K}^{(L)}(\omega,\mathbf{r})$ (7.17), shows that the time-reversed and Lorentz-adjoint currents and fields obey the same equations. Are we to conclude that if the currents are identical then so too are the fields? We shall address this question in Sec. 7.6, but here the discussion will be confined to media without sources, and first we compare eigenmodes in the source-free region.

Insertion of the plane-wave ansätze

$$\mathbf{e}_{\alpha}(\omega,\mathbf{r})\sim\exp(-i\mathbf{k}_{\alpha}\cdot\mathbf{r}) \tag{7.25}$$

$$\bar{\mathbf{e}}_\beta(\omega, \mathbf{r}) \sim \exp(i\mathbf{k}_\beta \cdot \mathbf{r}) \tag{7.26}$$

in the respective given and adjoint source-free equations, i.e. with $\mathbf{j}$ and $\bar{\mathbf{j}}$ equal zero in (7.20) and (7.23):

$$\left[i\omega\mathsf{K}(\omega, \mathbf{r}) + \mathsf{D}\right] \mathbf{e}(\omega, \mathbf{r}) = 0 \tag{7.27}$$

$$\left[i\omega\mathsf{K}^T(\omega, \mathbf{r}) - \mathsf{D}\right] \bar{\mathbf{e}}(\omega, \mathbf{r}) = 0 \tag{7.28}$$

yields two algebraic equations for the eigenmodes $\mathbf{e}_\alpha$ and $\bar{\mathbf{e}}_\beta$ propagating in the $\pm\hat{\mathbf{k}}$ direction (with $\mathbf{k}_\alpha = k_\alpha \hat{\mathbf{k}}$, $\bar{\mathbf{k}}_\beta = \bar{k}_\beta \hat{\mathbf{k}}$),

$$\left[\omega\mathsf{K}(\omega, \mathbf{r}) - k_\alpha \hat{\mathcal{K}}\right] \mathbf{e}_\alpha = 0 \tag{7.29}$$

$$\left[\omega\mathsf{K}^T(\omega, \mathbf{r}) - \bar{k}_\beta \hat{\mathcal{K}}\right] \bar{\mathbf{e}}_\beta = 0 \tag{7.30}$$

The symmetric matrix $\hat{\mathcal{K}}$ is given by $\mathcal{K} =: k\hat{\mathcal{K}}$ with $\mathcal{K}$ defined in (4.42). The eigenvalue equations

$$\det\left[\omega\mathsf{K}(\omega, \mathbf{r}) - k_\alpha \hat{\mathcal{K}}\right] = 0, \qquad \det\left[\omega\mathsf{K}^T(\omega, \mathbf{r}) - \bar{k}_\beta \hat{\mathcal{K}}\right] = 0 \tag{7.31}$$

with $\hat{\mathcal{K}} = \hat{\mathcal{K}}^T$, are identical, so that

$$k_\alpha = \bar{k}_\alpha \tag{7.32}$$

As to the given and adjoint eigenmodes, $\mathbf{e}_\alpha$ and $\bar{\mathbf{e}}_\beta$, we note that for progressive plane-wave solutions k_α and $\bar{k}_\alpha$ are real, as is $\hat{\mathcal{K}}$. Furthermore, $\mathcal{K}$ is hermitian, i.e. $\mathsf{K}^T = \mathsf{K}^\star$. Hence the entire matrix operator multiplying $\mathbf{e}_\alpha$ in (7.29) is the complex conjugate of that multiplying $\bar{\mathbf{e}}_\beta$ in (7.30). Consequently

$$\bar{\mathbf{e}}_\alpha = \mathbf{e}_\alpha^\star \tag{7.33}$$

Now the time-averaged Poynting vector $\langle \mathbf{S}_\alpha \rangle$ for the eigenmode α in loss-free media is given (aside from a factor 4) as in (2.51), by

$$\langle \mathbf{S}_\alpha \rangle = \mathbf{E}_\alpha \times \mathbf{H}_\alpha^\star + \mathbf{E}_\alpha^\star \times \mathbf{H}_\alpha \tag{7.34}$$

For the adjoint modes, given by (7.33), the time-averaged Poynting vector is unaltered,

$$\langle \bar{\mathbf{S}}_\alpha \rangle = \langle \mathbf{S}_\alpha \rangle \tag{7.35}$$

and therein lies the non-physicality of the adjoint eigenmodes: the direction of phase propagation as given by the plane-wave ansatz (7.26) has been reversed, while the direction of the Poynting vector (7.35) has not. 'Physicality' is restored in the Lorentz-adjoint eigenmodes, cf. (7.24), in which the sign of the magnetic wavefields, and hence of the Poynting vector, is changed by the matrix $\bar{\mathbf{I}}^{(6)}$:

$$\mathbf{e}_\alpha^{(L)} := \bar{\mathbf{I}}^{(6)}\bar{\mathbf{e}}_\alpha = \bar{\mathbf{I}}^{(6)}\mathbf{e}_\alpha^* = \mathbf{e}'_{-\alpha} = \mathcal{T}\mathbf{e}_\alpha \tag{7.36}$$

by virtue of (7.33), (7.13) and (7.7). In a lossy medium the Lorentz-adjoint eigenmodes would correspond to the *restricted* time-reversed eigenmodes, and we should then refrain from using overall complex conjugation if we wish to obtain physically meaningful results.

We note also, for later use, that the given and adjoint normalized eigenmodes, $\hat{\mathbf{e}}_\alpha$ and $\hat{\bar{\mathbf{e}}}_\alpha$, defined in (2.72), are similarly transformed under (restricted) time reversal:

$$\mathcal{T}\hat{\mathbf{e}}_\alpha := \hat{\mathbf{e}}'_{-\alpha} = \bar{\mathbf{I}}^{(6)}\hat{\bar{\mathbf{e}}}_\alpha, \qquad \mathcal{T}\hat{\bar{\mathbf{e}}}_\alpha := \hat{\bar{\mathbf{e}}}'_{-\alpha} = \bar{\mathbf{I}}^{(6)}\hat{\mathbf{e}}_\alpha \tag{7.37}$$

• The reader may well object to our conclusion that in a source-free medium the Lorentz-adjoint and time-reversed eigenmodes are identical, recalling that in Sec. 4.2.3 it was shown that the refractive-index surfaces for the *given* and Lorentz-adjoint eigenmodes were identical. Now the given and time-reversed eigenmodes are surely not always equal. However, that result was obtained for anisotropic (but not bianisotropic) media, in which the eigenvalues k_α satisfied a biquadratic equation, so that for every $\mathbf{k}_\alpha^{(L)} = -\mathbf{k}_\alpha$ there was also a solution $\mathbf{k}_\alpha^{(L)} = \mathbf{k}_\alpha$. This is not true in general for bianisotropic media.

7.2.3 Ray paths—the motion of wave packets

The result just found may be generalized to include the motion of wave packets. A wave packet may be considered to be a linear superposition of a frequency and an angular (directional) spectrum of plane waves or eigenmodes, which in free space for instance could be derived mathematically by a Fourier analysis of the wave packet in ω- and in $\mathbf{k}$-space. (The problematics of eigenmode decomposition in anisotropic media has been illustrated in Chap. 5, but can be performed in principle). If each eigenmode in the spectrum were time-reversed, it would represent a solution of Maxwell's equations in the time-reversed (Lorentz-adjoint) medium, and the superposition of such time-reversed eigenmodes would then yield a time-reversed wave packet propagating 'backwards' in the time-reversed medium. In an absorbing medium the attenuation of the wave packet would be the same for the given and the Lorentz-adjoint (restricted time-reversed) eigenmodes and wave packet.

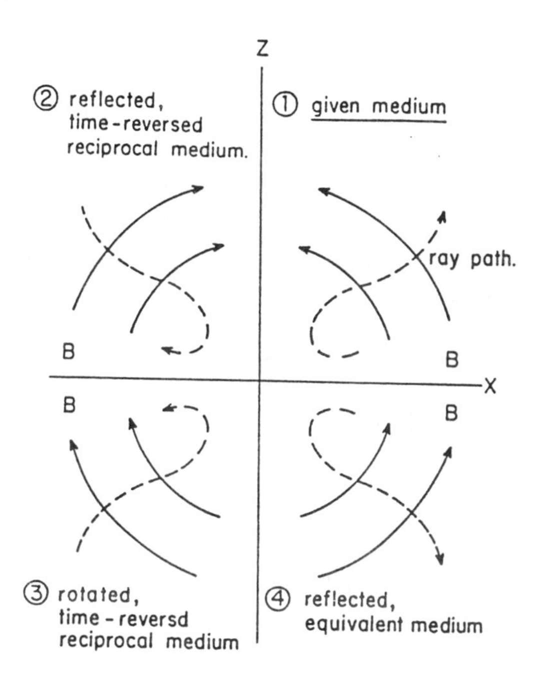

Figure 7.1: Ray paths in a spherically symmetric plasma immersed in an external dipole magnetic field.

The trajectory of the wave packet is the (directed) ***ray path***. If the ray path in the given medium is known, then the reversed ray path (i.e. the reversed trajectory) will also be a valid solution in the time-reversed medium. Spatial mappings of both given and time-reversed trajectories will also be valid solutions in the given and time-reversed mapped media. This is illustrated in Fig. 7.1 in which the medium is the same spherically symmetric plasma immersed in a dipole magnetic field, discussed in Sec. 4.4.1 and illustrated in Fig. 4.2. The given ray path in the first quadrant has been reflected or rotated, with or without time reversal, into the other quadrants, subject to the same spatial or temporal transformations as the media in the corresponding quadrants.

We could have derived this result with the aid of the Lorentz reciprocity theorem using a fixed transmitting (or receiving) antenna and a second, exploring probe antenna (receiving or transmitting) which could follow the given or mapped ray paths in each quadrant. This procedure is cumbersome compared with the intuitively simple result offered by the time-reversal analysis.

7.3 Scattering theorems and time reversal

7.3.1 Scattering from plane-stratified slabs

In Sec. 7.2.2 we discussed the transformation of eigenmodes under (restricted) time reversal. In order to rederive the scattering theorem (2.112) with the aid of time reversal, we must determine how the eigenmode *amplitudes* are transformed under this operation. We recall that the amplitudes, a_α and $\bar{a}_\alpha$, of the respective given and adjoint eigenmodes, $\mathbf{e}_\alpha$ and $\bar{\mathbf{e}}_\alpha$, were related to the normalized eigenmodes by (2.74), viz.

$$\mathbf{e}_\alpha = a_\alpha \hat{\mathbf{e}}_\alpha \,, \qquad \bar{\mathbf{e}}_\alpha = \bar{a}_\alpha \hat{\bar{\mathbf{e}}}_\alpha \tag{7.38}$$

Applying the relations (7.36) and (7.37) between the given and time-reversed eigenmodes to (7.38), we obtain

$$\bar{\mathsf{I}}^{(6)} \bar{\mathbf{e}}'_{-\alpha} = a_\alpha \bar{\mathsf{I}}^{(6)} \hat{\bar{\mathbf{e}}}'_{-\alpha} \tag{7.39}$$

or, upon premultiplication by $\bar{\mathsf{I}}^{(6)} = (\bar{\mathsf{I}}^{(6)})^{-1}$,

$$\bar{\mathbf{e}}'_{-\alpha} =: \bar{a}'_{-\alpha} \hat{\bar{\mathbf{e}}}'_{-\alpha} = a_\alpha \hat{\bar{\mathbf{e}}}'_{-\alpha} \tag{7.40}$$

Hence

$$\bar{a}'_{-\alpha} = a_\alpha \,, \qquad \text{and} \qquad a'_{-\alpha} = \bar{a}_\alpha \tag{7.41}$$

This means that in the treatment of eigenmode scattering from a plane-stratified slab (see Fig. 2.3) all incoming (outgoing) eigenmodes are (restricted) time reversed to the *adjoint* outgoing (incoming) amplitudes:

$$\begin{aligned} \boldsymbol{a}_{in} &= \bar{\boldsymbol{a}}'_{out} \,, & \boldsymbol{a}_{out} &= \bar{\boldsymbol{a}}'_{in} \\ \bar{\boldsymbol{a}}_{in} &= \boldsymbol{a}'_{out} \,, & \bar{\boldsymbol{a}}_{out} &= \boldsymbol{a}'_{in} \end{aligned} \tag{7.42}$$

The incoming- and outgoing-amplitude column matrices, $\boldsymbol{a}_{in}$ and $\boldsymbol{a}_{out}$, and their adjoints are defined as in (2.101) and (2.102).

The definition (2.103) of the scattering matrices S and $\bar{\mathsf{S}}$, viz.

$$\boldsymbol{a}_{out} =: \mathsf{S}\, \boldsymbol{a}_{in} \,, \qquad \bar{\boldsymbol{a}}_{out} =: \bar{\mathsf{S}}\, \bar{\boldsymbol{a}}_{in} \tag{7.43}$$

thus becomes, when the equalities in (7.42) are substituted,

$$\bar{\boldsymbol{a}}'_{in} = \mathsf{S} \bar{\boldsymbol{a}}'_{out} = \mathsf{S}\bar{\mathsf{S}}' \bar{\boldsymbol{a}}'_{in} \tag{7.44}$$

with $\bar{\boldsymbol{a}}'_{out} =: \bar{\mathsf{S}}' \bar{\boldsymbol{a}}'_{in}$. Hence

$$\mathsf{S}\bar{\mathsf{S}}' = \mathsf{I}^{(4)} = \bar{\mathsf{S}}'\mathsf{S} \tag{7.45}$$

since $\overline{\mathbf{S}}' = \mathbf{S}^{-1}$, and similarly

$$\overline{\mathbf{S}}\mathbf{S}' = \mathbf{I}^{(4)} = \mathbf{S}'\overline{\mathbf{S}} \tag{7.46}$$

Now the constancy of the bilinear concomitant vector and the biorthogonality of eigenmodes, (2.39), (2.78) and (2.105), led to the relation (2.108) linking the given and adjoint scattering matrices:

$$\overline{\mathbf{S}}^T\mathbf{S} = \mathbf{I}^{(4)} = \mathbf{S}^T\overline{\mathbf{S}} \tag{7.47}$$

Comparison with (7.46) yields

$$\mathbf{S}' = \mathbf{S}^T \tag{7.48}$$

i.e. the time-reversed scattering matrix is just the transpose of the given scattering matrix. To relate this to the scattering theorem, (2.112) or (6.126), we recall from Sec. 7.2.2 that the (restricted-) time-reversed and the Lorentz-adjoint eigenmodes are identical, and hence the time-reversed and Lorentz-adjoint scattering matrices that relate them are identical too:

$$\mathbf{S}' = \mathbf{S}^{(L)} \tag{7.49}$$

Furthermore, the scattering matrices were shown in Sec. 6.3.2 to be invariant under orthogonal mappings of the field vectors in the stratification plane, cf (6.115). All such invariant mappings of the Lorentz-adjoint scattering matrix $\mathbf{S}^{(L)}$ yield what was termed the conjugate scattering matrix $\mathbf{S}^c$, i.e. the scattering matrix in the conjugate medium. Thus

$$\mathbf{S}^c = \mathbf{S}^{(L)} = \mathbf{S}'$$

and (7.48) becomes

$$\mathbf{S}^c = \mathbf{S}^T \tag{7.50}$$

which is just the generalized scattering theorem given in (6.126).

It will no doubt have been noted that this proof could have been given without mention of time reversal, by use of the Lorentz-adjoint eigenmodes only. However, just as the use of time reversal in the analysis of reciprocal ray paths in the previous section provided an intuitively simpler solution, we believe that this approach in scattering problems provides clearer physical understanding of the mathematical procedure.

7.3.2 Eigenmode generalization of Kerns' scattering theorem

We now make use of the time-reversal approach to outline a proof of an eigenmode generalization of Kerns' scattering theorem, (2.8) and (2.9). Consider a scattering object imbedded in an anisotropic plane-stratified medium, rather than in free space, so that the incoming and outgoing angular spectra of plane waves consist of eigenmodes of the medium for each value of the transverse wave vector $\mathbf{k}_t$. A special case of this generalization is obtained when the plane-stratified medium is replaced by a homogeneous anisotropic medium.

The scattering object, as in Fig. 2.1 and with the notation of Sec. 2.1.2, is contained between two planes, $z = z^-$ and $z = z^+$, parallel to the stratification (if any). The scattering object and the medium between z^- and z^+ are assumed to be loss-free, i.e. absorb no energy, although the rest of the medium may be lossy. As before, the wave fields are assumed to be time harmonic, with $\exp(i\omega t)$ time dependence.

• First, we need to decompose the field $\mathbf{e}(\mathbf{r})$ into eigenmodes (four in all, cf. Sec. 2.2.3) for each value of $\mathbf{k}_t \equiv (k_x,\, k_y)$:

$$\mathbf{e}(x,y,z) = \frac{1}{4\pi^2}\int_{-\infty}^{\infty}\int_{-\infty}^{\infty} \underset{\sim}{a}_\beta(\mathbf{k}_t',z)\,\hat{\mathbf{e}}_\beta(\mathbf{k}_t')\,\exp[-i(k_x'x+k_y'y)]\,dk_x'\,dk_y' \quad (7.51)$$

with summation over the repeated index, $\beta = \pm 1, \pm 2$, implied; $\hat{\mathbf{e}}_\beta$ is a normalized eigenmode, as in (2.72), and $\underset{\sim}{a}_\beta(\mathbf{k}_t',z)$ is an amplitude density in transverse-$\mathbf{k}$ space, cf. eq. (2.4) and the discussion following it, where now

$$\begin{aligned}\underset{\sim}{a}_\alpha(\mathbf{k}_t,z) &= \hat{\hat{\mathbf{e}}}_\alpha^T(\mathbf{k}_t)\,\mathbf{U}_z\,\underset{\sim}{\mathbf{e}}(\mathbf{k}_t,z)\,\mathrm{sgn}(\alpha)\\ &= \int_{-\infty}^{\infty}\int_{-\infty}^{\infty}\hat{\hat{\mathbf{e}}}_\alpha^T(\mathbf{k}_t)\,\mathbf{U}_z\,\mathbf{e}(x,y,z)\,\mathrm{sgn}(\alpha)\,\exp[i(k_xx+k_yy)]\,dx\,dy \quad (7.52)\end{aligned}$$

in which $\hat{\hat{\mathbf{e}}}_\alpha(\mathbf{k}_t)$ is a normalized adjoint eigenmode (2.72) and $\mathbf{U}_z$ is defined in (2.25); $\underset{\sim}{\mathbf{e}}(\mathbf{k}_t,z)$ has been replaced in the second line by its inverse Fourier transform. The first equality in (7.52) is a generalized form of (2.77), which may be recovered if we write

$$\underset{\sim}{a}_\alpha(\mathbf{k}_t) = a_\alpha(\mathbf{k}_t)\delta(\mathbf{k}_t'-\mathbf{k}_t)\,, \qquad \underset{\sim}{\mathbf{e}}(\mathbf{k}_t) = \mathbf{e}(\mathbf{k}_t)\delta(\mathbf{k}_t'-\mathbf{k}_t)$$

If (7.51) is substituted in (7.52) we recover the conditions for orthogonality in $\mathbf{k}_t$-space and for modal biorthogonality (2.50) for any given value of $\mathbf{k}_t$:

$$\begin{aligned}\underset{\sim}{a}_\alpha(\mathbf{k}_t,z) &= \frac{1}{4\pi^2}\iiiint \hat{\hat{\mathbf{e}}}_\alpha^T(\mathbf{k}_t)\mathbf{U}_z\hat{\mathbf{e}}_\beta(\mathbf{k}_t')\,\mathrm{sgn}(\alpha)\\ &\qquad\cdot\exp[i\{(k_x-k_x')x+(k_y-k_y')y\}]\,\underset{\sim}{a}_\beta(\mathbf{k}_t',z)\,dk_x'\,dk_y'\,dx\,dy\\ &= \iint \delta_{\alpha\beta}\,\delta(\mathbf{k}_t-\mathbf{k}_t')\,\underset{\sim}{a}_\beta(\mathbf{k}_t',z)\,d^2k_t' \quad (7.53)\end{aligned}$$

(with $dk'_x\, dk'_y \to d^2k'_t$), in terms of the Kronecker delta $\delta_{\alpha\beta}$ and the Dirac delta function, cf. [72, eq. 2.46],

$$\frac{1}{4\pi^2}\iint \exp[i\{(k_x - k'_x)x + (k_y - k'_y)y\}]\, dx\, dy = \delta(\mathbf{k}_t - \mathbf{k}'_t) \tag{7.54}$$

• Second, we need to express the constancy of the energy flux across all planes in loss-free regions, and in particular across the planes $z = z^+$ and $z = z^-$ between which the scattering object is situated. This flux turns out to be the flux density of the mean Poynting vector $\langle S_z \rangle$ (2.52), summed over all eigenmodes and integrated over the entire transverse-**k** plane. This in turn equals the total flux of the bilinear concomitant vector P_z (2.78) across the planes $z = z^{\pm}$, since $P_z = \langle S_z \rangle$ (2.56) in loss-free regions. Furthermore, the constancy of $P_{z,\alpha}$ (2.49) for any eigenmode in plane-stratified media, whether loss-free or absorbing, implies that the total integrated flux of $P_{z,\alpha}$ has a constant value over the entire region $z \le z^-$ and $z \ge z^+$.

The total time-averaged energy flux across the planes $z = z^-$ and $z = z^+$ is given, with the aid of (2.52), (7.51), (7.25), (7.26) (7.54) and (2.50) by

$$\begin{aligned}
\iint \langle S_z \rangle\, dx\, dy &= \iint \bar{\mathbf{e}}^T(x,y)\mathbf{U}_z \mathbf{e}(x,y)\, dx\, dy \\
&= \frac{1}{4\pi^2}\iiiiiint \bar{a}_\alpha(\mathbf{k}_t)\, \hat{\bar{\mathbf{e}}}_\alpha^T(\mathbf{k}_t)\mathbf{U}_z\, a_\beta(\mathbf{k}'_t)\hat{\mathbf{e}}_\beta(\mathbf{k}'_t) \\
&\qquad \cdot \exp[i\{(k_x - k'_x)x + (k_y - k'_y)y\}]\, dx\, dy\, d^2k'_t\, d^2k_t \\
&= \iiiint \bar{a}_\alpha(\mathbf{k}_t) a_\beta(\mathbf{k}'_t)\, \{\hat{\bar{\mathbf{e}}}_\alpha^T(\mathbf{k}_t)\mathbf{U}_z \hat{\mathbf{e}}_\beta(\mathbf{k}_t)\}\, \delta(\mathbf{k}_t - \mathbf{k}'_t)\, d^2k'_t\, d^2k_t \\
&= \iint \delta_{\alpha\beta}\, \mathrm{sgn}(\alpha)\, \bar{a}_\alpha(\mathbf{k}_t)\, a_\beta(\mathbf{k}_t)\, d^2k_t \\
&= \iint \bar{a}_\alpha(\mathbf{k}_t)\, a_\alpha(\mathbf{k}_t)\, \mathrm{sgn}(\alpha)\, d^2k_t
\end{aligned} \tag{7.55}$$

with summation over repeated subscripts, α or β, implied. This is a straightforward generalization of Parseval's theorem [72, Sec. 14.5],in which the (time-averaged) energy flux density integrated over any plane has been shown to be equal to the energy flux density of all eigenmodes integrated over the $\mathbf{k}_t$-plane. We note that the expression in the last line of (7.55) is just the z-component of the concomitant vector P_z (2.78) integrated over the transverse-**k** plane, which is constant throughout the region $z \le z^-$ and $z \ge z^+$:

$$\iint \bar{a}_\alpha(\mathbf{k}_t)\, a_\alpha(\mathbf{k}_t)\, \mathrm{sgn}(\alpha)\, d^2k_t = \iint P_z(\mathbf{k}_t)\, d^2k_t = \mathrm{const} \tag{7.56}$$

• The next step is to discretize all integrations over the $\mathbf{k}_t$-plane by converting them into matrix products, and thereby enabling us to repeat the various steps in the proof of the scattering theorem for plane-stratified slabs given in the previous section.

We shall suppose that those transverse-$\mathbf{k}$ vectors that are physically relevant to the scattering problem occupy a finite area in the $\mathbf{k}_t$-plane, of the order of πk_{max}^2, where k_{max} is the largest value of $|k_\alpha|$ $(k_\alpha^2 = k_t^2 + k_z^2)$ encountered in the eigenmode decomposition (7.51). If we exclude the possibility of resonances, k_{max} will be finite. Values of $|\,\mathbf{k}_t\,|$ greater k_{max} will correspond to evanescent modes that will not contribute to the energy flux in (7.55) and may consequently be ignored in the application of (7.56).

Let us divide the physically relevant $\mathbf{k}_t$-plane into a very large number N of equal elementary areas, $d^2k_t \rightarrow \delta A_k$, which we shall number in some systematic way from 1 to N. For each elementary area, i.e. for each value of $\mathbf{k}_t$, there will be four corresponding eigenmodes, $\alpha = \pm 1, \pm 2$, as noted earlier, and hence there will be $4N$ different eigenmodes which we shall number as follows:

$$\underbrace{1, 2, \ldots N}_{\alpha=1} \quad \underbrace{N+1, N+2, \ldots 2N}_{\alpha=2} \quad \underbrace{-1, -2, \ldots -N}_{\alpha=-1} \quad \underbrace{-(N+1), -(N+2), \ldots -2N}_{\alpha=-2}$$

Eq. (7.56) now becomes

$$\iint P_z(\mathbf{k}_t)\, d^2k_t \rightarrow \delta A_k \sum_{i=-2N}^{2N} \underset{\sim}{\bar{a}}_i \underset{\sim}{a}_i \operatorname{sgn}(i) = \text{const} \tag{7.57}$$

where $\delta A_k := \delta k_x\, \delta k_y = \text{const}$. It is convenient at this stage to define elementary eigenmode amplitudes, a_i and $\bar{a}_i$,

$$a_i := (\delta A_k)^{1/2} \underset{\sim}{a}_i, \qquad \bar{a}_i := (\delta A_k)^{1/2} \underset{\sim}{\bar{a}}_i \tag{7.58}$$

to replace the amplitude densities, $\underset{\sim}{a}_i$ and $\underset{\sim}{\bar{a}}_i$, so that (7.57) takes the form

$$\sum_{i=-2N}^{2N} \bar{a}_i a_i \operatorname{sgn}(i) = \text{const} \tag{7.59}$$

Equating the integrated flux of P_z on $z = z^<$ with that on $z = z^>$, where $z^<$ and $z^>$ represent any value of z in the respective regions $z \leq z^-$ and $z \geq z^+$, we obtain, after regrouping terms as in (2.106),

$$\sum_{i=1}^{2N} \bar{a}_i(z^<) a_i(z^<) + \sum_{i=-1}^{-2N} \bar{a}_i(z^>) a_i(z^>) = \sum_{i=-1}^{-2N} \bar{a}_i(z^<) a_i(z^<) + \sum_{i=1}^{2N} \bar{a}_i(z^>) a_i(z^>) \tag{7.60}$$

We now introduce the $4N$-element column matrices, $\boldsymbol{a}_{in}$ and $\boldsymbol{a}_{out}$, (or $\bar{\boldsymbol{a}}_{in}$ and $\bar{\boldsymbol{a}}_{out}$), by analogy with (2.102), in terms of the $2N$-element columns $\boldsymbol{a}_+$ and $\boldsymbol{a}_-$, as in (2.101):

$$\boldsymbol{a}_{in} := \begin{bmatrix} \boldsymbol{a}_+(z^<) \\ \boldsymbol{a}_-(z^>) \end{bmatrix} := \begin{bmatrix} a_1(z^<) \\ \vdots \\ a_{2N}(z^<) \\ a_{-1}(z^>) \\ \vdots \\ a_{-2N}(z^>) \end{bmatrix}, \quad \boldsymbol{a}_{out} := \begin{bmatrix} \boldsymbol{a}_-(z^<) \\ \boldsymbol{a}_+(z^>) \end{bmatrix} := \begin{bmatrix} a_{-1}(z^<) \\ \vdots \\ a_{-2N}(z^<) \\ a_1(z^>) \\ \vdots \\ a_{2N}(z^>) \end{bmatrix} \tag{7.61}$$

with the adjoint column matrices $\bar{\boldsymbol{a}}_{in}$ and $\bar{\boldsymbol{a}}_{out}$ defined analogously. Flux conservation of P_z (7.60) then takes the compact form, as in (2.107),

$$\bar{\boldsymbol{a}}_{in}^T \boldsymbol{a}_{in} = \bar{\boldsymbol{a}}_{out}^T \boldsymbol{a}_{out} \tag{7.62}$$

Furthermore, $\boldsymbol{a}_{in}$, $\boldsymbol{a}_{out}$ and their adjoints are related by the $4N \times 4N$ scattering matrices S and $\overline{\mathsf{S}}$, as in (2.103) :

$$\boldsymbol{a}_{out} = \mathsf{S}\,\boldsymbol{a}_{in}, \qquad \bar{\boldsymbol{a}}_{out} = \overline{\mathsf{S}}\,\bar{\boldsymbol{a}}_{in} \tag{7.63}$$

or, written out in full, cf. (2.104),

$$\begin{bmatrix} \boldsymbol{a}_-(z^<) \\ \boldsymbol{a}_+(z^>) \end{bmatrix} = \begin{bmatrix} \mathsf{R}_+(z^<) & \mathsf{T}_-(z^<, z^>) \\ \mathsf{T}_+(z^>, z^<) & \mathsf{R}_-(z^>) \end{bmatrix} \begin{bmatrix} \boldsymbol{a}_+(z^<) \\ \boldsymbol{a}_-(z^>) \end{bmatrix} \tag{7.64}$$

in terms of the $2N \times 2N$ reflection and transmission matrices $\mathsf{R}_\pm$ and $\mathsf{T}_\pm$. Substitution of (7.63) in (7.62) yields

$$\bar{\boldsymbol{a}}_{in}^T \boldsymbol{a}_{in} = \bar{\boldsymbol{a}}_{in}^T \overline{\mathsf{S}}^T \mathsf{S}\,\boldsymbol{a}_{in}$$

or, as in (2.108),

$$\overline{\mathsf{S}}^T \mathsf{S} = \mathsf{I}^{(4N)} = \mathsf{S}\overline{\mathsf{S}}^T = \overline{\mathsf{S}}\mathsf{S}^T \tag{7.65}$$

since $\overline{\mathsf{S}}^T = \mathsf{S}^{-1}$.

• The final step in the proof involves time-reversal of the eigenmodes and their amplitudes, as in Sec. 7.3.1, so that with primes denoting time-reversed quantities we have, as in (7.42),

$$\begin{aligned} \boldsymbol{a}_{in} &= \bar{\boldsymbol{a}}'_{out}, & \boldsymbol{a}_{out} &= \bar{\boldsymbol{a}}'_{in} \\ \bar{\boldsymbol{a}}_{in} &= \boldsymbol{a}'_{out}, & \bar{\boldsymbol{a}}_{out} &= \boldsymbol{a}'_{in} \end{aligned} \tag{7.66}$$

Substitution into the second equation in (7.63) yields, as in (7.46),

$$\boldsymbol{a}'_{in} = \overline{\mathbf{S}}\,\boldsymbol{a}'_{out} = \overline{\mathbf{S}}\mathbf{S}'\boldsymbol{a}'_{in} \tag{7.67}$$

with $\boldsymbol{a}'_{out}$=:$\mathbf{S}'\boldsymbol{a}'_{in}$, and hence

$$\overline{\mathbf{S}}\mathbf{S}' = \mathbf{I}^{(4n)} \tag{7.68}$$

Comparison with (7.65) yields

$$\mathbf{S}' = \mathbf{S}^T \tag{7.69}$$

and since, as noted in Sec. 7.3.1,

$$\mathbf{S}' = \mathbf{S}^{(L)} = \mathbf{S}^c$$

we have finally

$$\mathbf{S}^{(L)} = \mathbf{S}^c = \mathbf{S}^T \tag{7.70}$$

The conjugate scattering matrix $\mathbf{S}^c$ operates on the amplitudes of eigenmodes in the conjugate medium, produced by orthogonal mappings in the transverse stratification plane of the Lorentz-adjoint eigenmodes that propagate in the Lorentz-adjoint medium, with the medium of the scattering object similarly mapped. The equality between the elements of the matrices in (7.70) gives typically

$$S^c_{ij} = S_{ji} \qquad \text{or} \qquad S^c_{\alpha\beta}(-\mathbf{k}^c_t, -\mathbf{k}^{c\prime}_t) = S_{\beta\alpha}(\mathbf{k}'_t, \mathbf{k}_t) \tag{7.71}$$

where $-\mathbf{k}^c_t$ and $-\mathbf{k}^{c\prime}_t$ are the conjugate (time-reversed orthogonal) mappings of $\mathbf{k}_t$ and $\mathbf{k}'_t$ respectively.

If we revert to the original integral, rather than matrix, representation we would have, as in (2.7), a scattering-density matrix element $\underset{\sim}{S}_{\alpha\beta}$ relating amplitude densities:

$$\underset{\sim}{a}^{out}_{\alpha}(\mathbf{k}_t) = \int \underset{\sim}{S}_{\alpha\beta}(\mathbf{k}_t, \mathbf{k}'_t)\, \underset{\sim}{a}^{in}_{\beta}(\mathbf{k}'_t)\, d^2k'_t$$

and the scattering theorem (7.71) for forward scattered eigenmodes, for instance, would have the form, cf. (2.9),

$$\underset{\sim}{S}_{\alpha\beta}(\mathbf{k}_t, \mathbf{k}'_t\,;\, z^>, z^<) = \underset{\sim}{S}^c_{\beta\alpha}(-\mathbf{k}^{c\prime}_t, -\mathbf{k}^c_t\,;\, z^<, z^>) \tag{7.72}$$

If the medium is homogeneous and loss-free, the theorem reduces to a straightforward generalization of Kerns' theorem, where the amplitude densities now refer to the eigenmodes of the anisotropic medium, and the reciprocal scattering situation requires that the constitutive tensors of both the scattering object and the surrounding medium be Lorentz adjoint.

If the medium surrounding the object is plane stratified, then in principle we have scattering both by the object and by the medium. A typical situation could involve, for instance, a scattering object having a scalar constitutive tensor, situated in the high ionospheric magnetoplasma, or below the ionosphere, satisfying the requirement in both cases that it be immersed in a locally loss-free medium. Such a medium, as discussed in Chaps. 2–4, would be self-conjugate under time reversal and reflection, and the scattering theorem, (7.71) and (7.72), would then require the scattering object in the conjugate scattering problem to be a reflection of the given object with respect to a magnetic meridian plane.

7.4 The compressible magnetoplasma

In our discussion of Lorentz reciprocity relating fields and their sources, we found that the (restricted-) time-reversed and Lorentz-adjoint equations were identical, and that both represented the behaviour of fields and currents in 'physically acceptable' media. The question of the physicality of the Lorentz-adjoint magnetoelectric medium, mentioned in Sec. 4.1.3, was problematic, but the problem could be resolved by a further reflection transformation of the medium. In this section we consider a 'warm', compressible electron magnetoplasma, which can support acoustic wave variables, p and $\boldsymbol{v}$, the pressure and macroscopic velocity of the electron gas, in addition to the electromagnetic field variables, $\mathbf{E}$ and $\mathbf{H}$. Whereas the electric field $\mathbf{E}$ is a polar vector even under time reversal, and the magnetic field $\mathbf{H}$ is an axial vector odd under time reversal, the velocity vector $\boldsymbol{v}$ is polar but odd under time reversal.

Another new feature will be that the Lorentz-adjoint medium and fields will not be uniquely defined; the question of the physicality of the time-reversed solutions will also have to be reexamined.

7.4.1 The Maxwell-Euler equations for a compressible magnetoplasma

We consider a warm electron magnetoplasma in which the free electrons, composing an 'electron gas', move freely in a background of heavy and essentially stationary ions. Very low frequency behaviour involving participation of heavy ions in the wave motion, and possibly entailing oscillatory motion of the external magnetic field lines, is not considered. Phenomena such as ion cyclotron whistlers, ion acoustic waves, magnetosonic waves, Alfvén shear waves etc. are thereby excluded. The electron gas is considered to be 'warm' rather than 'hot', implying that the electron thermal velocities are much smaller than the phase velocities of the electomagnetic-like waves that propagate in the plasma, so that the wave-particle interactions characteristic of a hot plasma may be ignored. The problem of radiation and waves in a compressible electron magnetoplasma has been treated by Deschamps and Kesler [48], by Wait [131, Chaps. 5 and 8] and Felsen and Marcuvitz [53, Secs. 1.1c and 1.3d], but the treatment here is based largely on ref. [15].

The governing equations, written out later in (7.77), include Maxwell's equations (2.21), in which the electric current term is separated into an external current source $\mathbf{J_e}$, and an internal current density $n_0 q\boldsymbol{v}$, due to the macroscopic, locally averaged electron velocity $\boldsymbol{v}$ in the plasma; $q{=}-|q|$ and n_0 are the electronic charge and the equilibrium number density respectively. (All equations are linearized, and so we have $n_0 q\boldsymbol{v}$ rather than $nq\boldsymbol{v}$, where n is the instantaneous density.)

Next, we have Euler's equation of motion [88, Sec. 2] governing the macroscopic motion of the electron 'fluid', with the electrons subject to electric and magnetic Lorentz forces, $n_0 q(\mathbf{E} + \boldsymbol{v} \times \mathbf{b})$ per unit volume, a pressure gradient force $-\boldsymbol{\nabla} p$ and an external force density $\mathbf{f}$ (gravity for example), cf. (1.1):

$$n_0 m \frac{\partial \boldsymbol{v}}{\partial t} = -\boldsymbol{\nabla} p + n_0 q\mathbf{E} - n_0 q\, \mathbf{b} \times \boldsymbol{v} + \mathbf{f} \tag{7.73}$$

The hydrodynamic time derivative in Euler's equation,

$$\frac{d\boldsymbol{v}}{dt} \equiv \frac{\partial \boldsymbol{v}}{\partial t} + (\boldsymbol{v} \cdot \boldsymbol{\nabla})\boldsymbol{v}$$

has been linearized — the second term on the right which is second order in $\boldsymbol{v}$ has been discarded; $\mathbf{b}$ is the external magnetic field and m is the electronic mass.

Finally, the changes in velocity and density (and through it of pressure) of the electron gas are related by the linearized equation of continuity:

$$\frac{\partial \rho}{\partial t} + \rho_0 \boldsymbol{\nabla} \cdot \boldsymbol{v} = \rho_0\, s \tag{7.74}$$

where $\rho_0{=}n_0 m$ is the equilibrium density, and $\rho_0\, s$, the source term on the right, represents the rate of electron mass production per unit volume (as for instance by ionizing radiation). With adiabatic (acoustic) pressure and density changes governed by

$$p\rho^{-\gamma} = \text{const}, \qquad \frac{1}{p_0}\frac{\partial p}{\partial t} = \frac{\gamma}{\rho_0}\frac{\partial \rho}{\partial t} \tag{7.75}$$

where $\gamma = 5/3$ is the 'monatomic' electron-gas specific heat ratio (Sec. 1.1), the equation of continuity, cf. (1.2), becomes

$$\frac{1}{\gamma p_0}\frac{\partial p}{\partial t} + \nabla \cdot \boldsymbol{v} = s \tag{7.76}$$

With harmonic time variation of all fields and sources, $\partial/\partial t \to i\omega$, we obtain finally the following Maxwell-Euler system of equations, as derived by Deschamps and Kesler [48, Sec. 2]:

$$\begin{bmatrix} i\omega\varepsilon_0 \mathsf{I}^{(3)} & -\nabla\times\mathsf{I}^{(3)} & n_0 q \mathsf{I}^{(3)} & 0 \\ \nabla\times\mathsf{I}^{(3)} & i\omega\mu_0\mathsf{I}^{(3)} & 0 & 0 \\ -n_0 q\mathsf{I}^{(3)} & 0 & n_0(i\omega m\mathsf{I}^{(3)} + q\mathrm{b}\times\mathsf{I}^{(3)}) & \nabla \\ 0 & 0 & \tilde{\nabla} & \frac{i\omega}{\gamma p_0} \end{bmatrix} \begin{bmatrix} \mathbf{E} \\ \mathbf{H} \\ \boldsymbol{v} \\ p \end{bmatrix} = - \begin{bmatrix} \mathbf{J}_{\mathrm{e}} \\ \mathbf{J}_{\mathrm{m}} \\ -\mathbf{f} \\ -s \end{bmatrix} \tag{7.77}$$

This may be written compactly, as in the Maxwell system (2.21), in terms of the 4-constituent, 10-component plasma fields and sources:

$$\mathsf{L}\mathrm{e} := [i\omega\mathsf{K}(\mathrm{b}) + \mathsf{D}]\,\mathrm{e} = -\mathrm{j} \tag{7.78}$$

where now

$$\mathrm{e} := \begin{bmatrix} \mathbf{E} \\ \mathbf{H} \\ \boldsymbol{v} \\ p \end{bmatrix}, \qquad \mathrm{j} := - \begin{bmatrix} \mathbf{J}_{\mathrm{e}} \\ \mathbf{J}_{\mathrm{m}} \\ -\mathbf{f} \\ -s \end{bmatrix} \tag{7.79}$$

$$\mathsf{K}(q) := \begin{bmatrix} \varepsilon_0\mathsf{I}^{(3)} & 0 & -\frac{i}{\omega}n_0 q\mathsf{I}^{(3)} & 0 \\ 0 & \mu_0\mathsf{I}^{(3)} & 0 & 0 \\ \frac{i}{\omega}n_0 q\mathsf{I}^{(3)} & 0 & n_0\left(m\mathsf{I}^{(3)} - \frac{i}{\omega}q\mathrm{b}\times\mathsf{I}^{(3)}\right) & 0 \\ 0 & 0 & 0 & \frac{1}{\gamma p_0} \end{bmatrix} = \tilde{\mathsf{K}}(-q) \tag{7.80}$$

$$\mathsf{D} := \begin{bmatrix} 0 & -\nabla\times\mathsf{I}^{(3)} & 0 & 0 \\ \nabla\times\mathsf{I}^{(3)} & 0 & 0 & 0 \\ 0 & 0 & 0 & \nabla \\ 0 & 0 & \tilde{\nabla} & 0 \end{bmatrix} = \mathsf{D}^T \tag{7.81}$$

In source-free media, assuming plane-wave solutions $\mathrm{e} \sim \exp[-ik\hat{\mathbf{k}}\cdot\mathbf{r}]$ to (7.78), we obtain an eigenmode equation of degree 6 in k, giving four predominantly electromagnetic, and two predominantly acoustic plane waves, three in the forward and three in the backward directions.

7.4.2 The Lorentz-adjoint system and Lorentz reciprocity

The Maxwell-Euler system formally adjoint to (7.78), cf. (4.29), is

$$\bar{\mathsf{L}}\bar{\mathrm{e}} := \left[i\omega\mathsf{K}^T(\mathrm{b},\mathbf{r}) - \mathsf{D}^T\right]\bar{\mathrm{e}}(\mathbf{r}) = -\bar{\mathrm{j}}(\mathbf{r}) \tag{7.82}$$

in which $\mathsf{K}^T(\mathrm{b})=\mathsf{K}(-\mathrm{b})$ (7.80), due to the antisymmetric term $\mathrm{b}\times\mathsf{I}^{(3)}$ which lies across the main diagonal. Application of the Lagrange identity to (7.78) and (7.82),

$$\bar{\mathrm{e}}^T\mathsf{L}\mathrm{e} - \mathrm{e}^T\bar{\mathsf{L}}\bar{\mathrm{e}} = \nabla\cdot\mathbf{P} \tag{7.83}$$

leads to the reciprocity relation

$$\begin{aligned} -\bar{\mathrm{e}}^T\mathrm{j} + \mathrm{e}^T\bar{\mathrm{j}} &= \bar{\mathrm{e}}^T\mathsf{D}\mathrm{e} + \mathrm{e}^T\mathsf{D}\bar{\mathrm{e}} \\ &= \nabla\cdot(\overline{\mathbf{E}}\times\mathbf{H} + \mathbf{E}\times\overline{\mathbf{H}} + \bar{p}v + p\bar{v}) \end{aligned} \tag{7.84}$$

with the bilinear concomitant vector $\mathbf{P}$ representing a Poynting-like energy flux density. Again, as pointed out in Sec. 4.2.2, the adjoint fields are unphysical in that the direction of energy flow is 'wrong' in relation to the direction of phase propagation. It is easily seen that the direction of energy flow is reversed, and the unphysical adjoint system (7.82) transformed into a physical Lorentz-adjoint Maxwell-Euler system by means of the operator $\bar{\mathsf{I}}^{(10)}$, where $\bar{\mathsf{I}}^{(10)}$ equals $\bar{\mathsf{I}}^{(10)}_{24}$ or $\bar{\mathsf{I}}^{(10)}_{23}$:

$$\bar{\mathsf{I}}^{(10)}_{23} := \begin{bmatrix} \mathsf{I}^{(3)} & \cdot & \cdot & \cdot \\ \cdot & -\mathsf{I}^{(3)} & \cdot & \cdot \\ \cdot & \cdot & -\mathsf{I}^{(3)} & \cdot \\ \cdot & \cdot & \cdot & 1 \end{bmatrix}, \qquad \bar{\mathsf{I}}^{(10)}_{24} := \begin{bmatrix} \mathsf{I}^{(3)} & \cdot & \cdot & \cdot \\ \cdot & -\mathsf{I}^{(3)} & \cdot & \cdot \\ \cdot & \cdot & \mathsf{I}^{(3)} & \cdot \\ \cdot & \cdot & \cdot & -1 \end{bmatrix} \tag{7.85}$$

which changes the sign of $\mathbf{H}$ and of v or p when operating on $\bar{\mathrm{e}}$. Thus, as in (4.32),

$$\mathsf{L}'\mathrm{e} := \bar{\mathsf{I}}^{(10)}\mathsf{L}\mathrm{e} := \bar{\mathsf{I}}^{(10)}[i\omega\mathsf{K} + \mathsf{D}]\,\mathrm{e}(\mathbf{r}) = -\bar{\mathsf{I}}^{(10)}\mathrm{j}(\mathbf{r}) \tag{7.86}$$

and the Lorentz-adjoint system formally adjoint to (7.86), cf. (4.33), becomes

$$\mathsf{L}^{(L)}\mathrm{e}^{(L)} \equiv \bar{\mathsf{L}}'\mathrm{e}^{(L)} = \left[i\omega\mathsf{K}^T - \mathsf{D}\right]\bar{\mathsf{I}}^{(10)}\mathrm{e}^{(L)}(\mathbf{r}) = -\bar{\mathsf{I}}^{(10)}\mathrm{j}^{(L)}(\mathbf{r}) \tag{7.87}$$

Premultiplication of (7.87) by $\bar{\mathsf{I}}^{(10)}$ yields the Maxwell-Euler system

$$\left[i\omega\mathsf{K}^{(L)} + \mathsf{D}\right]\mathrm{e}^{(L)}(\mathbf{r}) = -\mathrm{j}^{(L)}(\mathbf{r}), \qquad \mathsf{K}^{(L)} := \bar{\mathsf{I}}^{(10)}\mathsf{K}^T\bar{\mathsf{I}}^{(10)} \tag{7.88}$$

What about the physicality of $\mathsf{K}^{(L)}$? Let us denote the two possible forms of $\mathsf{K}^{(L)}$ by $\mathsf{K}_{23}^{(L)}$ and $\mathsf{K}_{24}^{(L)}$. We see by inspection that

$$\mathsf{K}_{24}^{(L)} := \bar{\mathsf{I}}_{24}^{(10)}\mathsf{K}^T\bar{\mathsf{I}}_{24}^{(10)} = \mathsf{K}(-q), \quad \text{but} \quad \mathsf{K}_{23}^{(L)} := \bar{\mathsf{I}}_{23}^{(10)}\mathsf{K}^T\bar{\mathsf{I}}_{23}^{(10)} = \mathsf{K}(-\mathrm{b}) \tag{7.89}$$

Now a magnetic field reversed electron plasma is of physical interest, whereas an electric charge conjugated positron plasma is perhaps physical but rather exotic. To obtain physically useful results we shall restrict attention to the magnetic field reversed Lorentz-adjoint media in which $\mathsf{K}^{(L)} \equiv \mathsf{K}_{23}^{(L)}$.

Application of the Lagrange identity to (7.86) and (7.87) with $\bar{\mathsf{I}}^{(10)} = \bar{\mathsf{I}}_{23}^{(10)}$ yields, cf. (4.35),

$$\tilde{\mathbf{e}}^{(L)}\mathsf{L}'\mathbf{e} - \tilde{\mathbf{e}}\bar{\mathsf{L}}'\mathbf{e}^{(L)} = \nabla\cdot\mathbf{P}^{(L)}, \quad \mathbf{P}^{(L)} = \mathbf{E}^{(L)}\times\mathbf{H} - \mathbf{E}\times\mathbf{H}^{(L)} + p^{(L)}\boldsymbol{v} - p\boldsymbol{v}^{(L)} \tag{7.90}$$

On integration over all space the divergence term vanishes, to give

$$\int \tilde{\mathbf{e}}^{(L)}\bar{\mathsf{I}}_{23}^{(10)}\mathbf{j}\, d^3r = \int \tilde{\mathbf{e}}\,\bar{\mathsf{I}}_{23}^{(10)}\mathbf{j}^{(L)}\, d^3r, \qquad \langle \mathbf{e}^{(L)}, \bar{\mathsf{I}}_{23}^{(10)}\mathbf{j}\rangle = \langle \mathbf{e}, \bar{\mathsf{I}}_{23}^{(10)}\mathbf{j}^{(L)}\rangle \tag{7.91}$$

or

$$\int \left(\mathbf{E}^{(L)}\cdot\mathbf{J}_{\mathrm{e}} - \mathbf{H}^{(L)}\cdot\mathbf{J}_{\mathrm{m}} + \boldsymbol{v}^{(L)}\cdot\mathbf{f} - p^{(L)}s\right) d^3r$$
$$= \int \left(\mathbf{E}\cdot\mathbf{J}_{\mathrm{e}}^{(L)} - \mathbf{H}\cdot\mathbf{J}_{\mathrm{m}}^{(L)} + \boldsymbol{v}\cdot\mathbf{f}^{(L)} - ps^{(L)}\right) d^3r \tag{7.92}$$

If the sources are purely electromagnetic, i.e. electric or magnetic currents, the reciprocity relation (7.92) reduces to that of a cold magnetoplasma, dicussed in previous chapters. If the only sources are mechanical transducers, that act as localized distributions of body force $\mathbf{f}$ and that generate or detect the velocity field $\boldsymbol{v}$, then (7.92) reduces to

$$\int \boldsymbol{v}^{(L)}\cdot\mathbf{f}\, d^3r = \int \boldsymbol{v}\cdot\mathbf{f}^{(L)}\, d^3r \tag{7.93}$$

The Maxwell-Euler system, (7.78) or (7.87), is readily shown to be invariant under orthogonal mappings [15], and such mappings of the Lorentz-adjoint system from a region V_1 to a region V_2 yield a conjugate Maxwell-Euler system in the (reciprocal) medium in V_2, cf. (6.74)–(6.76). The resultant reciprocity relations for purely acoustic sources, i.e. for mechanical transducers, are illustrated in Fig. 7.2. The reaction of an acoustic source $\mathbf{f}_a$ (through its velocity field $\boldsymbol{v}_a$) on a second source $\mathbf{f}_b$, equals the reaction of the mapped source $\mathbf{f}_{b'}$ (through its field $\boldsymbol{v}_{b'}$) in the conjugate medium on the mapped source $\mathbf{f}_{a'}$, cf. (4.89) and (4.91):

$$\langle \boldsymbol{v}_a, \mathbf{f}_b\rangle = \langle \boldsymbol{v}_{b'}, \mathbf{f}_{a'}\rangle \tag{7.94}$$

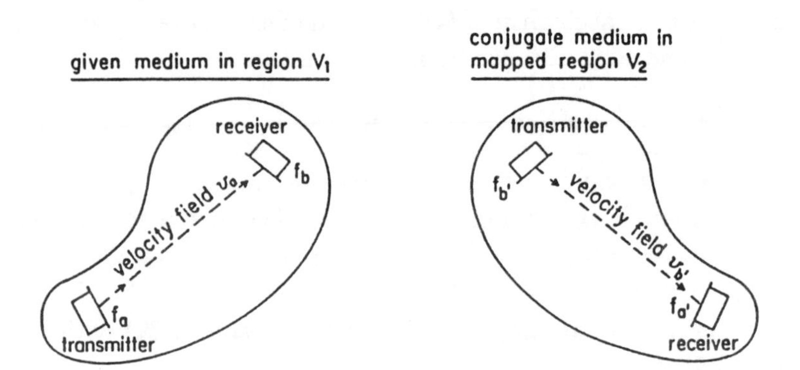

Figure 7.2: Reciprocity between given and mapped acoustic sources and fields in a compressible magnetoplasma.

7.4.3 The time-reversed Maxwell-Euler equations

Following the prescription developed in Sec. 7.1 for (restricted) time reversal of time-harmonic quantities, cf. (7.13), we may time-reverse the plasma field $\mathbf{e} \equiv \mathbf{e}(\omega)$ in (7.79) to give

$$\mathbf{e}' := \mathcal{T}\mathbf{e} := \begin{bmatrix} \mathbf{E}' \\ \mathbf{H}' \\ \boldsymbol{v}' \\ p' \end{bmatrix} = \begin{bmatrix} \mathbf{E}^\star \\ -\mathbf{H}^\star \\ -\boldsymbol{v}^\star \\ p^\star \end{bmatrix} = \bar{\mathsf{I}}_{23}^{(10)} \mathbf{e}^\star \tag{7.95}$$

recalling that $\mathbf{H}$ and $\boldsymbol{v}$ are odd under time reversal.

The source distribution, $\mathbf{j} \equiv \mathbf{j}(\omega)$ is similarly time-reversed,

$$\mathbf{j}' := \mathcal{T}\mathbf{j} := \begin{bmatrix} \mathbf{J}'_e \\ \mathbf{J}'_m \\ -\mathbf{f}' \\ -s' \end{bmatrix} = \begin{bmatrix} -\mathbf{J}^\star_e \\ \mathbf{J}^\star_m \\ -\mathbf{f}^\star \\ s^\star \end{bmatrix} = -\bar{\mathsf{I}}_{23}^{(10)} \mathbf{j}^\star \tag{7.96}$$

with $\mathbf{J}_e$ and s odd under time reversal.

We now take the complex conjugate of the Maxwell-Euler system (7.78) and premultiply by $-\bar{\mathsf{I}}_{23}^{(10)}$, using $\bar{\mathsf{I}}_{23}^{(10)} = \left[\bar{\mathsf{I}}_{23}^{(10)}\right]^{-1}$, to obtain

$$-\bar{\mathsf{I}}_{23}^{(10)} \left[-i\omega \mathsf{K}^\star(\mathrm{b}) + \mathsf{D}\right] \bar{\mathsf{I}}_{23}^{(10)} \bar{\mathsf{I}}_{23}^{(10)} \mathbf{e}^\star = \bar{\mathsf{I}}_{23}^{(10)} \mathbf{j}^\star \tag{7.97}$$

or, with the aid of (7.95), (7.96) and (7.81), this becomes

$$\left[i\omega \mathsf{K}'(\mathrm{b}) + \mathsf{D}\right] \mathbf{e}' = -\mathbf{j}' \tag{7.98}$$

in which we have identified the time-reversed constitutive tensor $\mathsf{K}'(\mathrm{b})$ as

$$\mathsf{K}'(\mathrm{b}) \equiv \bar{\mathsf{I}}_{23}^{(10)}\mathsf{K}^\star(\mathrm{b})\bar{\mathsf{I}}_{23}^{(10)} = \mathsf{K}(-\mathrm{b}) \tag{7.99}$$

and hence, not unexpectedly, the Maxwell-Euler equation is seen to be invariant under time reversal.

It is convenient at this stage to introduce a dissipation term explicitly into the Maxwell-Euler system through an effective collision frequency ν in the Euler equation of motion (7.73). The left-hand term, $n_0 m\partial \boldsymbol{v}/\partial t \rightarrow i\omega n_0 m\boldsymbol{v}$, will be modified by momentum loss to $(i\omega+\nu)n_0 m\boldsymbol{v}$, so that the third diagonal term in the matrix operator in (7.77) will have the form

$$-n_0\left[(i\omega+\nu)\mathsf{I}^{(3)} + q\mathrm{b}\times\mathsf{I}^{(3)}\right]$$

and will appear in $\mathsf{K}(\mathrm{b})$ (7.80) as

$$-n_0\left[m(1-i\frac{\nu}{\omega})\mathsf{I}^{(3)} - \frac{i}{\omega}q\mathrm{b}\times\mathsf{I}^{(3)}\right]$$

The (unrestricted) time reversal transformation will thus lead to

$$\mathsf{K}'(\nu,\mathrm{b}) = \mathsf{K}(-\nu,-\mathrm{b}) \tag{7.100}$$

in which the effect of a changing the sign of the collision frequency will be to convert an absorbing medium into an amplifying medium.

When the time-reversed loss-free tensor $\mathsf{K}'(\mathrm{b})$ (7.99) is compared with the Lorentz-adjoint tensor $\mathsf{K}_{23}^{(L)}$ (7.89) we find that

$$\mathsf{K}' = \mathsf{K}_{23}^{(L)}, \quad \text{i.e.} \quad \bar{\mathsf{I}}_{23}^{(10)}\mathsf{K}^\star\bar{\mathsf{I}}_{23}^{(10)} = \bar{\mathsf{I}}_{23}^{(10)}\mathsf{K}^T(\mathrm{b})\bar{\mathsf{I}}_{23}^{(10)} \tag{7.101}$$

as expected since K is hermitian, $\mathsf{K}^\star=\mathsf{K}^T$, but

$$\mathsf{K}' \neq \mathsf{K}_{24}^{(L)}$$

To summarize, we have found that when there is more than one possible Lorentz-adjoint system, the one that is physically meaningful is that given by the time-reversal transformation.

7.5 Isotropic chiral media

7.5.1 Phenomenological background

Till now we have restricted our attention to media in which the constsitutive tensor $\mathsf{K}(\omega,\mathbf{r})$, which related the fields $\mathbf{D}(\omega,\mathbf{r})$ and $\mathbf{B}(\omega,\mathbf{r})$ to $\mathbf{E}(\omega,\mathbf{r})$ and

$\mathbf{H}(\omega,\mathbf{r})$ in the frequency (Fourier-transform) domain, was frequency dependent (see the discussion in Sec, 7.1). This expressed the nonlocal dependence in the time domain of $\mathbf{D}(t,\mathbf{r})$ on $\mathbf{E}(t,\mathbf{r})$, for instance, in the form of a convolution integral as in (7.5). Such media are said to exhibit *temporal dispersion*. If, on the other hand, the relation between $\mathbf{D}(\mathbf{r})$ and $\mathbf{E}(\mathbf{r})$ involves *spatial* derivatives or integrals, then the dependence of $\mathbf{D}(\mathbf{r})$ on $\mathbf{E}(\mathbf{r})$ will be nonlocal in space, and the medium will be said to be *spatially dispersive*. In Fourier-transform $(\omega,\mathbf{k})$-space this becomes, in general, a point relation:

$$\mathbf{d}(\omega,\mathbf{k}) := \begin{bmatrix} \mathbf{D}(\omega,\mathbf{k}) \\ \mathbf{B}(\omega,\mathbf{k}) \end{bmatrix} = \mathsf{K}(\omega,\mathbf{k}) \begin{bmatrix} \mathbf{E}(\omega,\mathbf{k}) \\ \mathbf{H}(\omega,\mathbf{k}) \end{bmatrix} =: \mathsf{K}(\omega,\mathbf{k})\,\mathbf{e}(\omega,\mathbf{k}) \tag{7.102}$$

We shall complete our discussion of Lorentz-adjointness, reciprocity and time reversal in different types of media by considering an isotropic chiral medium as an example of a spatially dispersive medium. Objects, including molecules, that cannot be superposed on their mirror images are said to possess chirality or 'handedness'. Chiral media, when transparent, are *optically active*, and will rotate the plane of polarization of a monochromatic plane-polarized beam of light either to the left or to the right (the medium is then said to be ℓ-rotatory or r-rotatory). In optics the convention is to specify the sense of rotation with respect to the observer who is looking towards the source. Thus in a circularly-polarized left-handed mode, the electric wave-vector $\mathbf{E}$ at a fixed point in space rotates in an anti-clockwise sense when observed head-on. In the radio-wave and plasma literature the handedness of rotation of circular (elliptic) polarization, is generally specified with respect to an observer who is looking along the direction of propagation $\hat{\mathbf{k}}$, cf. Budden [33, Sec. 4.3]. In this section we adopt the optical convention. An unambiguous characterization of the handedness of a wave is its *helicity* [72, Sec. 7.2], which specifies the direction (+ or −) of the angular momentum vector of the wave with respect to $\hat{\mathbf{k}}$. Thus a left-handed wave is said to have 'positive helicity'.

The optical activity of a medium may stem from the chirality of its crystalline structure, or from the chirality of molecules that are distributed with random orientation in an isotropic host medium. An example of the first type is a quartz (SiO_2) crystal in which the silicon or oxygen atoms lie along helices about the optic axis, with either a left- or a right-handed screw sense [90, Sec. 5.5.4]. Examples of the second type are aqueous solutions of sugar — sucrose or dextrose (d-glucose) which are optically r-rotatory, or levulose (fructose) which is ℓ-rotatory. Naturally occurring amino acids, with a single exception (glycine), are all ℓ-rotatory [70, Sec. 8.10.2].

The *specific rotation* of a sample, which specifies the angle of rotation of a plane-polarized beam of monochromatic light per unit length of the medium, is a function of the wavelength (frequency) of the light. This may be demonstrated by passing a plane-polarized microwave beam through a box containing arbitrarily oriented copper helices [125]. It is found that the specific rotation follows a ***Drude equation*** [49, Sec. 6.5, eq. (36)] of the form

$$\alpha = \frac{A}{\lambda^2 - \lambda_0^2} \tag{7.103}$$

where A is a constant and λ_0 is a characteristic resonant wavelength. α is seen to change sign as a resonant wavelength is crossed, and more than one resonant frequency may be present and affect the optical behaviour in a given frequency region. If an absorption term is inserted in (7.103), the sharp resonance broadens into an absorption band, and (7.103) then gives a reasonably good description of the behaviour of an optical system on both sides of the absorption band [93, Sec. 2.2]. Certain substances,the so-called ***cholesteric liquid crystals***, have a helical molecular structure of very small pitch and exhibit extremely large specific rotations, of the order of $40,000°/\text{mm}$ [70, Sec. 8.10.1].

It is found experimentally that both left and right circularly polarized light can propagate independently in a chiral medium, i.e. they are eigenmodes of the medium. A plane-polarized incident beam may be decomposed into two oppositely rotating circular modes, each travelling with a different phase velocity. The slower mode, say the left-handed mode, will have the smaller wavelength, and hence in its passage through the medium will undergo a larger specific rotation α (7.103) than the other mode, and so the medium will be left-handed. The characteristic behaviour of the real and imaginary components of the modal refractive indices, n_L and n_R, in and around an absorption band, described by A. Cotton in 1896, constitute the ***Cotton effect***, and can be represented by a superposition of two Drude-type equations (7.103), including absorption, for each of the two modes, [93, Sec.2.2].

7.5.2 Eigenmodes in the chiral medium

The constitutive relations that have been proposed, cf. [109, eq. 3], for isotropic chiral media are

$$\begin{aligned} \mathbf{D}(\mathbf{k}) &= \varepsilon[\mathbf{E} - i\beta(k)\,\mathbf{k}\times\mathbf{E}] \\ \mathbf{B}(\mathbf{k}) &= \mu[\mathbf{H} - i\beta(k)\,\mathbf{k}\times\mathbf{H}] \end{aligned} \tag{7.104}$$

where the small chirality constant β ($\beta k << 1$) is a measure of the optical activity, i.e. of the specific rotatory power, of the medium. Since $\mathbf{k}\times\mathbf{E}$ is an axial vector and $\mathbf{k}\times\mathbf{H}$ is polar, (see Sec. 6.1.4), it is evident that β must be a ***pseudoscalar*** that changes sign in a reflection or inversion mapping. The pseudoscalar character of β is also evident from the constitutive tensor K (7.19) for chiral media, which will be derived presently, in view of the discussion in Sec. 6.1.6.

We now substitute these constitutive relations into the source-free ($\mathbf{j}$=0) Maxwell system (2.21), assume a plane-wave ansatz

$$\mathbf{e} \sim \exp(-i\mathbf{k}\cdot\mathbf{r}), \qquad \mathbf{k} = k_\alpha \hat{\mathbf{k}} \tag{7.105}$$

and look for solutions (eigenvalues) k_α for a given direction of propagation, $\hat{\mathbf{k}}$. The ansatz (7.105) converts the differential matrix operator D into an algebraic matrix operator, cf. (4.42) and (4.46),

$$\mathsf{D} \to -ik_\alpha \begin{bmatrix} 0 & -\hat{\mathbf{k}}\times\mathsf{I}^{(3)} \\ \hat{\mathbf{k}}\times\mathsf{I}^{(3)} & 0 \end{bmatrix} =: -ik_\alpha\hat{\mathcal{K}}, \quad \hat{\mathcal{K}} = \hat{\mathcal{K}}^T \tag{7.106}$$

and we obtain, cf. (4.43),

$$\left[i\omega(\mathsf{K}_0 - i\beta k_\alpha\hat{\mathcal{C}}) - ik_\alpha\hat{\mathcal{K}}\right]\mathbf{e}_\alpha(\hat{\mathbf{k}}) = 0 \tag{7.107}$$

where

$$\mathsf{K}_0 := \begin{bmatrix} \varepsilon\mathsf{I}^{(3)} & 0 \\ 0 & \mu\mathsf{I}^{(3)} \end{bmatrix}, \qquad \hat{\mathcal{C}} := \begin{bmatrix} \varepsilon\hat{\mathbf{k}}\times\mathsf{I}^{(3)} & 0 \\ 0 & \mu\hat{\mathbf{k}}\times\mathsf{I}^{(3)} \end{bmatrix} = -\hat{\mathcal{C}}^T \tag{7.108}$$

$\hat{\mathcal{C}}$ may be termed the 'chirality operator'.

Premultiplying (7.107) by $[\hat{\mathbf{k}}^T, \hat{\mathbf{k}}^T]$, we find $\hat{\mathbf{k}}\cdot\mathbf{E}_\alpha = \hat{\mathbf{k}}\cdot\mathbf{H}_\alpha = 0$; in other words, the wave vectors $\mathbf{E}_\alpha$ and $\mathbf{H}_\alpha$, as expected, lie in a plane perpendicular to $\hat{\mathbf{k}}$. Hence (7.107) is a relation between the transverse components of $\mathbf{E}$ and $\mathbf{H}$. Let us take the z-axis along $\hat{\mathbf{k}}$, and express the wave fields in terms of the rotating circular basis vectors:

$$\hat{\epsilon}_\pm := \frac{1}{\sqrt{2}}(\hat{\mathbf{x}} \pm i\hat{\mathbf{y}}), \quad (\hat{\epsilon}_+)^*\cdot\hat{\epsilon}_+ = \hat{\epsilon}_-\cdot\hat{\epsilon}_+ = 1, \quad (\hat{\epsilon}_+)^*\cdot\hat{\epsilon}_- = 0 \tag{7.109}$$

The components $E^\pm$ and $H^\pm$ of the wave fields in this basis are

$$E^\pm = (\hat{\epsilon}_\pm)^*\cdot\mathbf{E} = \hat{\epsilon}_\mp\cdot\mathbf{E}, \qquad H^\pm = (\hat{\epsilon}_\pm)^*\cdot\mathbf{H} = \hat{\epsilon}_\mp\cdot\mathbf{H} \tag{7.110}$$

where

$$\mathbf{E} = \hat{\epsilon}_+E^+ + \hat{\epsilon}_-E^- =: \mathbf{E}_+ + \mathbf{E}_- \quad \text{and} \quad \mathbf{H} = \hat{\epsilon}_+H^+ + \hat{\epsilon}_-H^- =: \mathbf{H}_+ + \mathbf{H}_- \tag{7.111}$$

Furthermore, with $\hat{\mathbf{k}}=\hat{\mathbf{z}}$, we have

$$\hat{\mathbf{k}} \times \hat{\epsilon}_{\pm} = \frac{1}{\sqrt{2}}(\hat{\mathbf{y}} \mp i\hat{\mathbf{x}}) = \frac{\mp i}{\sqrt{2}}(\hat{\mathbf{x}} \pm i\hat{\mathbf{y}}) = \mp i\hat{\epsilon}_{\pm} \tag{7.112}$$

so that

$$\left[\hat{\mathbf{k}} \times \mathsf{I}^{(2)}\right] \cdot \left\{ \begin{matrix} \mathbf{E}_+ \\ \mathbf{E}_- \end{matrix} \right\} = -i \left\{ \begin{matrix} \mathbf{E}_+ \\ -\mathbf{E}_- \end{matrix} \right\}, \quad \left[\hat{\mathbf{k}} \times \mathsf{I}^{(2)}\right] \cdot \left\{ \begin{matrix} \mathbf{H}_+ \\ \mathbf{H}_- \end{matrix} \right\} = -i \left\{ \begin{matrix} \mathbf{H}_+ \\ -\mathbf{H}_- \end{matrix} \right\} \tag{7.113}$$

and (7.107) then separates into two independent pairs of equations,

$$\begin{gathered} \begin{bmatrix} i\omega\varepsilon(1-\beta k_+) & k_+ \\ -k_+ & i\omega\mu(1-\beta k_+) \end{bmatrix} \begin{bmatrix} E^+ \\ H^+ \end{bmatrix} = 0 \\ \begin{bmatrix} i\omega\varepsilon(1+\beta k_-) & -k_- \\ k_- & i\omega\mu(1+\beta k_-) \end{bmatrix} \begin{bmatrix} E^- \\ H^- \end{bmatrix} = 0 \end{gathered} \tag{7.114}$$

The eigenvalues, $k=k_{\pm}$, associated with the wave fields $(\mathbf{E}_+,\mathbf{H}_+)$ and $(\mathbf{E}_-,\mathbf{H}_-)$ respectively, are obtained by equating the determinants of the 2×2 matrices in (7.114) to zero, and we get

$$k_{\pm}^2 = \omega^2 \varepsilon\mu(1 \mp \beta k_{\pm})^2 \tag{7.115}$$

There are two pairs of solutions, which we denote $k_{\pm}^{(+)}$ and $k_{\pm}^{(-)}$, corresponding to propagation in the positive- or negative-z directions,

$$k_{\pm}^{(+)} = -k_{\mp}^{(-)} = \frac{\omega\sqrt{\varepsilon\mu}}{1 \pm \omega\sqrt{\varepsilon\mu}\beta} \tag{7.116}$$

In terms of the eigenmodes k_α in (7.107) this reads

$$k_\alpha = k_{\pm}^{(+)}, \qquad k_{-\alpha} = k_{\mp}^{(-)}, \qquad \alpha = 1\,,\,2$$

When propagating in the positive-z direction, the wave field $(\mathbf{E}_+^{(+)},\mathbf{H}_+^{(+)})$, associated with the eigenvalue $k_+^{(+)}$, is right-handed (in the optical convention), rotating in a clockwise sense in a given plane z=const when viewed head on. When propagation is in the negative-z direction, it is the wave field $(\mathbf{E}_-^{(-)},\mathbf{H}_-^{(-)})$, associated with the eigenvalue $k_-^{(-)}$, that is right-handed. Thus in a fixed plane, transverse to $\hat{\mathbf{k}}$, they rotate in opposite directions, but when viewed head on from opposite directions they have the same 'handedness'. Compare this with the circularly polarized modes that propagate in a magnetoplasma, parallel or anti-parallel to the external magnetic field $\mathbf{b}$. The labelling of the eigenmodes there depends only on the sense of rotation of the wave-field vectors with respect to the external magnetic field, and not on their sense of rotation with respect to the direction of propagation $\hat{\mathbf{k}}$.

Let us consider only the positive eigenmode solutions of (7.115), $k_\pm = k_\pm^{(+)}$, bearing in mind that the negative-going eigenmodes are related to them by (7.116). The wavelengths $\lambda_\pm$ of the two eigenmodes are, with (7.116),

$$\lambda_\pm = \frac{2\pi}{k_\pm} = \frac{2\pi}{\omega\sqrt{\varepsilon\mu}} \pm 2\pi\beta \tag{7.117}$$

so that $2\pi\beta$ is seen to be the increase or decrease of the wavelength due to the chirality of the medium. In a distance Δz the k_- wave will rotate through an angle $k_-\Delta z$ to the left, the k_+ wave will rotate through $k_+\Delta z$ to the right, and a plane-polarized wave will rotate through $\Delta\varphi$ to the left or right, according as $\beta \gtrless 0$:

$$\Delta\varphi = \frac{k_- - k_+}{2}\Delta z = \frac{\omega^2\varepsilon\mu\beta}{1-\omega^2\varepsilon\mu\beta^2}\Delta z \approx \beta\omega^2\varepsilon\mu\,\Delta z \tag{7.118}$$

with the aid of (7.116).

7.5.3 The Lorentz-adjoint system and Lorentz reciprocity

With the purpose of deriving a Lorentz-adjoint Maxwell system, we note that the constitutive relation (7.104) may be derived from the relationship [86, eq. 2],

$$\mathbf{d} := \begin{bmatrix} \mathbf{D} \\ \mathbf{B} \end{bmatrix} = \begin{bmatrix} \varepsilon(\mathsf{I}^{(3)} + \beta\nabla\times\mathsf{I}^{(3)}) & 0 \\ 0 & \mu(\mathsf{I}^{(3)} + \beta\nabla\times\mathsf{I}^{(3)}) \end{bmatrix} \begin{bmatrix} \mathbf{E} \\ \mathbf{H} \end{bmatrix} =: [\mathsf{K}_0 + \beta\mathsf{D}_0]\,\mathbf{e} \tag{7.119}$$

with

$$\mathsf{K}_0 := \begin{bmatrix} \varepsilon\mathsf{I}^{(3)} & 0 \\ 0 & \mu\mathsf{I}^{(3)} \end{bmatrix}, \qquad \mathsf{D}_0 := \begin{bmatrix} \varepsilon\nabla\times\mathsf{I}^{(3)} & 0 \\ 0 & \mu\nabla\times\mathsf{I}^{(3)} \end{bmatrix} = -\mathsf{D}_0^T \tag{7.120}$$

when a plane-wave ansatz (7.105) is assumed.

The Maxwell system (2.21), with $\exp(i\omega t)$ time dependence, then becomes

$$\mathsf{L}\,\mathbf{e} := [i\omega(\mathsf{K}_0 + \beta\mathsf{D}_0) + \mathsf{D}]\,\mathbf{e} = -\mathbf{j} \tag{7.121}$$

with $\mathsf{D} = \mathsf{D}^T$ defined in (2.22). Following the procedure adopted in Sec. 4.2.2, we write

$$\mathsf{L}'\mathbf{e} := \bar{\mathsf{I}}\,\mathsf{L}\,\mathbf{e} := \bar{\mathsf{I}}\,[i\omega(\mathsf{K}_0 + \beta\mathsf{D}_0) + \mathsf{D}]\,\mathbf{e} = -\bar{\mathsf{I}}\,\mathbf{j} \tag{7.122}$$

and the equation formally adjoint to (7.122) is then

$$\begin{aligned} \mathsf{L}^{(L)}\mathbf{e}^{(L)} &\equiv \bar{\mathsf{L}}'\mathbf{e}^{(L)} := \left[i\omega(\mathsf{K}_0^T - \beta\mathsf{D}_0^T) - \mathsf{D}^T\right]\bar{\mathsf{I}}^T\mathbf{e}^{(L)} \\ &= \left[i\omega(\mathsf{K}_0 + \beta\mathsf{D}_0) - \mathsf{D}\right]\bar{\mathsf{I}}\,\mathbf{e}^{(L)} = -\bar{\mathsf{I}}\,\mathbf{j}^{(L)} \end{aligned} \tag{7.123}$$

with the aid of (7.120) and the substitutions $\mathsf{D}^T{=}\mathsf{D}$ (2.22) and $\bar{\mathsf{I}}{=}\bar{\mathsf{I}}^T$ (2.81). If (7.123) is premultiplied by $\bar{\mathsf{I}}$, the Maxwell system (7.121) is recovered with the field $\mathbf{e}$ and current $\mathbf{j}$ replaced by $\mathbf{e}^{(L)}$ and $\mathbf{j}^{(L)}$ respectively. This means that *the isotropic chiral medium is Lorentz self-adjoint or self-reciprocal.* Application of the Lagrange identity to (7.122) and (7.123) yields, cf. (4.35),

$$\tilde{\mathbf{e}}^{(L)}\mathsf{L}'\mathbf{e} - \tilde{\mathbf{e}}\bar{\mathsf{L}}'\mathbf{e}^{(L)} = \nabla\cdot\mathbf{P}^{(L)}$$

$$\mathbf{P}^{(L)} = (\mathbf{E}^{(L)}\times\mathbf{H} - \mathbf{E}\times\mathbf{H}^{(L)}) + i\omega\beta(\varepsilon\mathbf{E}\times\mathbf{E}^{(L)} - \mu\mathbf{H}\times\mathbf{H}^{(L)}) \tag{7.124}$$

To show that $\mathbf{P}^{(L)}$ vanishes at infinity, where the given and Lorentz-adjoint eigenmodes are superpositions of outgoing, circularly polarized eigenmodes we note, with (7.109), that

$$\hat{\boldsymbol{\epsilon}}_\pm\times\hat{\boldsymbol{\epsilon}}_\mp = \mp i\hat{\mathbf{z}}\,, \qquad \hat{\boldsymbol{\epsilon}}_\pm\times\hat{\boldsymbol{\epsilon}}_\pm = 0 \tag{7.125}$$

so that if the wave fields $\mathbf{e}$ and $\mathbf{e}^{(L)}$ in (7.124) are decomposed into circularly polarized components, as in (7.110), the concomitant vector $\mathbf{P}^{(L)}$ in (7.124) becomes

$$\begin{aligned} \mathbf{P}^{(L)} = i\hat{\mathbf{z}}&\left[(E^{(L)-}H^+ - E^{(L)+}H^-) - (E^-H^{(L)+} - E^+H^{(L)-})\right] \\ &-\omega\beta\hat{\mathbf{z}}\left[\varepsilon(E^{(L)+}E^- - E^{(L)-}E^+) - \mu(H^{(L)+}H^- - H^{(L)-}H^+)\right] \end{aligned} \tag{7.126}$$

From (7.114), (7.116) and (7.123) it is easily shown that

$$\sqrt{\varepsilon}E^\pm = \pm i\sqrt{\mu}H^\pm \qquad \sqrt{\varepsilon}E^{(L)\pm} = \pm i\sqrt{\mu}H^{(L)\pm} \tag{7.127}$$

and when this is substituted into (7.126) we find that $\mathbf{P}^{(L)}$ vanishes. This is true of course only at large distances from the sources, where the given and Lorentz-adjoint modes propagate radially outwards, so that their wave fields at any point may be expressed in terms of the same circular rotating basis vectors $\hat{\boldsymbol{\epsilon}}_\pm$. Integration of (7.124) over all space then gives, with the aid of (7.121), (7.123) and (7.124), with $\mathbf{P}^{(L)}{=}0$ in the far field,

$$\int(\tilde{\mathbf{e}}\,\bar{\mathsf{I}}\,\mathbf{j}^{(L)} - \tilde{\mathbf{e}}^{(L)}\bar{\mathsf{I}}\,\mathbf{j})\,d^3r = 0 \tag{7.128}$$

This is just the Lorentz reciprocity theorem, cf. (4.51), in an isotropic, chiral medium.

If the theorem is applied to two independent current sources, $\mathbf{j}_a$ and $\mathbf{j}_b$, and the fields they generate, $\mathbf{e}_a$ and $\mathbf{e}_b$, (7.128) may be written in the form

$$\langle \mathbf{e}_a \,,\, \bar{\mathbf{I}}\mathbf{j}_b \rangle = \langle \mathbf{e}_b \,,\, \bar{\mathbf{I}}\mathbf{j}_a \rangle \qquad \text{or} \qquad \langle a\,,\, b\rangle = \langle b\,,\, a\rangle \tag{7.129}$$

in which we have dropped the superscript (L), since the medium is Lorentz self-adjoint. This straightforward form of the Lorentz reciprocity theorem has been derived by Lakhtakia et al. [86].

7.5.4 The eigenmode formulation of the Lorentz reciprocity theorem

The formally adjoint and Lorentz-adjoint eigenmodes

In this section we consider the Lorentz reciprocity problem when the reaction of one localized current distibution (or antenna) on another is mediated by a single eigenmode, or by several eigenmodes, which propagate from one source to the other. For this purpose we must be able to identify which Lorentz-adjoint eigenvalues (and eigenmodes) correspond to the given eigenvalues, $k_+^{(+)}$ and $k_-^{(+)}$ (7.116). We do this systematically by constructing first the system that is *formally adjoint* to the given system (7.121) in a source-free (j=0) medium:

$$\bar{\mathsf{L}}\bar{\mathrm{e}} := \left[i\omega(\mathsf{K}_0^T - \beta\mathsf{D}_0^T) - \mathsf{D}^T\right]\bar{\mathrm{e}} = \left[i\omega(\mathsf{K}_0 + \beta\mathsf{D}_0) - \mathsf{D}\right]\bar{\mathrm{e}} = 0 \tag{7.130}$$

with $\mathsf{K}_0^T = \mathsf{K}^T$, $\mathsf{D}_0^T = -\mathsf{D}_0$, $\mathsf{D}^T = \mathsf{D}$, cf. (7.120) and (2.22). Application of the plane-wave ansatz

$$\bar{\mathrm{e}}_\alpha \sim \exp(i\bar{k}_\alpha z) \tag{7.131}$$

to (7.130) yields with the aid of (7.106), (7.108) and (7.120), and with $\nabla \to i\bar{k}_\alpha \hat{\mathbf{k}}$,

$$\left[i\omega(\mathsf{K}_0 + i\beta\bar{k}_\alpha\hat{\mathcal{C}}) - i\bar{k}_\alpha\hat{\mathcal{K}}\right]\bar{\mathrm{e}}_\alpha = 0 \tag{7.132}$$

This is the same as the given eigenmode equation (7.107) if we let $\beta \to -\beta$ and consequently, instead of (7.115), we obtain

$$\bar{k}_\pm^2 = \omega^2\varepsilon\mu(1 \pm \beta\bar{k}_\pm)^2 \tag{7.133}$$

where $\bar{k}_\pm$ is associated with the adjoint fields $(\overline{E}^\pm, \overline{H}^\pm)$. Comparison of (7.133) with (7.115), and (7.132) with (7.107), yields

$$\bar{k}_\pm^2 = k_\mp^2\,, \qquad \bar{k}_\pm = k_\mp \tag{7.134}$$

(This is analogous to the relation $\bar{q}_\alpha = q_\alpha$ (2.46) in a magnetoplasma — although $\bar{q}_\alpha$ was there associated with a differently polarized eigenmode that propagated in a magnetic field reversed medium.)

The direction of phase propagation has been reversed in the adjoint mode [compare the plane-wave ansatz in the given and in the adjoint mode, (7.105) and (7.131)], but the ratio of $\mathbf{H}_\pm$ to $\mathbf{E}_\pm$, and consequently the direction of the Poynting vector, has remained unchanged. 'Physicality' is restored to the Lorentz-adjoint modes (see the discussion in Sec. 3.4.1) by the matrix operator $\bar{\mathsf{I}}$ (2.81) which reverses the magnetic wave fields, and thereby the direction of energy flow, and hence

$$k_\pm^{(+)} = -k_\mp^{(L)(-)} \tag{7.135}$$

where $k_\pm$ and $\mathbf{k}_\pm^{(L)}$ are associated with the respective fields

$$\mathbf{e}_\pm := \begin{bmatrix} \mathbf{E}_\pm \\ \mathbf{H}_\pm \end{bmatrix} \qquad \text{and} \qquad \mathbf{e}_\pm^{(L)} := \begin{bmatrix} \mathbf{E}_\pm^{(L)} \\ \mathbf{H}_\pm^{(L)} \end{bmatrix}$$

Let us clarify what we have just proved. The eigenvalue Lorentz-adjoint to the given eigenvalue $k_\pm^{(+)}$ is $k_\mp^{(L)(-)}$ which, by (7.135) and (7.116), yields

$$k_\mp^{(L)(-)} = k_\mp^{(-)} \tag{7.136}$$

This was to be expected since the medium is Lorentz self-adjoint. The mode that is Lorentz-adjoint to the given mode has the same handedness as the given mode but propagates in the opposite direction. For the sake of clarity in the ensuing discussion it will be convenient to rename the modes according to their handedness, ℓ or r. Thus

$$k_+^{(+)} \equiv k_r^{(+)}, \qquad k_-^{(+)} \equiv k_\ell^{(+)}, \qquad k_+^{(-)} \equiv k_\ell^{(-)}, \qquad k_-^{(-)} \equiv k_r^{(-)} \tag{7.137}$$

so that (7.116) and (7.136) take the simple form:

$$k_{r,\ell}^{(+)} = -k_{r,\ell}^{(-)}, \qquad k_{r,\ell}^{(L)(\pm)} = k_{r,\ell}^{(\pm)} \tag{7.138}$$

and the eigenvalue, Lorentz-adjoint to the given eigenvalue $k_{r,\ell}^{(+)}$, is consequently $-k_{r,\ell}^{(L)(-)} = -k_{r,\ell}^{(-)}$.

Eigenmode linkage between two antennas

Suppose that the antennas a and b, i.e. the localized current distributions $\mathbf{j}_a$ and $\mathbf{j}_b$, radiate and receive only left-handed eigenmodes. In the case of radiowaves this could be achieved by using crossed dipole antennas in which the input or induced currents were phase shifted by 90° [42, Sec. 5.2]. It will suit our purposes better to use an optical model. We suppose that the current sources are enclosed by filters that transmit only circularly polarized light, say left handed, *from either side*. [Cholesteric liquid crystals, for instance, have just such a property [65, Chap. 14]: in a narrow band of wavelengths centred at a certain wavelength λ_0, where the specific rotation changes sign and becomes zero, light of a given circular polarization, say left handed, is completely transmitted, and the other (right-handed) mode is totally reflected. Alternatively, one could conceive an optically active material in which one mode is transmitted and the other is absorbed. Technical details are not essential for our thought experiments as long as the setup is feasible in principle!]

The Lorentz reciprocity theorem (7.129) then takes the form

$$\langle \mathbf{e}_\ell^a, \bar{\mathbf{I}}\mathbf{j}_b \rangle = \langle \mathbf{e}_\ell^b, \bar{\mathbf{I}}\mathbf{j}_a \rangle \tag{7.139}$$

where $\mathbf{e}_\ell^a$ and $\mathbf{e}_\ell^b$ are the left-handed modal wave-field components generated by the currents $\mathbf{j}_a$ and $\mathbf{j}_b$. This may be written symbolically as

$$\langle a\,(\overset{+}{T}_{11})\,b \rangle = \langle b\,(\overset{-}{T}_{11})\,a \rangle \tag{7.140}$$

and represents a short-hand form of the statement: the reaction of the field of the source a, via the transmission channel $\overset{+}{T}_{11}$, on the source b, equals the reaction of the source b, via the reverse transmission channel $\overset{-}{T}_{11}$, on the source a. The left-hand mode has been denoted mode 1, the right-hand mode will be denoted mode 2.

If the two sources are encompassed by right circularly polarized filters we obtain similarly

$$\langle a\,(\overset{+}{T}_{22})\,b \rangle = \langle b\,(\overset{-}{T}_{22})\,a \rangle \tag{7.141}$$

If, on the other hand, a is enclosed by a left-handed filter and b by a right-handed one, the reaction between the two sources will differ from zero provided that the medium is non-homogeneous, so that intermode coupling occurs in the intervening medium. The reciprocity theorem then takes the form

$$\langle a\,(\overset{+}{T}_{12})\,b \rangle = \langle b\,(\overset{-}{T}_{21})\,a \rangle \tag{7.142}$$

Suppose next that a is encompassed by a left-handed filter whereas b has no encompassing filter. We could describe the result symbolically as follows:

$$\langle a\,(\overset{+}{T}_{11}, \overset{+}{T}_{12})\,b \rangle = \langle b \begin{pmatrix} \overset{-}{T}_{11} \\ \overset{-}{T}_{21} \end{pmatrix} a \rangle \tag{7.143}$$

Finally, we remove all filters, i.e. all transmission channels are open, and the result may be written symbolically as

$$\langle a \begin{pmatrix} \overset{+}{T}_{11} & \overset{+}{T}_{12} \\ \overset{+}{T}_{21} & \overset{+}{T}_{22} \end{pmatrix} b\rangle = \langle a\,, b\rangle = \langle b\,, a\rangle = \langle b \begin{pmatrix} \overset{-}{T}_{11} & \overset{-}{T}_{21} \\ \overset{-}{T}_{12} & \overset{-}{T}_{22} \end{pmatrix} a\rangle \tag{7.144}$$

or

$$\langle a\,(\mathbf{T}_{+})\,b\rangle = \langle a\,, b\rangle = \langle b\,, a\rangle = \langle b\,(\tilde{\mathbf{T}}_{-})\,a\rangle \tag{7.145}$$

In summary, the Lorentz reciprocity theorem may in general be written in the form

$$\langle a\,(\mathbf{T}_{+})\,b\rangle = \langle b\,(\tilde{\mathbf{T}}_{-})\,a\rangle \tag{7.146}$$

with $\mathbf{T}_{\pm}$ defined by (7.144) and (7.145). If all eigenmode transmission channels are open then $\mathbf{T}_{\pm}$ may be dropped in the formulation. However, if only part of the eigenmodes are transmitted or received because of the antenna structure, or because of filtering, then the appropriate matrix elements of $\mathbf{T}_{\pm}$ in (7.146) will describe the resultant reciprocal interaction.

The two antennas could of course be linked by more than a single ray path for each type of polarization. If there were n different left-handed, and m different right-handed ray paths, then $\mathbf{T}_{\pm}$ would simply be a $(n+m)\times(n+m)$ square matrix to express the contributions of the $n+m$ different 'modes'.

Application to gyrotropic or other media

Suppose that the sources a and b above are immersed in a gyrotropic medium, such as a magnetoplasma. Suppose also, for simplicity, that the two sources are linked by one of the external magnetic field lines $\mathbf{b}$, so that the eigenmodes mediating the reaction between a and b are circularly polarized. Let the two sources be enclosed by the left-handed filters described above.

The reaction of a on b via the left-handed transmission channel is given by $\langle a\,(\overset{+}{T}_{11})\,b\rangle$, as before. In the reciprocal, Lorentz-adjoint medium the external magnetic field is reversed in direction, and so the backward propagating mode remains left-handed. It will consequently pass through both filters and the reaction will be given by $\langle b\,(\overset{-}{T}_{11})\,a\rangle$, as in (7.140).

By similar arguments we may show that (7.146) is the eigenmode generalization of Lorentz reciprocity in any medium and its Lorentz-adjoint counterpart.

7.5.5 Time reversal and reflection mapping of eigenmodes in the chiral medium

The time-reversed modes

Intuitively we would expect that under time reversal ('running the film backwards'), a right- or left-handed mode would reverse its direction of propagation but remain right- or left-handed. Let us check whether this is borne out by the mathematical formalism developed in Secs. 7.1 and 7.2.

We take the complex conjugate of the Maxwell system (7.121) and premultiply it by $-\bar{\mathsf{I}}$, with $\bar{\mathsf{I}}=\bar{\mathsf{I}}^{-1}$ (2.81), to give

$$-\bar{\mathsf{I}}\left[-i\omega(\mathsf{K}_0^* + \beta\mathsf{D}_0^*) + \mathsf{D}\right]\bar{\mathsf{I}}\,\bar{\mathsf{I}}\,\mathbf{e}^* = \bar{\mathsf{I}}\mathbf{j}^* \tag{7.147}$$

If β and the scalars ε and μ in K_0 and D_0 are assumed to be real, and with time-reversed fields and currents, $\mathbf{e}'$ and $\mathbf{j}'$ (7.13), given by

$$\mathbf{e}' = \bar{\mathsf{I}}\,\mathbf{e}^*, \qquad \mathbf{j}' = -\bar{\mathsf{I}}\mathbf{j}^* \tag{7.148}$$

(7.147) becomes, with $\bar{\mathsf{I}}\,\mathsf{D}\,\bar{\mathsf{I}}=-\mathsf{D}$, (2.28) and (2.81), and with $\bar{\mathsf{I}}\,\mathsf{D}_0\,\bar{\mathsf{I}}=\mathsf{D}_0$ (7.120),

$$[i\omega(\mathsf{K}_0 + \beta\mathsf{D}_0) + \mathsf{D}]\,\mathbf{e}' = -\mathbf{j}' \tag{7.149}$$

with $\bar{\mathsf{I}}\,\mathsf{K}_0^*\bar{\mathsf{I}}=\mathsf{K}_0$. We have retrieved the original Maxwell system (7.121), and in a source-free region ($\mathbf{j}$=0) we may expect to obtain the same eigenmodes as in the original system, but ordered differently. Note that the rotating basis vectors (7.109) are interchanged by complex conjugation,

$$\hat{\boldsymbol{\epsilon}}_\pm^* = \hat{\boldsymbol{\epsilon}}_\mp$$

and $\bar{\mathsf{I}}$ operating on an eigenmode $\mathbf{e}_\alpha$ reverses its direction of propagation. Hence (7.148) yields

$$k'^{(+)}_\pm = -k^{(-)}_\mp \tag{7.150}$$

which is exactly the same as the relation (7.135) that links the Lorentz-adjoint to the given eigenmodes, with the time-reversed eigenmodes preserving the handedness of the given eigenmodes, as anticipated.

We noted in Sec. 7.2.1 that the Lorentz-adjoint and time-reversed constitutive tensors coincide, provided that the tensor is hermitian. Now the constitutive relations in $\mathbf{k}$-space are given by (7.104). Maxwell's equations for a monochromatic plane wave, $\mathbf{e}\sim\exp i(\omega t - \mathbf{k}\cdot\mathbf{r})$, have the form

$$\mathbf{k}\times\mathbf{E} = \omega\mathbf{B}, \qquad \mathbf{k}\times\mathbf{H} = -\omega\mathbf{D} \tag{7.151}$$

When substituted into (7.104) these give the *biisotropic* constitutive tensor (7.19) for an isotropic chiral medium, which is indeed hermitian.

Eigenmodes in a reflected chiral medium

Here too we would expect r-rotatory media to be mapped by reflection into ℓ-rotatory media (right-handed helices become left-handed), and left- and right-handed eigenmodes to be interchanged.

Let us apply a reflection mapping, cf. (6.58) and (6.60), to the Maxwell system (7.107),

$$\boldsymbol{\Gamma}\left[i\omega(\mathsf{K}_0(\mathbf{r}) - \beta k\widehat{\mathcal{C}}) - ik\widehat{\mathcal{K}}\right]\boldsymbol{\Gamma}^T\boldsymbol{\Gamma}\mathrm{e}(\mathbf{r}) = 0 \tag{7.152}$$

with

$$\boldsymbol{\Gamma} := \begin{bmatrix} \gamma & 0 \\ 0 & (\det\gamma)\gamma \end{bmatrix}, \qquad \boldsymbol{\Gamma}^T = \boldsymbol{\Gamma}^{-1}$$

as in (6.46) and (6.47), with $\det \boldsymbol{\Gamma} = \det\gamma = -1$ in a reflection mapping. Using primes to denote reflected quantities in this subsection we have, with (6.21),

$$\boldsymbol{\Gamma}\mathrm{e}(\mathbf{r}) = \mathrm{e}'(\mathbf{r}')\,, \qquad \gamma(\hat{\mathbf{k}}\times\mathsf{I}^{(3)})\gamma^T = (\det\gamma)\hat{\mathbf{k}}'\times\mathsf{I}^{(3)} = -\hat{\mathbf{k}}'\times\mathsf{I}^{(3)}$$

$$\boldsymbol{\Gamma}\mathsf{K}_0(\mathbf{r})\boldsymbol{\Gamma}^T = \mathsf{K}_0(\mathbf{r}')\,, \qquad \gamma\mathbf{r} = \mathbf{r}'$$

$$\boldsymbol{\Gamma}\widehat{\mathcal{C}}\boldsymbol{\Gamma}^T = -\begin{bmatrix} \varepsilon\,\hat{\mathbf{k}}'\times\mathsf{I}^{(3)} & 0 \\ 0 & \mu\,\hat{\mathbf{k}}'\times\mathsf{I}^{(3)} \end{bmatrix} = -\widehat{\mathcal{C}}'$$

$$\boldsymbol{\Gamma}\widehat{\mathcal{K}}\boldsymbol{\Gamma}^T = \begin{bmatrix} 0 & -\hat{\mathbf{k}}'\times\mathsf{I}^{(3)} \\ \hat{\mathbf{k}}'\times\mathsf{I}^{(3)} & 0 \end{bmatrix} = \widehat{\mathcal{K}}'$$

cf. (6.45), (6.46), (6.53), (7.106) and (7.108). Thus (7.152) becomes

$$\left[i\omega(\mathsf{K}_0(\mathbf{r}') + \beta k\widehat{\mathcal{C}}') - ik\widehat{\mathcal{K}}'\right]\mathrm{e}'(\mathbf{r}') = 0 \tag{7.153}$$

and we have retrieved the original system (7.107), but with the chirality constant β reversed in sign. This confirms that β is a pseudo-scalar, as pointed out in Sec. 7.2.2.

To examine the behaviour of an eigenfield such as $\mathbf{E}_\pm := E^\pm\hat{\boldsymbol{\epsilon}}_\pm$ (7.110) under reflection (here the direction of propagation is in the z-direction and $\hat{\boldsymbol{\epsilon}}_\pm$ is given by (7.109)), we consider for simplicity a reflection mapping with respect to the y=0 plane. Then, with $\mathsf{q} = -\mathsf{q}_2 \equiv -\mathsf{q}_y$ (6.7), we obtain

$$\mathsf{q}\,\hat{\boldsymbol{\epsilon}}_\pm = \hat{\boldsymbol{\epsilon}}_\mp \tag{7.154}$$

so that an eigenmode $(\mathbf{E}_\pm\,, \mathbf{H}_\pm)$ is mapped into $(\mathbf{E}'_\mp\,, \mathbf{H}'_\mp)$, and the sense of rotation is reversed, as expected.

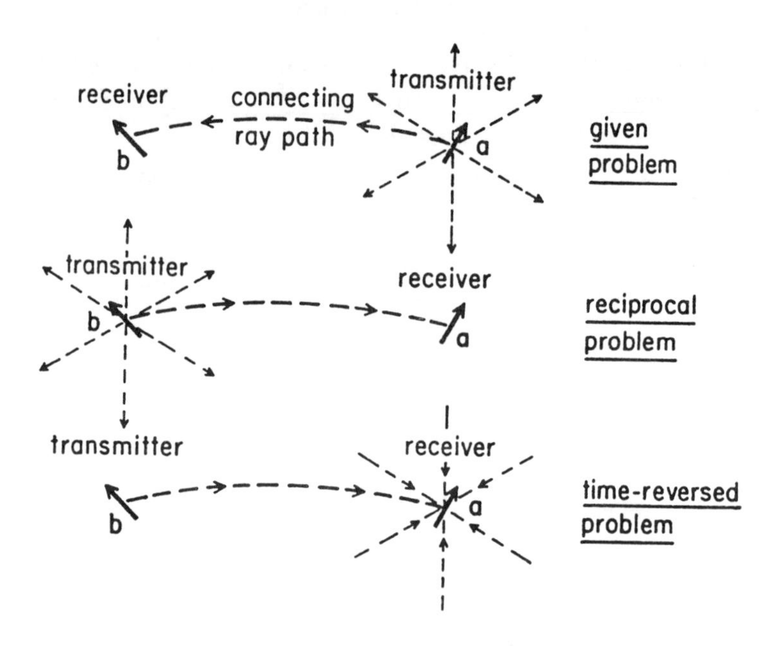

Figure 7.3: Reciprocity and time reversal with two antennas.

7.6 Time reversal and causality

It was seen in Sec. 7.2.2 that the time-reversed and Lorentz-adjoint inhomogeneous Maxwell equations, (7.21) and (7.24), with $\mathbf{K}'=\mathbf{K}^{(L)}$(7.18), were identical. We then raised the question whether the solutions were also identical. In the case of eigenmodes and ray paths, i.e. solutions in source-free media, the answer was in the affirmative. We now reexamine the question when the media contain sources.

Suppose that the sources are currents that flow in two antennas, a and b, and we consider the Lorentz-reciprocity problem. An input current $I^{(a)}(0)$ in antenna a produces an outgoing wave field which propagates in the given medium and induces an open-circuit voltage $V_{o.c.}^{(b)}$ in antenna b. In the reciprocal problem an input current $I^{(b)}(0)$ in antenna b emits an outgoing wave field which propagates in the Lorentz-adjoint (i.e. time-reversed) medium and induces an open-circuit voltage $V_{o.c.}^{(a)}$ in antenna a. We saw in Sec. 4.5.2, cf. (4.108), that $V_{o.c.}^{(b)}=V_{o.c.}^{(a)}$ when $I^{(a)}(0)=I^{(b)}(0)$.

Now if in the two cases we restrict our attention to a comparison of the wave fields only, it is evident that the fields in the second case are not a time-reversed copy of the fields in the first case — except for the ray (or rays) connecting the two antennas which, in view of the results of Sec. 7.2.3, is just that! This is illustrated in Fig. 7.3. It is evident that the field structures at both the receiver and transmitter are 'non-physical' in the time-reversed problem. There is no outgoing wave field at the transmitter b, *except along the ray path (or paths) linking it to the receiver a*. On the other hand there is an incoming wave field, converging from infinity in all directions on the receiver a. Time reversal has simply transformed the 'retarded', causal solutions into 'advanced', non-causal (and non-physical) solutions.

Does this mean that the time-reversal procedure gives invalid or incorrect results when sources are present? Not necessarily! In fact, it is clear that the only possible interaction between the transmitting and receiving antennas, in both the given and the Lorentz-adjoint problems, is mediated by the ray (or rays) which emanates from one antenna and terminates on the other. But the spatial configuration of the ray, and even the attenuation of the wave field along it, is identical in both the given and in the Lorentz-adjoint (i.e. in the restricted time-reversed) problems. This is the physical basis of the Lorentz reciprocity theorem, and in this sense, i.e. the linking of the receiving and transmitting antennas by the same ray path in both directions, one may say that Lorentz reciprocity is an expression of the invariance of Maxwell's equations under time reversal. (In a different context, Budden has performed the reciprocity calculations for antennas linked by the same ray path [33, Secs. 14.13-14.15]; see also [35].)

In connection with the above discussion one recalls the result found by Wheeler and Feynman [134], summarized in ref. [96, Sec. 21.12] in which it is shown that even if it is assumed that advanced potentials and fields have the same physical validity as the corresponding retarded quantities, and if both are included symmetrically in the solutions of Maxwell's equations, the net interaction between any two charges will be given by the retarded, causal fields only.

Appendix A

A.1 The spectral resolution of a matrix

The tensors associated with a cold magnetoplasma, such as the conductivity or dielectric tensors, are characterized by a single real axis of symmetry, the direction $\hat{\mathbf{b}}$ of the external magnetic field. They belong to a class of tensors, named *cyclotonics* by Gibbs [58, Sec. 133], having the following structure

$$\mathbf{A} := \lambda_1 \hat{\boldsymbol{a}}_1 \bar{\boldsymbol{a}}^{1^T} + \alpha(\hat{\boldsymbol{a}}_2 \bar{\boldsymbol{a}}^{2^T} + \hat{\boldsymbol{a}}_3 \bar{\boldsymbol{a}}^{3^T}) + \beta(\hat{\boldsymbol{a}}_3 \bar{\boldsymbol{a}}^{2^T} - \hat{\boldsymbol{a}}_2 \bar{\boldsymbol{a}}^{3^T})$$

where $\hat{\boldsymbol{a}}_1$, $\hat{\boldsymbol{a}}_2$ and $\hat{\boldsymbol{a}}_3$ represent a set of linearly independent, but not necessarily orthogonal normalized base vectors, and $\bar{\boldsymbol{a}}^1$, $\bar{\boldsymbol{a}}^2$ and $\bar{\boldsymbol{a}}^3$ represent the reciprocal set of base vectors [135]. The eigenvalues are

$$\lambda_1, \qquad \lambda_2 = \alpha + i\beta, \qquad \lambda_3 = \alpha - i\beta$$

For any matrix $\mathbf{A}$ having three distinct eigenvalues, λ_1, λ_2 and λ_3, as above for example, we may construct the (complete) system of three orthogonal *projectors* $\mathcal{P}_1$, $\mathcal{P}_2$ and $\mathcal{P}_3$ of the matrix $\mathbf{A}$ (see for instance [26, Sec. 3.6]),

$$\mathcal{P}_1 = \frac{[\lambda_2 \mathbf{I} - \mathbf{A}][\lambda_3 \mathbf{I} - \mathbf{A}]}{(\lambda_2 - \lambda_1)(\lambda_3 - \lambda_1)}, \quad \mathcal{P}_2 = \frac{[\lambda_3 \mathbf{I} - \mathbf{A}][\lambda_1 \mathbf{I} - \mathbf{A}]}{(\lambda_3 - \lambda_2)(\lambda_1 - \lambda_2)}, \quad \mathcal{P}_3 = \frac{[\lambda_1 \mathbf{I} - \mathbf{A}][\lambda_2 \mathbf{I} - \mathbf{A}]}{(\lambda_1 - \lambda_3)(\lambda_2 - \lambda_3)} \tag{A.1}$$

Now the pairs of matrices in each of the numerators above commute so that, with the characteristic equation for the matrix $\mathbf{A}$,

$$[\lambda_i \mathbf{I} - \mathbf{A}]\hat{\mathbf{u}}_i = 0, \qquad i = 1, 2 \text{ or } 3 \tag{A.2}$$

where $\hat{\mathbf{u}}_i$ denotes the (normalized) eigenvectors of $\mathbf{A}$, we find that

$$\mathcal{P}_i \hat{\mathbf{u}}_j = \delta_{ij} \hat{\mathbf{u}}_j \tag{A.3}$$

Since any vector $\mathbf{u}$ may be expressed as a linear combination of the eigenvectors $\hat{\mathbf{u}}_i$,

$$\mathbf{u} = \gamma_1\hat{\mathbf{u}}_1 + \gamma_2\hat{\mathbf{u}}_2 + \gamma_3\hat{\mathbf{u}}_3 \tag{A.4}$$

we obtain, with the aid of (A.3),

$$\mathcal{P}_i\mathbf{u} = \mathcal{P}_i\sum_j \gamma_j\hat{\mathbf{u}}_j = \gamma_i\hat{\mathbf{u}}_i \tag{A.5}$$

so that $\mathcal{P}_i$ is seen to project an arbitrary vector onto the (normalized) eigenvector $\hat{\mathbf{u}}_i$, and hence its name.

Applying $\mathcal{P}_i$ again to (A.5), we find

$$\mathcal{P}_i{}^2\mathbf{u} = \gamma_i\hat{\mathbf{u}}_i = \mathcal{P}_i\mathbf{u}$$

so that

$$\mathcal{P}_i{}^2 = \mathcal{P}_i \tag{A.6}$$

(matrices possessing this property are said to be *idempotent*) whereas

$$\mathcal{P}_i\mathcal{P}_j\mathbf{u} = \gamma_j\mathcal{P}_i\hat{\mathbf{u}}_j = 0$$

because of (A.3). The projectors are thus orthogonal

$$\mathcal{P}_i\mathcal{P}_j = \delta_{ij}\mathcal{P}_j \tag{A.7}$$

Furthermore, by virtue of (A.3) and (A.4),

$$[\mathcal{P}_1 + \mathcal{P}_2 + \mathcal{P}_3]\mathbf{u} = \gamma_1\hat{\mathbf{u}}_1 + \gamma_2\hat{\mathbf{u}}_2 + \gamma_3\hat{\mathbf{u}}_3 = \mathbf{u}$$

so that, with $\mathbf{I}$ denoting the unit matrix,

$$\mathcal{P}_1 + \mathcal{P}_2 + \mathcal{P}_3 = \mathbf{I} \tag{A.8}$$

which is the completeness condition for the three projectors.

Now apply the matrix $\mathbf{A}$ to an arbitrary vector $\mathbf{u}$. Using (A.4), (A.2) and (A.3) we get

$$\begin{aligned}
\mathbf{Au} &= \mathbf{A}\sum_i \gamma_i\hat{\mathbf{u}}_i = \sum_i \gamma_i\lambda_i\hat{\mathbf{u}}_i = \sum_i \gamma_i\lambda_i\mathcal{P}_i\hat{\mathbf{u}}_i \\
&= \sum_i \lambda_i\mathcal{P}_i\sum_j \gamma_j\hat{\mathbf{u}}_j \qquad \text{[in view of (A.3)]} \\
&= [\lambda_1\mathcal{P}_1 + \lambda_2\mathcal{P}_2 + \lambda_3\mathcal{P}_3]\mathbf{u}
\end{aligned} \tag{A.9}$$

yielding the ***spectral resolution*** of the matrix $\mathbf{A}$:

$$\mathbf{A} = \lambda_1 \mathcal{P}_1 + \lambda_2 \mathcal{P}_2 + \lambda_3 \mathcal{P}_3 \tag{A.10}$$

in terms of the eigenvalue spectrum λ_i and the projectors $\mathcal{P}_i$.

The reciprocal tensor $\mathbf{A}^{-1}$ may now be derived with the aid of (A.7), (A.8) and (A.10). Consider the expression

$$\begin{aligned} \mathbf{A}\left[\frac{1}{\lambda_1}\mathcal{P}_1 + \frac{1}{\lambda_2}\mathcal{P}_2 + \frac{1}{\lambda_3}\mathcal{P}_3\right] &= \left[\frac{1}{\lambda_1}\mathcal{P}_1 + \frac{1}{\lambda_2}\mathcal{P}_2 + \frac{1}{\lambda_3}\mathcal{P}_3\right]\mathbf{A} \\ &= [\mathcal{P}_1 + \mathcal{P}_2 + \mathcal{P}_3] = \mathbf{I} \end{aligned}$$

which yields

$$\mathbf{A}^{-1} = \frac{1}{\lambda_1}\mathcal{P}_1 + \frac{1}{\lambda_2}\mathcal{P}_2 + \frac{1}{\lambda_3}\mathcal{P}_3 \tag{A.11}$$

We note finally that the matrix

$$\mathbf{U} := [\hat{\mathbf{u}}_1 \; \hat{\mathbf{u}}_2 \; \hat{\mathbf{u}}_3] \tag{A.12}$$

formed by the juxtaposition of the three linearly independent eigenvectors $\hat{\mathbf{u}}_i$ is non-singular, and the reciprocal matrix $\overline{\mathbf{U}}$ may be formed:

$$\overline{\mathbf{U}} := \mathbf{U}^{-1} \equiv \begin{bmatrix} \bar{\mathbf{u}}^{1^T} \\ \bar{\mathbf{u}}^{2^T} \\ \bar{\mathbf{u}}^{3^T} \end{bmatrix} \tag{A.13}$$

where $\bar{\mathbf{u}}^{i^T}$ ($i = 1, 2, 3$) denotes any of the three rows of $\overline{\mathbf{U}}$. Since $\overline{\mathbf{U}}\mathbf{U}=\mathbf{I}$, we have (in vector or matrix notation),

$$\bar{\mathbf{u}}^i \cdot \hat{\mathbf{u}}_j \equiv \bar{\mathbf{u}}^{i^T}\hat{\mathbf{u}}_j = \delta_{ij} \tag{A.14}$$

thereby determining the system of reciprocal eigenvectors $\bar{\mathbf{u}}^i$ that are biorthogonal to the eigenvectors $\hat{\mathbf{u}}_j$.

The matrix $\mathbf{U}$ diagonalizes $\mathbf{A}$ (this property may be derived with aid of (A.2))

$$\mathbf{A}\mathbf{U} = \mathbf{U}\mathit{\Lambda}, \qquad \mathit{\Lambda} := \begin{bmatrix} \lambda_1 & 0 & 0 \\ 0 & \lambda_2 & 0 \\ 0 & 0 & \lambda_3 \end{bmatrix}$$

or

$$\mathbf{U}^{-1}\mathbf{A} = \mathit{\Lambda}\mathbf{U}^{-1}, \qquad \bar{\mathbf{u}}^{i^T}\mathbf{A} = \lambda_i \bar{\mathbf{u}}^{i^T}$$

so that

$$\mathbf{A}^T\bar{\mathbf{u}}^i = \lambda_i \bar{\mathbf{u}}^i \tag{A.15}$$

The reciprocal eigenvectors are thus seen to be eigenvectors of the transposed matrix $\mathbf{A}^T$, and are often referred to as the *left eigenvectors* of $\mathbf{A}$. In view of (A.14) it is clear that the projectors may be expressed in terms of the given and reciprocal sets of eigenvectors,

$$\mathcal{P}_i = \hat{\mathbf{u}}_i \bar{\mathbf{u}}^{iT} \tag{A.16}$$

A.2 Application to gyrotropic tensors

The various tensors associated with a magnetoplasma (such as the mobility tensor $\boldsymbol{\mu}_s$, the resistivity tensor $\boldsymbol{\eta}_s$, the conductivity tensor $\boldsymbol{\sigma}$, the dielectric tensor $\boldsymbol{\varepsilon}$) are gyrotropic, i.e. they have a single symmetry axis $\hat{\mathbf{b}}$, the direction of the (axial) magnetic-field vector, and are a special case of the cyclotonic tensors discussed above. Their general form, cf. Chen [36, Sec. 1.10], is

$$\begin{aligned} \mathbf{A}(\hat{\mathbf{b}}) &= a_\perp(\mathbf{I} - \hat{\mathbf{b}}\hat{\mathbf{b}}^T) + ia_\times(\hat{\mathbf{b}} \times \mathbf{I}) + a_\parallel \hat{\mathbf{b}}\hat{\mathbf{b}}^T \\ &= \mathbf{A}^T(-\hat{\mathbf{b}}) \end{aligned} \tag{A.17}$$

and the three orthogonal projectors are given by

$$\mathcal{P}_{\pm 1} := \frac{1}{2}(\mathbf{I} - \hat{\mathbf{b}}\hat{\mathbf{b}}^T \pm i\hat{\mathbf{b}} \times \mathbf{I}), \qquad \mathcal{P}_0 := \hat{\mathbf{b}}\hat{\mathbf{b}}^T \tag{A.18}$$

The completeness relation for the projectors

$$\sum_{j=-1}^{1} \mathcal{P}_j = \mathbf{I} \tag{A.19}$$

is clearly satisfied, and their orthogonality,

$$\mathcal{P}_j \mathcal{P}_k = \delta_{jk} \mathcal{P}_k \tag{A.20}$$

cf. (A.7), follows from the expansion

$$(\hat{\mathbf{b}} \times \mathbf{I})(\hat{\mathbf{b}} \times \mathbf{I}) = \hat{\mathbf{b}} \times (\hat{\mathbf{b}} \times \mathbf{I}) = \hat{\mathbf{b}}\hat{\mathbf{b}}^T - \mathbf{I}$$

This can be verified by post-multiplication with an arbitrary vector $\mathbf{v}$ so that, in mixed matrix and vector notation,

$$(\hat{\mathbf{b}} \times \mathbf{I})(\hat{\mathbf{b}} \times \mathbf{I})\mathbf{v} = \hat{\mathbf{b}} \times (\hat{\mathbf{b}} \times \mathbf{v}) = \hat{\mathbf{b}}\hat{\mathbf{b}} \cdot \mathbf{v} - \mathbf{v} = (\hat{\mathbf{b}}\hat{\mathbf{b}}^T - \mathbf{I})\mathbf{v}$$

The linear combinations

$$\mathcal{P}_1 + \mathcal{P}_{-1} = \mathbf{I} - \hat{\mathbf{b}}\hat{\mathbf{b}}^T, \qquad \mathcal{P}_1 - \mathcal{P}_{-1} = i\hat{\mathbf{b}} \times \mathbf{I} \tag{A.21}$$

of the two projectors have the following interpretation. If applied to an arbitrary vector the first, $(\mathsf{I} - \hat{\mathbf{b}}\hat{\mathbf{b}}^T)$, projects it onto the plane perpendicular to $\hat{\mathbf{b}}$, whereas the second, $(\hat{\mathbf{b}} \times \mathsf{I})$, does the same but in addition rotates it through an angle $\pi/2$ about $\hat{\mathbf{b}}$.

Finally, the spectral resolution of the matrix (A.10)

$$\begin{aligned} \mathsf{A}(\hat{\mathbf{b}}) = &= \lambda_1 \mathcal{P}_1 + \lambda_{-1} \mathcal{P}_{-1} + \lambda_0 \mathcal{P}_0 \\ &\equiv a_\perp(\mathcal{P}_1 + \mathcal{P}_{-1}) + a_\times(\mathcal{P}_1 - \mathcal{P}_{-1}) + a_\parallel \mathcal{P}_0 \end{aligned} \tag{A.22}$$

yields the eigenvalues

$$\lambda_1 = a_\perp + a_\times, \qquad \lambda_{-1} = a_\perp - a_\times, \qquad \lambda_0 = a_\parallel \tag{A.23}$$

and hence, with the aid of (A.11), the inverse matrix is determined. With the coefficients $a_\perp$, $a_\parallel$ and $a_\times$ expressed in terms of the eigenvalues:

$$a_\perp = \frac{1}{2}(\lambda_1 + \lambda_{-1}), \qquad a_\times = \frac{1}{2}(\lambda_1 - \lambda_{-1}), \qquad a_\parallel = \lambda_0 \tag{A.24}$$

the general form of the gyrotropic tensor (A.17) becomes

$$\mathsf{A}(\hat{\mathbf{b}}) = \frac{1}{2}(\lambda_1 + \lambda_{-1})(\mathsf{I} - \hat{\mathbf{b}}\hat{\mathbf{b}}^T) + \frac{i}{2}(\lambda_1 - \lambda_{-1})(\hat{\mathbf{b}} \times \mathsf{I}) + \lambda_0 \hat{\mathbf{b}}\hat{\mathbf{b}}^T \tag{A.25}$$

Bibliography

[1] Allis W.P., Buchsbaum S.J. and Bers A. (1963) *Waves in Anisotropic Plasmas*, M.I.T. Press, Cambridge, Mass.

[2] Altman C. (1971) Reciprocity relations for electromagnetic wave propagation through birefringent stratified media, *Proc. International Symp. on E.M. Wave Theory, Tbilisi, 9-15 Sept. 1971*, pp.790–793.

[3] Altman C. and Cory H. (1969a) The simple thin film optical method in electromagnetic wave propagation, *Radio Sci.* **4**, 449–457.

[4] Altman C. and Cory H. (1969b) The generalized thin film optical method in electromagnetic wave propagation, *Radio Sci.* **4**, 459–470.

[5] Altman C. and Cory H. (1970) Multiple reflection processes in the D- and E-regions at low and very low frequencies, *J. Atmos. Terr. Phys.* **32**, 1439–1455.

[6] Altman C. and Fijalkow E. (1970) Coupling processes in the D and E regions at low and very low frequencies. II. Frequencies greater than critical: a study of the nighttime ionosphere, *J. Atmos. Terr. Phys.* **32**, 1475–1488.

[7] Altman C. and Postan A. (1971) A reciprocity theorem for magnetoionic modes, *Radio Sci.* **6**, 483–487.

[8] Altman C. and Schatzberg A. (1981) A reciprocity theorem relating currents and fields in the presence of a plane-stratified ionosphere, *J. Atmos. Terr. Phys.* **43**, 339–344.

[9] Altman C. and Schatzberg A. (1982) Reciprocity and equivalence in reciprocal and non-reciprocal media through reflection transformations of the current distributions, *Appl. Phys.* **B28**, 327–333.

[10] Altman C., Schatzberg A. and Suchy K. (1981) Symmetries and scattering relations in plane-stratified anisotropic, gyrotropic and bianisotopic media, *Appl. Phys.* **B26**, 147–153.

[11] Altman C., Schatzberg A. and Suchy K. (1984) Symmetry transformations and time reversal of currents and fields in bounded (bi)anisotropic media, *IEEE Trans. Antennas Propagat.* *AP*-**32**, 1204–1210.

[12] Altman C. and Suchy K. (1979a) Eigenmode scattering relations for plane-stratified gyrotropic media, *Appl. Phys.* **19**, 213–219.

[13] Altman C. and Suchy K. (1979b) Generalization of a scattering theorem for plane-stratified gyrotropic media, *Appl. Phys.* **19**, 337–343.

[14] Altman C. and Suchy K. (1980) Penetrating and non-penetrating mode propagation in the ionosphere in the light of eigenmode scattering relations, *J. Atmos. Terr. Phys.* **42**, 161–165.

[15] Altman C. and Suchy K. (1986) Spatial mapping and time reversal in magnetized compressible media, *IEEE Trans. Antennas Propagat.* *AP*-**34**, 1294–1299.

[16] Appleton E.V. (1928) The influence of the earth's magnetic field on wireless transmission, *Proc. Union Radio Sci. Int.* **1**, 2–3. [Paper presented at the Washington Assembly of U.R.S.I., 1927.]

[17] Appleton E.V. (1932) Wireless studies of the ionosphere, *J. Instn. Elect. Engrs* **71**, 642–650.

[18] Arnaud J.A. (1973) Biorthogonality relations for bianisotropic media, *J. Opt. Soc. Am.* **63**, 238–240.

[19] Bahar E. and Agrawal B.S. (1976) Vertically polarized waves in inhomogeneous media with critical coupling regions, energy conservation and reciprocity relationships, *Radio Sci.* **11**, 885–896.

[20] Bahar E. and Agrawal B.S. (1979) Generalized characteristic functions applied to propagation in bounded inhomogeneous anisotropic media—reciprocity and energy relationships, *J. Atmos. Terr. Phys.* **41**, 565–578.

[21] Barron D.W. and Budden K.G. (1959) The numerical solution of differential equations governing the reflexion of long radio waves from the ionosphere. III, *Proc. R. Soc. Lond. A* **249**, 387–401.

[22] Booker H.G. (1936) Oblique propagation of electromagnetic waves in a slowly varying non-isotropic medium, *Proc. R. Soc. Lond. A* **155**, 235–257.

[23] Bossy L. (1971) La détermination précise du matrizant pour la résolution des equations de Maxwell dans une ionosphère stratifiée, *Ann. Soc. Sci. de Bruxelles* **85**, 29–38.

[24] Bossy L. (1979) Wave propagation in stratified anisotropic media, *J. Geophys.* **46**, 1–14.

[25] Bossy L. and Claes H. (1974) "A numerical method for the integration of the equations governing the propagation of E.M. waves in a plane-stratified magnetoionic medium" in *ELF-VLF Radio Wave Propagation*, ed. J.A.Holtet, Reidel, Dordrecht, pp. 111–116.

[26] Bradbury T.C. (1968) *Theoretical Mechanics*, Wiley, New York.

[27] Bremmer H. (1951) The W.K.B. approximation as the first term of a geometric-optical series, *Commun. Pure Appl. Math.* **4**, 105–115.

[28] Bremmer H. (1958) "Propagation of Electromagnetic Waves" in *Encyclopedia of Physics* Vol. 16, *Electric Fields and Waves*, pp. 423–639, ed. J.Bartels, Springer, Berlin.

[29] Budden K.G. (1954) A reciprocity theorem on the propagation of radio waves via the ionosphere, *Proc. Camb. Phil. Soc.* **50**, 604–613.

[30] Budden K.G. (1955a) The numerical solution of differential equations governing the reflection of long radio waves from the ionosphere, *Proc. R. Soc. Lond. A* **227**, 516–537.

[31] Budden K.G. (1955b) The numerical solution of differential equations governing the reflection of long radio waves from the ionosphere. II, *Phil. Trans. R. Soc. Lond. A* **248**, 45–72.

[32] Budden K.G. (1966) *Radio Waves in the Ionosphere*, 2nd ed., Cambridge University Press.

[33] Budden K.G. (1985) *The Propagation of Radio Waves*, Cambridge University Press.

[34] Budden K.G. and Jones D. (1987) The theory of radio windows in planetary magnetospheres, *Proc. R. Soc. Lond. A* **412**, 1–23.

[35] Budden K.G. and Jull G.W. (1964) Reciprocity and nonreciprocity with magnetoionic rays, *Can. J. Phys.* **42**, 113–130.

[36] Chen H.C. (1983) *Theory of Electomagnetic Waves: A Coordinate-Free Approach*, McGraw-Hill, New York.

[37] Cheng D.K. and Kong J.A. (1968) Covariant description of bianisotropic media, *Proc IEEE*, **56**, 248–251.

[38] Clemmow P.C. (1966) *The Plane Wave Spectrum Representation of Electromagnetic Fields*, Pergamon, London.

[39] Clemmow P.C. and Heading J. (1954) Coupled forms of the differential equations governing radio propagation in the ionosphere, *Proc. Camb. Phil. Soc.* **50**, 319–333.

[40] Cohen M.H. (1955) Reciprocity theorem for anisotropic media, *Proc. IRE* **43**, 103.

[41] Collin R.E. (1973) On the incompleteness of E and H modes in wave-guides, *Can. J. Phys.* **51**, 1135–1140.

[42] Collin R.E. (1985) *Antennas and Radiowave Propagation*, McGraw-Hill, New York.

[43] Collin R.E. and Zucker F.J. (eds.) (1969) *Antenna Theory*, part I, McGraw-Hill, New York.

[44] Cory H. (1967) Analysis of reflection and transmission processes for very low frequency electromagnetic waves in the D-region of the ionosphere, D.Sc. thesis, Technion (Haifa).

[45] Cory H., Quemada D. and Vigneron J. (1970) Analyse théorique des mesure de champs électromagnétiques effectuées au sol, en fusée ou en satellite. 1., *Ann. Géophys.* **26**, 665–669.

[46] Courant R. and Hilbert D. (1962) *Methods of Mathematical Physics*, Vol.II, Interscience, New York.

[47] Dällenbach W. (1942) Der Reziprozitätssatz des elektromagnetischen Feldes, *Arch. Elektotech.* **36**, 153–165.

[48] Deschamps G.D. and Kesler O.B. (1967) Radiation of an antenna in a compressible magnetoplasma, *Radio Sci.* **2**, 757–767.

[49] Drude P. (1959) *The Theory of Optics*, Dover, New York. First published in 1900 as *Lehrbuch der Optik*, Hirzel-Verlag, Leipzig.

[50] Dzyaloshinskii I.E. (1960) On the magneto-electrical effect in antiferromagnets, *Sov. Phys. JETP* **10**, 628–629.

[51] Eckersley T.L. (1927) Short-wave wireless telegraphy, *I.E.E. Wireless Proceedings* **2**, 85–129.

[52] Everitt W.L. (1937) *Communication Engineering*, 2nd ed., McGraw-Hill, New York.

[53] Felsen L.B. and Marcuvitz N. (1973) *Radiation and Scattering of Waves*, Prentice Hall, Englewood Cliffs, N.J.

[54] Fijalkow E., Altman C. and Cory H. (1973) A full-wave study of ion-cyclotron whistlers in the ionosphere, *J. Atmos. Terr. Phys.* **35**, 317–337.

[55] Fryer G.J. and Frazer L.N. (1984) Seismic waves in stratified anisotropic media, *Geophys. J. R. astr. Soc.* **78**, 691–710.

[56] Gans R. (1915) Fortpflanzung des Lichts durch ein inhomogenes Medium, *Ann. Phys. Lpz.* **47**, 709–736.

[57] Gantmacher F.R. (1959) *The Theory of Matrices*, Vols. I and II, Chelsea Publishing Co., New York.

[58] Gibbs J.W. (1960) *Vector Analysis*, ed. E.B.Wilson, Dover, New York. Published originally by Scribner, New York (1901).

[59] Goldstein H. (1980) *Classical Mechanics*, 2nd ed., Addison-Wesley, Reading, Mass.

[60] Goldstein S. (1928) The influence of the earth's magnetic field on electric transmission in the upper atmosphere, *Proc. R. Soc. Lond. A* **121**, 260–285.

[61] Hamermesh M. (1962) *Group Theory*, Addison-Wesley, Reading, Mass.

[62] Harrington R.F. (1961) *Time-Harmonic Electromagnetic Fields*, McGraw Hill, New York.

[63] Harrington R.F. and Villeneuve A.T. (1958) Reciprocity relations for gyrotropic media, *IRE Trans. Microwave Theory Tech.* **MTT-6**, 308–310.

[64] Hartree D.R. (1931) The propagation of electromagnetic waves in a refracting medium in a magnetic field, *Proc. Camb. Phil. Soc.* **27**, 143–162.

[65] Hartshorne N.H. and Stuart A. (1970) *Crystals and the Polarising Microscope*, 4th ed., Edward Arnold, London.

[66] Heading J. (1971) Identities between reflexion and transmission coefficients and electric field components for certain anisotropic modes of oblique propagation, *J. Plasma Phys.* **6**, 257–270.

[67] Heading J. (1973) Generalized reciprocity relations and conditions relating to solutions of n second-order partial differential equations, *Q. J. Mech. Appl. Math.* **26**, 515–528.

[68] Heading J. (1975) A six-fold generalization of ionospheric reciprocity, *Q. J. Mech. Appl. Math.* **28**, 185–206.

[69] Heading J. (1978) General theorems relating to wave propagation governed by self-adjoint and Hermitian self-adjoint differential operators of order 2n, *Proc. R. Soc. Lond. A* **360**, 279–300.

[70] Hecht E. and Zajac A. (1974) *Optics*, Addison-Wesley, Reading.

[71] Inoue Y. and Horowitz S. (1966) Numerical solution of full-wave equation with mode coupling, *Radio Sci.* **1**, 957–970.

[72] Jackson J.D. (1975) *Classical Electrodynamics*, 2nd ed., Wiley, New York.

[73] Johler J.R and Harper J.D. (1962) Reflection and transmission of radio waves at a continuously stratified plasma with arbitrary magnetic induction, *J. Res. NBS* **64D**, 81–99.

[74] Jones D. (1969) The effect of the latitudinal variation of the terrestrial magnetic field strength on ion cyclotron whistlers, *J. Atmos. Terr. Phys.* **31**, 971–981.

[75] Jones D. (1970) "The theory of the effect of collisions on ion cyclotron whistlers", in *Plasma Waves in Space and in the Laboratory*, Vol. 2, pp. 471–485, ed. J.O.Thomas and B.J.Landmark, Edinburgh University Press.

[76] Jones D. (1976) Source of terrestrial non-thermal radiation, *Nature, Lond.* **260**, 686–689.

[77] Jordan E.C. and Balmain K.G. (1968) *Electromagnetic Waves and Radiating Systems*, 2nd ed., Prentice-Hall, New York.

[78] Keller H.B. and Keller J.B. (1962) Exponential-like solutions of systems of ordinary differential equations, *J. Soc. Indust. Appl. Math.* **10**, 246–259.

[79] Kelso J.M., Nearhoof H.J., Nertney R.J. and Waynick A.H. (1951) The polarisation of vertically incident long radio waves, *Ann. Géophys.* **7**, 215–244.

[80] Kennett B.L.N. (1974) Reflections, rays and reverberations, *Bull. Seism. Soc. Am.* **64**, 1685–1696.

[81] Kerns D.M. (1976) Plane-wave scattering-matrix theory of antennas and antenna-antenna interactions: formulation and applications, *J. Res. NBS—B. Mathematical Sciences* **80B**, 5–51.

[82] Kong J.A. (1972) Theorems of bianisotropic media, *Proc. IEEE* **60**, 1036–1046.

[83] Kong J.A. (1975) *Theory of Electromagnetic Waves*, Wiley, New York.

[84] Kong J.A. and Cheng D.K. (1970) Modified reciprocity theorem for bianisotropic media, *Proc. IEE* **117**, 349–350.

[85] Lacoume J. L. (1967) Etude de la propagation des ondes électromagnétiques dans un magnétoplasma stratifié, *Ann. de Géophys.* **23**, 477–493.

[86] Lakhtakia A., Varadan V.V. and Varadan V.K. (1988) Field equations, Huygen's principle, integral equations and theorems for radiation and scattering of electromagnetic waves in isotropic chiral media, *J. Opt. Soc. Am. A* **5**, 175–184.

[87] Landau L. and Lifshitz E. (1975) *The Classical Theory of Fields*, 4th ed., Pergamon, Oxford.

[88] Landau L. and Lifshitz E. (1987) *Fluid Mechanics*, 2nd ed., Pergamon, Oxford.

[89] Lassen H. (1927) Über den Einfluß des Erdmagnetfeldes auf die Fortpflanzung der elektrischen Wellen der drahtlosen Telegraphie in der Atmosphäre, *Elektrische Nachrichtentechnik* **4**, 324-334.

[90] Lipson S.G. and Lipson H. (1981) *Optical Physics*, 2nd ed., Cambridge University Press.

[91] Lorentz H.A. (1895-1896) The theorem of Poynting concerning the energy in the electromagnetic field and two general propositions concerning the propagation of light, *Versl. Kon. Akad. Wetensch. Amsterdam* **4**, 176. English translation in *Collected Papers*, Vol.III, (1936) pp.1–11, Martinus Nijhoff, The Hague.

[92] Marcuse D. (1972) *Light Transmission Optics*, Van Nostrand Reinhold, New York.

[93] Mislow K. (1966) *Introduction to Stereochemistry*, Benjamin, New York.

[94] Morse P.M. and Feshbach H. (1953) *Methods of Theoretical Physics*, part 1, McGraw-Hill, New York.

[95] Nagano I., Mambo M. and Hutatsuishi G. (1975) Numerical calculation of electromagnetic waves in an anisotropic multilayered medium, *Radio Sci.* **10**, 611–617.

[96] Panofsky W.K.H. and Phillips M. (1962) *Classical Electricity and Magnetism* 2nd ed., Addison-Wesley, Reading, Mass.

[97] Pathak P.H. (1983) On the eigenfunction expansion of electromagnetic dyadic Green's functions, *IEEE Trans. Antennas Propagat.* **AP-31**, 837–846.

[98] Pitteway M.L.V. (1965) The numerical calculation of wave fields, reflection coefficients and polarisations for long radio waves in the lower ionosphere, *Phil. Trans. R. Soc. Lond. A* **257**, 219–241.

[99] Pitteway M.L.V. and Horowitz S. (1969) Numerical solutions of differential equations with slowly varying coefficients, *Alta Frequenza* **38**, (Numero Speciale), 280–281.

[100] Pitteway M.L.V. and Jespersen J.L. (1966) A numerical study of the excitation, internal reflection and limiting polarization of whistler waves in the lower ionosphere, *J. Atmos. Terr. Phys.* **28**, 17–43.

[101] Post E.J. (1962) *Formal Structure of Electromagnetics*, North-Holland, Amsterdam.

[102] Price G.H. (1964) Propagation of electromagnetic waves through a continuously varying stratified anisotropic medium, *Radio Sci. J. Res. NBS* **68D**, 407–418.

[103] Rahmat-Samii Y. (1975) On the question of computation of the dyadic Green's function at the source region in waveguides and cavities, *IEEE Trans. Microwave Theory Tech.* **MTT-23**, 762–765.

[104] Ratcliffe J.A. (1959) *The Magneto-ionic Theory and its Applications to the Ionosphere*, Cambridge University Press.

[105] Rawer K. and Suchy K. (1967) "Radio Observations of the Ionosphere" in *Encyclopedia of Physics* Vol. 49/2, *Geophysics* III/2, pp. 1–546, ed. S.Flügge, Springer, Heidelberg.

[106] Rumsey V.H. (1954) Reaction concept in electromagnetic theory, *Phys. Rev.* **94**, 1483–1491.

[107] Schatzberg A. (1980) Reciprocity and scattering relations in electromagnetic wave propagation in plane-stratified and other structured media, M.Sc. thesis, Technion (Haifa).

[108] Schatzberg A. and Altman C. (1981) Reciprocity relations between currents and fields in plane-stratified magnetoplasmas, *J. Plasma Phys.* **26**, 333–344.

[109] Silverman M.P. (1986) Reflection and refraction at the surface of a chiral medium: comparison of gyrotropic constitutive relations invariant or noninvariant under a duality transformation, *J. Opt. Soc. Am. A* **3**, 830–837.

[110] Sommerfeld A. (1925) Das Reziprozitäts-theorem der drahtlosen Telegraphie, *Jb. drahtl. Telegr.* **26**, 93–98.

[111] Sommerfeld A. (1949) *Partial Differential Equations in Physics*, Academic Press, New York.

[112] Spiegel M.R. (1964) *Complex Variables*, Schaum, New York.

[113] Stix T.H. (1962) *The Theory of Plasma Waves*, McGraw-Hill, New York.

[114] Storey L.R.O. (1953) An investigation of whistling atmospherics, *Phil. Trans. R. Soc. Lond. A* **246**, 113–141.

[115] Stratton J.A. (1941) *Electomagnetic Theory*, McGraw-Hill, New York.

[116] Suchy K. (1964) Neue Methoden in der Kinetischen Theorie verdünnter Gase, *Ergebnisse der exakten Naturwissenschaften* **35**, 103–294.

[117] Suchy K. (1984) "Transport Coefficients and Collision Frequencies for Aeronomic Plasmas" in *Encyclopedia of Physics* Vol. 49/7, *Geophysics* III/7, pp. 57–221, ed. K.Rawer, Springer, Heidelberg.

[118] Suchy K. and Altman C. (1975a) The Maxwell field, its adjoint field and the 'conjugate' field in anisotropic absorbing media, *J.Plasma Phys.* **13**, 299–316.

[119] Suchy K. and Altman C. (1975b) Reflection and transmission theorems for characteristic waves in stratified anisotropic absorbing media, *J. Plasma Phys.* **13**, 437–449.

[120] Suchy K. and Altman C. (1989) Relations between eigenmode scattering matrices in curved-stratified cold magnetoplasmas, *J. Atmos. Terr. Phys.* **51**, 707–714.

[121] Suchy K., Altman C. and Schatzberg A. (1985) Orthogonal mappings of time-harmonic electromagnetic fields in inhomogeneous (bi)anisotropic media, *Radio Sci.* **20**, 149–160.

[122] Tai C.T. (1961) On the transposed radiating systems in an anisotropic medium, *IRE Trans. Antennas Propagat.* **9**, 502.

[123] Tai C.T. (1973) On the eigenfunction expansion of dyadic Green's functions, *Proc. IEEE* **61**, 480–481.

[124] Tellegen B.D.H. (1948) The gyrator, a new electric network element, *Philips Res. Reps.* **3**, 81–101.

[125] Tinoco I. and Freeman M.P. (1957) The optical activity of oriented copper helices, *J. Phys. Chem.* **61**, 1196–1200.

[126] Tsuruda K. (1973) Penetration and reflection of VLF waves through the ionosphere: Full wave calculations with ground effect, *J. Atmos. Terr. Phys.* **35**, 1377–1405.

[127] van Bladel J. (1984) *Relativity and Engineering*, Springer-Verlag, Berlin.

[128] Volland H. (1962a) The propagation of plane electromagnetic waves in a horizontally stratified ionosphere, *J. Atmos. Terr. Phys.* **24**, 853–856.

[129] Volland H. (1962b) Die Streumatrix der Ionosphäre, *Archiv Elektr. Übertragung* **16**, 328–334.

[130] Volland H. (1962c) Kopplung und Reflexion von elektromagnetischen Wellen in der Ionosphäre, *Archiv Elektr. Übertragung* **16**, 515–524.

[131] Wait J.R. (1968) *Electromagnetics and Plasmas*, Holt, Rinehart and Winston, New York.

[132] Weiglhofer W. (1988) Symbolic derivation of the electrodynamic Green's tensor in an isotopic medium, *Am. J. Phys.* **56**, 1095–1096.

[133] Weiglhofer W. (1989) A simple and straightforward derivation of the dyadic Green's function of an isotropic chiral medium, *Archiv für Elektronik und Ubertragungstechnik* **43**, 51–52.

[134] Wheeler J.A. and Feynman R.P. (1945) Interaction with the absorber as the mechanism of radiation, *Revs. Mod. Phys.* **17**, 157–181.

[135] Wills A.P. (1958) *Vector Analysis with an Introduction to Tensor Analysis*, Dover, New York. First published in 1931.

[136] Yaghjian A.D. (1980) Electric dyadic Green's functions in source regions, *Proc. IEEE* **68**, 248–263.

Notation and Symbols

N.1 Type styles and notation

Three basic type styles are used, 'roman', 'sans serif' and *'italic'*, the first two also in bold variants: **bold roman**, and **bold sans serif**. Greek and calligraphic letters, both in ordinary and bold type, are used too: μ, $\boldsymbol{\mu}$, $\mathcal{T}$, $\boldsymbol{\mathcal{T}}$.

• Components of vectors and scalar quantities are denoted by italic letters (E_x, k_α, a, q).

• Vectors, column and row matrices are denoted by bold roman letters ($\mathbf{E}$, $\mathbf{e}$). Bold Greek letters are used for unit vectors ($\hat{\boldsymbol{\xi}}$, $\hat{\boldsymbol{\eta}}$, $\hat{\boldsymbol{\epsilon}}_\pm$). Exceptions are $\boldsymbol{a}$, $\boldsymbol{v}$, $\boldsymbol{l}$, $\boldsymbol{\ell}$, $\boldsymbol{\phi}$.

• Tensors and matrix operators (but not row and column matrices) are denoted by bold sans serif letters (q, K, D, L), by bold Greek ($\boldsymbol{\chi}$, $\boldsymbol{\lambda}$, $\boldsymbol{\Gamma}$), and by bold calligraphic letters ($\boldsymbol{\mathcal{K}}$, $\boldsymbol{\mathcal{M}}$, $\boldsymbol{\mathcal{T}}$).

• Square brackets [...] and parentheses (...). The equation

$$\mathbf{e} = \begin{bmatrix} \mathbf{E} \\ \mathbf{H} \end{bmatrix} = [E_x\ E_y\ E_z\ H_x\ H_y\ H_z]^T$$

is understood to mean that $\mathbf{e}$ equals a column matrix formed from $\mathbf{E}$ and $\mathbf{H}$, and equals the transpose of a 6-element row matrix, whereas

$$\mathbf{e} = (\mathbf{E},\ \mathbf{H},\ \boldsymbol{v},\ p)$$

states that the constituents of the vector $\mathbf{e}$ are the three vectors $\mathbf{E}$, $\mathbf{H}$, $\boldsymbol{v}$ and the scalar p.

• The transpose of a matrix is denoted by a superscript 'T' or by a tilde ($\sim$), according to typographical convenience. Thus $\tilde{\mathsf{G}}^c$ is the transpose of G^c, while $\bar{\mathsf{G}}^T$ is the transpose of $\bar{\mathsf{G}}$.

• Primes ($\mathbf{e}'$, $\mathbf{j}'$, K') are generally used to denote *spatially mapped* (rotated, reflected, inverted) quantities; but in Chap. 7 they denote *time-reversed* quantities.

• Overbars ($\bar{q}_\alpha$, $\bar{\mathbf{e}}$, $\overline{\mathbf{E}}$, $\overline{\mathsf{K}}$) denote adjoint (non-physical) quantities that satisfy a non-physical, adjoint equation. Adjoint vectors are related to the corresponding physical quantities by a biorthogonality relationship.

• The superscript (L) ($a^{(L)}$, $\mathbf{e}^{(L)}$, $\mathbf{K}^{(L)}$) denotes 'Lorentz-adjoint' quantities. These obey Maxwell's equations, or the Maxwell-Euler equations, in a (restricted) time-reversed medium, ('restricted' in the sense that dissipative processes are not time reversed). These are physical quantities, and in the case of the electromagnetic field, a reversal of sign (direction) of the adjoint magnetic field yields the Lorentz-adjoint field.

• The superscript c ($q^c_{,}$ $\mathbf{e}^c$, $\mathbf{K}^c$, $\mathbf{S}^c$) denotes 'conjugate' quantities. In Chap. 2 this concept is used for quantities that obey Maxwell's equations in a reflected, time-reversed medium, that is identical with the original medium. Later the concept is extended to all spatially mapped, time-reversed quantities, that are just spatial mappings of the corresponding Lorentz-adjoint quantities, and obey Maxwell's equations in the mapped time-reversed medium.

N.2 List of symbols

a_α	eigenmode amplitude, $\alpha{=}\pm1,\pm2$	27
$\underset{\sim}{a}^\pm_\alpha$	$:=\eta_\alpha^{1/2}\underset{\sim}{A}^\pm_\alpha$, normalized amplitude density	56
da	column of 'primary' eigenmode amplitudes generated by current element $\mathbf{j}(\mathbf{k}_t,z')\,dz'$	177
$\boldsymbol{a}$	$:=(a_1,a_2,a_{-1},a_{-2})$, column of eigenmode amplitudes	27,48
$\boldsymbol{a}_{in}$, $\boldsymbol{a}_{out}$	column of incoming/outgoing eigenmode amplitudes	36,81
$\boldsymbol{a}_+$, $\boldsymbol{a}_-$	$:=(a_1,a_2)$ or (a_{-1},a_{-2}), 2-element amplitude columns	36,81
$\langle a\,,\,b\rangle$	reaction of field of a on source b	148
$\underset{\sim}{A}^\pm_\alpha$	spectral amplitude density in $\mathbf{k}_t$-space ($\alpha{=}\pm1,\pm2$)	56
$d\mathsf{A}_\pm(z')$	column of multiply-reflected wave amplitudes at z'	178
$\hat{b}_x,\hat{b}_y,\hat{b}_z$	direction cosines of $\mathbf{b}$ in (x,y,z) frame	78
$\hat{b}_{x'},\hat{b}_{y'},\hat{b}_{z'}$	direction cosines of $\mathbf{b}$ in (x',y',z') frame	12
$\mathbf{b}$	external (Earth's) magnetic induction field	6
$\mathbf{B}$	magnetic induction field	10
c	speed of light in vacuo	6
C	$:=S-P$, plasma parameter	12
C	$:=\omega\mathsf{K}-k_x\mathsf{U}_x-k_y\mathsf{U}_y$;	21
	also times $1/k_0$: $\rightarrow c\mathsf{K}-s_x\mathsf{U}_x-s_y\mathsf{U}_y$	65
$\mathsf{C}_t,\mathsf{C}_{1,2},\mathsf{C}_z$	partitioned transverse and normal constituents of C	174
$\hat{\mathcal{C}}$	6×6 chirality operator in optically active media	244
$\mathbf{d}$	downgoing wave-field component of $\mathbf{g}$ ($=\mathbf{u}+\mathbf{d}$)	32
$\mathbf{d}_n$	wave field of downgoing non-penetrating mode	34
D	$:=(R-L)/2$, plasma parameter	11
$\mathbf{D}$	electric displacement field	10

$\mathcal{D}$	2×2 downgoing amplitude matrix for modes in $\mathbf{g}^{(1)}$, $\mathbf{g}^{(2)}$	33
D	$:=\begin{bmatrix} 0 & -\nabla\times\mathsf{I}^{(3)} \\ \nabla\times\mathsf{I}^{(3)} & 0 \end{bmatrix}$, Maxwell differential operator;	64
	also 10×10 Maxwell-Euler differential operator	237
$\mathsf{D}_t, \mathsf{D}_w$	tangential and normal parts of differential operator D	85
$\mathbf{e}$	$:=(E_x,E_y,E_z;H_x,H_y,H_z)$, 6-component wave field;	64
	used also for transverse components (E_x,E_y,H_x,H_y);	45
	$:=(\mathbf{E},\mathbf{H},\boldsymbol{v},p)$, 10-component wave field in compressible magnetoplasma	237
$\mathbf{e}^{(4)}$	$:=(E_x,E_y,\mathcal{H}_x,\mathcal{H}_y)$, 4-component transverse wave field	25
$\mathbf{e}_\alpha$	characteristic wave fields or eigenmodes ($\alpha=\pm1,\pm2$)	47,65
$\bar{\mathbf{e}}$, $\bar{\mathbf{e}}_\alpha$	6-component adjoint wave field (eigenmode)	66,68
$\hat{\mathbf{e}}_\alpha$, $\hat{\bar{\mathbf{e}}}_\alpha$	$:=(1,\pm\rho_\alpha,0;\mp Y_0\rho_\alpha,Y_0,0)/[2Y_0\,q(1-\rho_\alpha{}^2)]^{1/2}$, given and adjoint normalized eigenvectors in free space	167
$\mathbf{e}'$	spatially mapped 6-component wave field;	145,196
	(in Chap. 7) time-reversed 6-component wave field	221
$\mathbf{e}^{(L)}$	$:=\bar{\mathsf{I}}\,\bar{\mathbf{e}}=(\mathbf{E}^{(L)},\mathbf{H}^{(L)})$, Lorentz-adjoint wave field	118
$\mathbf{e}^a_\ell$, $\mathbf{e}^a_r$	left/right-handed wave fields generated by current $\mathbf{j}_a$	250
E_z	component of $\mathbf{E}$ normal to stratification	23
$\mathbf{E}$, $\overline{\mathbf{E}}$	given/adjoint electric wave field	6,67
$\mathbf{E}_t$	projection of $\mathbf{E}$ on stratification plane	23
$\mathbf{E}_p^\pm$, $\mathbf{E}_n^\pm$	electric fields, up(+) or downgoing(−), in $\mathbf{u}_p$, $\mathbf{u}_n$, $\mathbf{d}_p$, $\mathbf{d}_n$	34
$\mathbf{E}^{in}$, $\mathbf{E}^{out}$	incoming/outgoing (scattered) electric wave fields	55,56
$\mathbf{E}_\parallel$, $\mathbf{E}_\perp$	projection of $\mathbf{E}$ on/normal to relative velocity vector $\boldsymbol{v}$	124
$\overline{\mathbf{E}}_\alpha$, $\overline{\mathbf{H}}_\alpha$	adjoint eigenmode wave fields;	72
	(also) reciprocal (left) wave-field eigenvectors	17
$\mathbf{E}_k$, $\mathbf{H}_k$	Fourier-analyzed wave fields in $\mathbf{k}$-space	162
E	$:=[\hat{\mathbf{e}}_1\ \hat{\mathbf{e}}_2\ \hat{\mathbf{e}}_{-1}\ \hat{\mathbf{e}}_{-2}]$, 4×6 normalized eigenmode matrix	48,81
$\overline{\mathsf{E}}$	4×6 normalized adjoint eigenmode matrix	80
E_+, E_-	$:=[\hat{\mathbf{e}}_{\pm1}\ \hat{\mathbf{e}}_{\pm2}]$, 2×6 normalized eigenmode matrices	81
$\mathbf{f}$, $\mathbf{f}_s$	force density on particles (of species s)	6
f_α	modal amplitude with rapidly varying phase removed	48
$\mathbf{f}$	$:=(f_1,f_2,f_{-1},f_{-2})$, column of eigenmode amplitudes	48
$\mathbf{f}_\alpha$	$:=(E_x,E_y,H_x,H_y)_\alpha$, 4-component eigenmode	91
F	$:=[\hat{\mathbf{f}}_1\ \hat{\mathbf{f}}_2\ \hat{\mathbf{f}}_{-1}\ \hat{\mathbf{f}}_{-2}]$, 4×4 normalized eigenmode matrix	91
$\mathsf{F}_\pm(\mathbf{k}_t,z,z')$	transfer matrices relating primary to resultant multiply-reflected amplitudes, $d\mathsf{A}(z)=:\mathsf{F}_\pm(\mathbf{k}_t,,z,z')da(z')$	180
$\mathbf{g}_u,\mathbf{g}_v,\mathbf{g}_w$	basis vectors in u,v,w frame, $\mathrm{dr}=:\mathbf{g}_u du+\mathbf{g}_v dv+\mathbf{g}_w dw$	85
$\mathbf{g}^u,\mathbf{g}^v,\mathbf{g}^w$	basis vectors reciprocal to $\mathbf{g}_u,\mathbf{g}_v,\mathbf{g}_w$, i.e. $\mathbf{g}^i\mathbf{g}_j=\delta_{ij}$	85

$\mathbf{g}_\alpha$	$:=(E_x, -E_y, \mathcal{H}_x, \mathcal{H}_y)_\alpha$, 4-element transverse eigenvector	26
$\mathbf{g}^{(1)}$, $\mathbf{g}^{(2)}$	independent solutions of coupled wave equations	31,32
$G(\tau)$	$:=(1/2\pi)\int_{-\infty}^{\infty}\chi(\omega)\exp(i\omega\tau)\,d\omega$, susceptibility kernel	219
G	$:=[\hat{\mathbf{g}}_1\ \hat{\mathbf{g}}_2\ \hat{\mathbf{g}}_{-1}\ \hat{\mathbf{g}}_{-2}]$, 4×4 normalized eigenmode matrix	26
G_+, G_-	$:=[\hat{\mathbf{g}}_1\ \hat{\mathbf{g}}_2]$, $[\hat{\mathbf{g}}_{-1}\ \hat{\mathbf{g}}_{-2}]$, 2×4 normalized eigenmode matrices	36
G_{in}, G_{out}	4×4 normalized incoming/outgoing eigenmode matrices	37
$\mathsf{G}(\mathbf{r},\mathbf{r}')$	dyadic Green's function, $\mathbf{e}(\mathbf{r})=:\int\mathsf{G}(\mathbf{r},\mathbf{r}')\,\mathbf{j}(\mathbf{r}')\,d^3r'$	169
$\bar{\mathsf{G}}$, G^c, $\mathsf{G}^{(L)}$	adjoint/conjugate/Lorentz-adjoint Green's functions	205/6
$\mathsf{G}(\mathbf{k}_t, z, z')$	dyadic Green's function in $\mathbf{k}_t$-space	169
$\mathsf{G}_z^0(\mathbf{k}_t,z,z')$	source term in Green's function $\mathsf{G}(\mathbf{k}_t,z,z')$	176
$h(z-z')$	unit step function	172
H_z	component of $\mathbf{H}$ normal to stratification	23
$\mathbf{H}$, $\overline{\mathbf{H}}$	given/adjoint magnetic wave fields	7,67
$\mathbf{H}_t$	projection of $\mathbf{H}$ on stratification plane	23
$\mathsf{H}(z-z')$	diagonal 4×4 unit step matrix	172
$\mathcal{H}$	$:=\sqrt{\mu_0/\varepsilon_0}\,\mathbf{H}$, normalized magnetic wave field	19
I	inclination of Earth's magnetic field	12
I_i, $I(s)$, $I(0)$	current in antenna i/ at point s/ at input terminals	151/2
$I_{s.c.}$	short-circuit current at receiving antenna terminals	153
I, $\mathsf{I}^{(n)}$	unit $(n\times n)$ tensor	8,63
$\bar{\mathsf{I}}$, $\bar{\mathsf{I}}^{(6)}$	Poynting-vector reversing operator; diagonal matrix with elements $\mathsf{I}^{(3)}$ and $-\mathsf{I}^{(3)}$	77
$\bar{\mathsf{I}}_{23}^{(10)}$	10×10 unit matrix with $(\mathsf{I}^{(3)}, -\mathsf{I}^{(3)}, -\mathsf{I}^{(3)}, 1)$ along diagonal	238
$\bar{\mathsf{I}}_{24}^{(10)}$	10×10 unit matrix with $(\mathsf{I}^{(3)}, -\mathsf{I}^{(3)}, \mathsf{I}^{(3)}, -1)$ along diagonal	238
$\mathbf{j}$	$:=(\mathbf{J}_\mathrm{e}, \mathbf{J}_\mathrm{m})$, 6-component current density;	20
	used also for 3-component electric current density;	8
	$:=(\mathbf{J}_\mathrm{e}, \mathbf{J}_\mathrm{m}, -\mathbf{f}, -s)$, 10-component current density in compressible magnetoplasma	237
$\mathbf{j}'$	spatially mapped 6-component current density;	145,196
	(in Chap. 7) time-reversed 6-component current density	221
$\bar{\mathbf{j}}$	adjoint 6-component current density	130
$\mathbf{j}^{(L)}$	$:=\bar{\mathsf{I}}\bar{\mathbf{j}}$, Lorentz-adjoint current density	131
$\mathbf{J}_\perp^\pm$	$:=[\mathsf{I}-\hat{\mathbf{k}}_0^\pm\tilde{\hat{\mathbf{k}}}_0^\pm]\mathbf{J}$, projection of current density $\mathbf{J}$ $(=\mathbf{J}_\mathrm{e})$ on plane perpendicular to wave vector $\hat{\mathbf{k}}_0^\pm$	165
$\mathbf{J}_\mathrm{e}$, $\mathbf{J}_\mathrm{m}$	electric/equivalent magnetic current density	20
$\mathbf{J}_{\mathrm{e}t}$, $\mathbf{J}_{\mathrm{m}t}$	projection of current densities on stratification plane	24
J	$:=\begin{bmatrix} 0 & \mathsf{I}^{(2)} \\ \mathsf{I}^{(2)} & 0 \end{bmatrix}$, interchanges up/downgoing modes	95

Symbol	Meaning	Page
$\bar{\mathsf{J}}$	$:=\bar{\mathsf{I}}^{(4)}\mathsf{J}$	103
k_0	$:=\omega\sqrt{\varepsilon_0\mu_0}=\omega/c$	19
$\mathbf{k}$	propagation (wave) vector	7
$\mathbf{k}_t$	projection of $\mathbf{k}$ on stratification plane	20
$\mathbf{k}_t'$, $\mathbf{k}_t''$	values of $\mathbf{k}_t$ for incident/scattered waves (Fig. 2.1)	55,57
$k_+^{(\pm)}$, $k_-^{(\pm)}$	$:=k_r^{(\pm)}$, $k_\ell^{(\pm)}$, wave vectors for circularly polarized (r/ℓ) propagation in positive/negative $(\pm)$ direction	245,249
K	$:=\begin{bmatrix} \varepsilon & \xi \\ \eta & \mu \end{bmatrix}$, 6×6 constitutive tensor;	21,63
	(also) 10×10 constitutive tensor for compressible plasma	237
K'	$:=\Gamma\mathsf{K}\Gamma^T$, mapped K (rotated, reflected, inverted);	213
	(in Chap. 7) time-reversed constitutive tensor	221
$\mathsf{K}^{(L)}$	$:=\bar{\mathsf{I}}\mathsf{K}^T\bar{\mathsf{I}}$, Lorentz-adjoint constitutive tensor	118
$\mathcal{K}$	$:=\begin{bmatrix} 0 & -\mathbf{k}\times\mathsf{I} \\ \mathbf{k}\times\mathsf{I} & 0 \end{bmatrix}=:k\hat{\mathcal{K}}$, derived from curl operator D	133
$\ell_{eff}^{(trans)}$, $\ell_{eff}^{(rec)}$	effective length of transmitting/receiving antenna	155
$\tilde{\ell}_1, \tilde{\ell}_2, \tilde{\ell}_3$	orthonormal row vectors of mapping matrix $\boldsymbol{\lambda}$	186
$d\boldsymbol{l}_i$, $\Delta\boldsymbol{l}_i$	elementary length of antenna	151
L, R, P	eigenvalues of plasma permittivity tensor	10,11
L	$[i\omega\mathsf{K}+\mathsf{D}]$, the Maxwell operator	117
	$\to [ik_0\mathsf{C}+\mathsf{U}_z\,d/dz]$ in stratified media	65
$\bar{\mathsf{L}}$	$:=[i\omega\mathsf{K}^T-\mathsf{D}^T]$, adjoint Maxwell operator;	117
	$\to [ik_0\mathsf{C}^T-\mathsf{U}_z\,d/dz]$ in stratified media	66
$\mathsf{L}^{(L)}$, $\bar{\mathsf{L}}'$	$:=[i\omega\mathsf{K}^T-\mathsf{D}^T]\bar{\mathsf{I}}^T$, Lorentz-adjoint Maxwell operator	131
m, m_s	electron mass, particle mass of species s	6,236
M	matrizant relating wave fields at two levels	46
$\bar{\mathsf{M}}$, M^c	adjoint/conjugate matrizants	114
$\mathcal{M}$	matrizant relating slowly varying amplitudes at two levels	49
$\mathbf{n}$	$:=c\mathbf{k}/\omega$, refractive index vector	13
$\mathbf{n}_t$	projection of $\mathbf{n}$ on stratification plane	22
n	$:=\|\mathbf{n}\|$, refractive index	13
n_β, $\bar{n}_\beta$	given and adjoint modal refractive indices $(\beta=\pm1,\pm2)$	68
n_0, n_{s0}	(equilibrium) particle number density (plasma species s)	6,236
$\mathsf{N}^{(4)}$	4×4 propagation matrix in coupled wave equations	25
p, p_s	small pressure perturbation (in plasma species s)	6,236
$\langle P_{z^\mp,\alpha}\rangle$	time-averaged energy flux across surfaces $z=z^\mp$	56
P_z, $\mathbf{P}$	(z-component of) bilinear concomitant vector, $\mathbf{P}=\overline{\mathbf{E}}\times\mathbf{H}+\mathbf{E}\times\overline{\mathbf{H}}$ for Maxwell system	66,67; 118

Symbol	Meaning	Page
$\mathbf{P}^{(L)}$	$:=\mathbf{E}^{(L)}\times\mathbf{H}-\mathbf{E}\times\mathbf{H}^{(L)}$, Lorentz-adjoint concomitant vector	118
$\mathbf{P}_{\alpha\beta}$	$:=\mathbf{E}_{\alpha}\times\overline{\mathbf{H}}_{\beta}+\overline{\mathbf{E}}_{\beta}\times\mathbf{H}_{\alpha}$, bilinear concomitant vector	86
$\mathbf{P}$	(also) polarization vector of medium	222
P, P^{ν}	propagator (for νth layer) relating wave amplitudes, $a(z^{+})=:\mathsf{P}(z^{+},z^{-})a(z^{-})$	37,115
$\overrightarrow{\mathsf{P}}$, $\overleftarrow{\mathsf{P}}$	propagators, $\overrightarrow{\mathsf{P}}:=\mathsf{P}(z^{+},z^{-})$, $\overleftarrow{\mathsf{P}}:=\mathsf{P}(z^{-},z^{+})$, $z^{+}>z^{-}$	116
$\mathcal{P}_0$, $\mathcal{P}_{\pm 1}$	orthogonal projectors of a 3×3 tensor	9,256
q, q_s	charge of particle (of species s)	6,236
q, q_{α}	$:=n_z$, component of refractive index vector normal to stratification; eigenvalue of Maxwell operator for stratified media, root of Booker quartic equation	22,26,65
	$:=(1-s_x{}^2-s_y{}^2)^{1/2}$, poles of n in complex Fourier space	163
q	$:=\pm\mathsf{q}_n$, n=1,2,3, reflection/rotation operator with respect to x_n=0 plane/x_n axis; $\pm\mathsf{q}_0$ is the identity/inversion operator	187
q_x, q_y, q_z	3×3 diagonal reflection matrices with elements (–1,1,1), (1,–1,1), (1,1,–1)	77,187
Q_i	$:=\begin{bmatrix} \mathsf{q}_i & 0 \\ 0 & -\mathsf{q}_i \end{bmatrix}$, $(i=x,y,z)$, 6×6 reflection matrix	78,141
$\overline{\mathsf{Q}}_i$, Q_i^c	$:=\mathsf{Q}_i\bar{\mathsf{I}}$, $(i=x,y,z)$, adjoint/conjugate reflection matrix	78,126
$\mathsf{Q}_{(4)}^c$	4×4 equivalent of Q_y	95
Q	$:=\begin{bmatrix} \mathsf{q} & 0 \\ 0 & (\det\mathsf{q})\mathsf{q} \end{bmatrix}$, 6×6 diagonal mapping operator;	191
	(also) diagonal matrix with eigenvalues q_{α} as elements	27
$\mathbf{r}$, $\mathbf{r}'$	given/mapped position vector, $\mathbf{r}'=\mathsf{q}\mathbf{r}$, $\lambda\mathbf{r}$ or $\gamma\mathbf{r}$;	187/8,195
	(in Chap. 5) observation/source point	163
r_+, r_-	interface reflection matrix for up/downgoing incidence	36
R_n	reflection coefficient for non-penetrating mode	34
R, $\mathsf{R}_{\pm}$	2×2 reflection matrix (up/downgoing incidence) for:	
	linear free-space modes, ${}_{\parallel}R_{\parallel}$, ${}_{\parallel}R_{\perp}$, ${}_{\perp}R_{\parallel}$, ${}_{\perp}R_{\perp}$;	33,96
	left-, right-handed free-space modes, ${}_{l}R_{l}$, ${}_{l}R_{r}$, ${}_{r}R_{l}$, ${}_{r}R_{r}$;	97
	eigenmodes in the medium, $R^{\pm}_{\alpha,\beta}$ (α,β=1 or 2)	38,59
$\mathsf{R}_{\pm}^{(L)}$, $\mathsf{R}_{\pm}^{c}$	Lorentz-adjoint/conjugate reflection matrices	58,83
$\mathcal{R}$	reflection operator	74
s	plasma species, ($s=e$) refers to an electron plasma;	6
	$:=\sin\theta$, θ is angle of incidence on stratified medium;	25
	$\rho_0\, s$, source term, electron production rate in plasma	236
s_x, s_y	components of $\mathbf{k}_t/k_0$	61

Symbol	Meaning	Page
S	$:=(R+L)/2$, plasma parameter	11
$\langle S_z\rangle$, $\langle \mathbf{S}\rangle$	(z-component of) time-averaged Poynting vector	35,69
$\mathbf{S}$	$=:\begin{bmatrix} \mathbf{R}_+ & \mathbf{T}_- \\ \mathbf{T}_+ & \mathbf{R}_- \end{bmatrix}$, scattering matrix, $\boldsymbol{a}_{out}=:\mathbf{S}\boldsymbol{a}_{in}$	38
$\overline{\mathbf{S}}, \mathbf{S}^c, \mathbf{S}^{(L)}$	adjoint/conjugate/Lorentz-adjoint scattering matrices	81,83,58
$\underset{\sim}{\mathbf{S}}$, $\underset{\sim}{S}_{\alpha,\beta}$	scattering-density matrix (element of)	57
$\mathbf{t}_+$, $\mathbf{t}_-$	interface transmission matrices, up/downgoing incidence	36
$T^{\pm}_{\alpha,\beta}$	$\alpha,\beta=1,2$, symbolic modal transmission channels	250
T^+_{LR}, T^+_{RR}	elements of $\mathbf{T}_+$ for right-/left-handed modes in medium	42
$\mathbf{T}$, $\mathbf{T}_{\pm}$	2×2 transmission matrix (up/downgoing) in terms of:	
	linear free-space modes, ${}_{\parallel}T_{\parallel}$, ${}_{\parallel}T_{\perp}$, ${}_{\perp}T_{\parallel}$, ${}_{\perp}T_{\perp}$;	96
	left-, right-handed free-space modes, ${}_lT_l$, ${}_lT_r$, ${}_rT_l$, ${}_rT_r$;	97
	eigenmodes in the medium, $T^{\pm}_{\alpha,\beta}$ (α,β=1 or 2)	38,59
$\mathbf{T}^{(L)}_{\pm}$, $\mathbf{T}^c_{\pm}$	Lorentz-adjoint/conjugate transmission matrices	58,83
$\mathbf{T}$	4×4 propagation matrix in coupled wave equations	26
$\mathcal{T}$	time-reversal operator	75
$\mathcal{T}$	propagation matrix relating $\mathbf{f}'$ and $\mathbf{f}$ (analogous to $\mathbf{T}$)	49
u, v, w	orthogonal curvilinear coordinates on curved surface	84
$\mathbf{u}$	upgoing wave-field component of $\mathbf{g}$ ($=\mathbf{u}+\mathbf{d}$)	32
$\mathbf{u}_p, \mathbf{u}_n$	upgoing penetrating/non-penetrating modes	34
$\mathbf{U}_x$, $\mathbf{U}_y$, $\mathbf{U}_z$	$:=\begin{bmatrix} 0 & -\hat{\mathbf{x}}_i\times\mathbf{I} \\ \hat{\mathbf{x}}_i\times\mathbf{I} & 0 \end{bmatrix}$, $i=1,2,3$, $x_1,x_2,x_3 \rightarrow x,y,z$	21
$\mathbf{U}_w$	as in $\mathbf{U}_{x_i}$ with $\hat{\mathbf{x}}_i \rightarrow \mathbf{g}^w$	85
$\mathbf{U}^{(4)}_z$, $\mathbf{U}_t$	4×4 form of $\mathbf{U}_z$ with null rows and columns removed	91,174
$\mathcal{U}$	2×2 upgoing amplitude matrix for modes in $\mathbf{g}^{(1)}$, $\mathbf{g}^{(2)}$	33
$\boldsymbol{v}$, $\boldsymbol{v}_s$	mean particle velocity (of species s) in a plasma	6,236
$\boldsymbol{v}$	velocity of medium in laboratory frame	124
$V(0)$	voltage at antenna input terminals	152
$V^{(i)}_{o.c.}$	open-circuit voltage at terminals of antenna i	153
x, y, z	cartesian frame with $\hat{\mathbf{z}}$ vertically up and $\hat{\mathbf{y}}$ normal to magnetic meridian plane (Fig. 2.2);	61,75
	also used (Chap. 1) with $\hat{\mathbf{z}}:=\hat{\mathbf{b}}$, $\hat{\mathbf{y}}:=\hat{\mathbf{b}}\times\mathbf{n}/\lvert\hat{\mathbf{b}}\times\mathbf{n}\rvert$	11,13
x', y', z'	auxiliary frame with $\hat{\mathbf{z}}'$ vertically up and $\hat{\mathbf{y}}'$ normal to plane of incidence (Fig. 3.1);	91,95
	also used (Chap. 1) with $\hat{\mathbf{z}}'$ vertically up and $\hat{\mathbf{y}}'$ normal to magnetic meridian plane	12
x'_c, y'_c, z'_c	conjugate frame with incidence plane $\hat{\mathbf{x}}', \hat{\mathbf{z}}'$ ($\angle\phi$ to $\hat{\mathbf{b}}, \hat{\mathbf{z}}'$ plane) rotated through $(\pi-\phi)$ about $\hat{\mathbf{z}}'$ (Fig. 3.1)	91,95

X, X_s	$:=\omega_p^2/\omega^2$, ω_{ps}^2/ω^2, plasma parameter	11,14
Y_0	$:=\sqrt{\varepsilon_0/\mu_0}=1/Z_0$, admittance of free space	71
Y, Y_s	$:=\omega_c/\omega$, ω_{cs}/ω, plasma parameter	11,14
δz	thickness of elementary layer	37
Z, Z_s	$:=\nu/\omega$, ν_s/ω, plasma parameter	11,14
Z_0	$:=\sqrt{\mu_0/\varepsilon_0}$, impedance of free space	24
Z_A	$:=V(0)/I(0)$, antenna input impedance	152
Z_{0s}	$:=V(0)/I(s)$, antenna transfer impedance	152
$\mathbf{Z}_s$	$\hat{\mathbf{n}}\times\mathbf{E}=:\mathbf{Z}_s\mathbf{H}$, $\mathbf{Z}_s\hat{\mathbf{n}}=0$, surface impedance tensor	137
α	specific rotation of linearly polarized light in medium	242
α,β	$(=\pm1,\pm2)$, eigenmode numbers	16,65
β	chirality constant in optically active medium	243
γ	specific-heat ratio	7
γ	$:=\gamma_\pm=\pm\exp[\phi\times\mathbf{I}]$, coordinate-free rotation operator without/with inversion	192,195
Γ	$:=\begin{bmatrix}\gamma & 0\\ 0 & (\det\gamma)\gamma\end{bmatrix}$, 6×6 coordinate-free mapping operator	196
$\bar{\Gamma}$	$:=\Gamma\bar{\mathbf{I}}=\bar{\mathbf{I}}\Gamma$, 6×6 adjoint mapping operator	200
Γ_D, Γ	the diagonal/off-diagonal parts of $\mathbf{T}^{-1}\mathbf{T}'$	48
$\boldsymbol{\Delta}^\nu$	4×4 diagonal phase matrix with elements $\boldsymbol{\Delta}_\pm^\nu$	37
$\boldsymbol{\Delta}_\pm$, $\boldsymbol{\Delta}_\pm^\nu$	2×2 diagonal phase matrices with elements $\exp(-ik_0q_{\pm1}\delta z)$ and $\exp(-ik_0q_{\pm2}\delta z)$	37,39
∇	$:=\partial/\partial\mathbf{r}=\mathbf{g}^u\partial/\partial u+\mathbf{g}^v\partial/\partial v+\mathbf{g}^w\partial/\partial w$, differential operator	84
∇_t	$:=\mathbf{g}^u\partial/\partial u+\mathbf{g}^w\partial/\partial w$, tangential differential operator	84
∇_w	$:=\mathbf{g}^w\partial/\partial w$, normal differential operator	84
ε', μ'	transformed values of ε, μ in $\mathbf{K}$ for moving media	124
ε_0	permittivity of free space	10
ε_L, ε_R, ε_P	eigenvalues of plasma permittivity tensor	10,11
$\boldsymbol{\varepsilon}$	electric permittivity tensor, constituent of $\mathbf{K}$	10,63
$\boldsymbol{\varepsilon}_{tt}$, $\vec{\boldsymbol{\varepsilon}}_{tz}$, $\vec{\boldsymbol{\varepsilon}}_{zt}^T$, ε_{zz}	constituents of partitioned permittivity tensor $\boldsymbol{\varepsilon}$	23
$\hat{\epsilon}_\parallel$, $\hat{\epsilon}_{\pm1}$	$:=\mathbf{k}_t/\lvert\mathbf{k}_t\rvert$, basis vectors for $k_z>0$ or $k_z<0$	56
$\hat{\epsilon}_\perp$, $\hat{\epsilon}_{\pm2}$	$:=\hat{\mathbf{z}}\times\hat{\epsilon}_{\pm1}$, basis vectors	56
$\hat{\epsilon}_\pm$	$:=(\hat{\mathbf{x}}\pm i\hat{\mathbf{y}})/\sqrt{2}$, rotating circular basis vectors	244
$\hat{\epsilon}_\alpha$	$:=(\hat{\xi}^\pm+\rho_\alpha\hat{\eta})/(1-\rho_\alpha{}^2)^{1/2}$, $\alpha=\pm1,\pm2$, elliptic basis vectors	166
$\hat{\bar{\epsilon}}_\alpha$	$:=(\hat{\xi}^\pm-\rho_\alpha\hat{\eta})/(1-\rho_\alpha{}^2)^{1/2}$, adjoint elliptic basis vectors	166
$\eta_{s,0}$, $\eta_{s,\pm1}$	eigenvalues of resistivity tensor for species s	9
$\eta_{\pm1}$, $\eta_{\pm2}$	$:=\omega\varepsilon_0/\lvert k_z\rvert$, $\lvert k_z\rvert/\omega\mu_0$, characteristic wave admittances	56

$\boldsymbol{\eta}_s$	resistivity tensor for plasma species s	8
$\boldsymbol{\eta}$	off-diagonal coupling tensor in $\mathbf{K}$	63
θ	$:=\arccos(\hat{\mathbf{b}}\cdot\hat{\mathbf{n}})$;	13
	$:=\arccos(\hat{\mathbf{n}}\cdot\hat{\mathbf{z}})$, angle of incidence	25
κ	$:=k_z$, wave-vector component normal to stratification	22
λ	wavelength of light in medium	243
$\boldsymbol{\lambda}$	orthonormal mapping operator in cartesian frame	186
$\boldsymbol{\Lambda}$	$:=\begin{bmatrix}\boldsymbol{\lambda} & 0\\ 0 & (\det\boldsymbol{\lambda})\boldsymbol{\lambda}\end{bmatrix}$, 6×6 cartesian mapping operator	191
μ_0	permeability of free space	10
$\mu_{sL}, \mu_{sR}, \mu_{sP}$	eigenvalues of mobility tensor for plasma species s	10
$\boldsymbol{\mu}$	magnetic permeability tensor, constituent of $\mathbf{K}$	63
$\boldsymbol{\mu}_s$	mobility tensor for plasma species s	8
ν, ν_s	collision frequency (of plasma species s) with neutrals	6,16
$\hat{\boldsymbol{\nu}}$	unit normal to boundary curve	86
ξ, η, ς	cartesian frame with $\hat{\varsigma}:=\hat{\mathbf{n}}$, $\hat{\eta}:=\hat{\mathbf{b}}\times\hat{\mathbf{n}}/\lvert\hat{\mathbf{b}}\times\hat{\mathbf{n}}\rvert$	18
ξ	relativistic coupling term in moving media	124
$\boldsymbol{\xi}$	off-diagonal coupling tensor in $\mathbf{K}$	63
ρ_0	$:=n_0 m$, equilibrium density of plasma	236
ρ, ρ_α	$:=(E_\eta/E_\xi)_\alpha$, transverse (modal) wave polarization	18,19
σ, σ_α	$:=(E_\varsigma/E_\xi)_\alpha$, longitudinal (modal) wave polarization	19
$\sigma_R, \sigma_L, \sigma_P$	eigenvalues of plasma conductivity tensor	10
$\boldsymbol{\sigma}$	conductivity tensor	8,9
τ_p	transmission coefficient for penetrating mode	35
$\boldsymbol{\tau}$	$:=-ik_0\mathbf{T}$, form of propagation matrix $\mathbf{T}$	46
$\boldsymbol{\phi}$	$:=\varphi\hat{\boldsymbol{\phi}}$, angle φ times unit vector along rotation axis	192
$\chi(\omega)$	scalar susceptibilty, $\mathbf{D}(\omega)=:\varepsilon_0\{1+\chi(\omega)\}\mathbf{E}(\omega)$	219
$\boldsymbol{\chi}$	$:=-i\boldsymbol{\sigma}/\varepsilon_0\omega$, susceptibility tensor	112
ω	(angular) wave frequency	7
ω_c, ω_{cs}	$:=\lvert q_s\mathbf{b}\rvert/m_s$, gyrofrequency (of plasma species s)	7,14
ω_p, ω_{ps}	$:=(q_s{}^2 n_{s0}/m_s\varepsilon_0)^{1/2}$, plasma frequency (of species s)	11,14

Index

DEVELOPMENTS IN ELECTROMAGNETIC THEORY AND APPLICATIONS

1. J. Caldwell and R. Bradley (eds.): *Industrial Electromagnetics Modelling*. 1983
ISBN 90-247-2889-4
2. H.G. Booker: *Cold Plasma Waves*. 1984 ISBN 90-247-2977-7
3. J. Lekner: *Theory of Reflection of Electromagnetic and Particle Waves*. 1987
ISBN 90-247-3418-5
4. A.E. Lifschitz: *Magnetohydrodynamics and Spectral Theory*. 1989
ISBN 90-247-3713-3
5. H.C.K. Rawer: *Waves in Ionized Media*. (forthcoming) ISBN 0-7923-0775-5
6. G. Franceschetti and O.M. Bucci: *The Efficient Representation of Scattered Fields*. (forthcoming) ISBN 0-7923-0776-3
7. K.I. Hopcraft and P.R. Smith: *An Introduction to Electromagnetic Inverse Scattering*. (forthcoming) ISBN 0-7923-0777-1
8. A.P. Shivarova, A.D. Boardman and Yu.M. Aliev: *Non-linear Plasma and Optical Waveguides*. (forthcoming) ISBN 0-7923-0778-X
9. C. Altman and K. Suchy: *Reciprocity, Spatial Mapping and Time Reversal in Electromagnetics*. 1991 ISBN 0-7923-1339-9

KLUWER ACADEMIC PUBLISHERS – DORDRECHT / BOSTON / LONDON